Proceedings of the Ringberg Workshop
New Trends in
HERA Physics
2003

Proceedings of the Ringberg Workshop

NEW TRENDS IN HERA PHYSICS 2003

edited by

G. Grindhammer
Max-Planck Institute for Physics
Munich, Germany

B. A. Kniehl
Hamburg University, Germany

G. Kramer
Hamburg University, Germany

W. Ochs
Max-Planck Institute for Physics
Munich, Germany

 **World Scientific**

NEW JERSEY • LONDON • SINGAPORE • BEIJING • SHANGHAI • HONG KONG • TAIPEI • CHENNAI

Published by

World Scientific Publishing Co. Pte. Ltd.
5 Toh Tuck Link, Singapore 596224
USA office: Suite 202, 1060 Main Street, River Edge, NJ 07661
UK office: 57 Shelton Street, Covent Garden, London WC2H 9HE

British Library Cataloguing-in-Publication Data
A catalogue record for this book is available from the British Library.

Cover picture courtesy of Luftbildverlag Hans Bertram GmbH, Munich.

NEW TRENDS IN HERA PHYSICS 2003
Proceedings of the Ringberg Workshop

Copyright © 2004 by World Scientific Publishing Co. Pte. Ltd.

All rights reserved. This book, or parts thereof, may not be reproduced in any form or by any means, electronic or mechanical, including photocopying, recording or any information storage and retrieval system now known or to be invented, without written permission from the Publisher.

For photocopying of material in this volume, please pay a copying fee through the Copyright Clearance Center, Inc., 222 Rosewood Drive, Danvers, MA 01923, USA. In this case permission to photocopy is not required from the publisher.

ISBN 981-238-835-4

Printed in Singapore by World Scientific Printers (S) Pte Ltd

PREFACE

The international workshop entitled *New Trends in HERA Physics 2003* took place from 28 September to 3 October 2003 at Ringberg Castle, which overlooks Lake Tegernsee in the foothills of the Bavarian Alps, one of the most picturesque locations to be found in the whole of Germany. The castle was built during the first half of the twentieth century by Duke Luitpold in Bavaria (Herzog Luitpold in Bayern), a member of the Wittelsbach family which ruled Bavaria over 800 years, and his friend Friedrich Attenhuber, an all-round artist, architect, and interior decorator. The castle is entirely their creation, from the massive Renaissance-inspired exterior right down to the fittings and furniture, which, in every detail, were designed by Attenhuber himself and executed by native craftsmen. Attenhuber also painted every single picture exhibited in the castle. He found his models in the farmhouses around Lake Tegernsee. The castle embodies all trends of art and styles which dominated the first half of last century, combined with local Alpine originality and the individual creative power of its constructors. According to the Duke's last will, the castle passed into the hands of the Max Planck Society after his death, in 1973. The castle was then transformed into a conference venue, where scientists can exchange their latest ideas and discuss problems with their colleagues from all over the world in beautiful surroundings and in a relaxed mountain atmosphere, high above the daily business activities.

This was the fourth event in a series of Ringberg workshops on HERA physics, which was started in 1997 and was continued in 1999 and 2001. In fact, at the end of these workshops, many participants expressed the opinion that this was a successful endeavour to bring theorists and experimentalists together in order to interpret the latest HERA data, and that it would be useful to organize a follow-up workshop in the same spirit.

On the occasion of the 2003 Ringberg workshop, forty-one experts of elementary-particle physics, both theorists and experimentalists, from twenty-three universities and research institutions in fourteen countries congregated to present their latest results on the various aspects of HERA physics. Specifically, there were sixteen theorists and twenty-five experimentalists, the latter representing the H1, HERMES, and ZEUS collaborations at HERA. The topics included: proton structure function; polarized ep scattering; photoproduction of hadrons, jets, and prompt photons; final states in deep inelastic scattering, with special emphasis on hadrons, jets, resonances, and gluonic mesons; heavy-flavour and charmonium production; elastic and diffractive ep scattering; new physics at HERA; and plans for HERA-II and beyond. We hope that the high-energy-physics community will benefit from these proceedings, in which the ongoing efforts in understanding the nature of the strong interactions, with particular emphasis on HERA physics, are documented.

We wish to thank all our friends and colleagues who have contributed to these proceedings. We are indebted to the workshop secretary, Mrs. Rosita Jurgeleit, for her assistance before, during, and after the workshop and to Dr. Annette Holtkamp for her technical assistance in the editorial work. The local costs at Ringberg Castle and the costs for the publication and dissemination of these proceedings were covered in equal parts by the Deutsches Elektronen-Synchrotron at Hamburg and the Max-Planck-Institut für Physik at Munich, which we gratefully acknowledge.

Hamburg, January 2004   GÜNTER GRINDHAMMER
                                            BERND A. KNIEHL
                                            GUSTAV KRAMER
                                            WOLFGANG OCHS

CONTENTS

Preface v

1 Proton Structure

1.1 B. Portheault: High Q^2 Structure Functions and
    Parton Distributions 3
1.2 D. Kcira: Structure Functions and the Transition Region
    from Photoproduction to Deep Inelastic Scattering 15
1.3 S. Moch, J.A.M. Vermaseren and A. Vogt: Deep Inelastic
    Structure Functions at NNLO and Beyond 26
1.4 V. Chekelian: Gluon Density and Strong Coupling from
    Structure Functions and Measurement of F_L 34
1.5 S. Alekhin: Power corrections in the Deep
    Inelastic Scattering 45
1.6 J. Pumplin: The Art of Global Fitting for
    Parton Distributions 55

2 Spin Physics

2.1 M. Tytgat: New Results from HERMES 67
2.2 D. Boer: Theoretical Aspects of Spin Physics 77
2.3 U. Stösslein: Effects of $e^{\pm}$ Polarization on Final States
    at HERA 87

3 Production of Hadrons, Jets and Photons

3.1 J. Cvach: Photoproduction of Jets and Prompt Photons
    at HERA 101
3.2 G. Heinrich: Photoproduction of Isolated Photons, Single
    Hadrons and Jets at NLO 114
3.3 C. Glasman: Jet Production in Deep Inelastic ep Scattering
    at HERA 126
3.4 L. Jönsson: Forward Jet and Particle Production at
    Low x 136
3.5 M. Maniatis: Single Hadron Production in Deep
    Inelastic Scattering 150
3.6 F. Corriveau: Photo- and Electroproduction of Single
    Hadrons and Resonances 158
3.7 P. Minkowski and W. Ochs: Gluonic Meson Production . 169

4 Heavy-Flavour Production

4.1 *S. Padhi:* Open Charm and Beauty Production 183
4.2 *I. Schienbein:* Open Heavy-Flavour Photoproduction at NLO 197
4.3 *P. Thompson:* Experimental Results on Exclusive Vector Mesons and Heavy Quarkonium 207

5 Elastic and Diffractive *ep* Scattering

5.1 *S. Levonian:* Diffractive Structure Functions and QCD Fits 219
5.2 *N. Vlasov:* Diffractive Jet and Charm Production . . . 231
5.3 *R. B. Peschanski:* Beyond BFKL 241
5.4 *O. Nachtmann:* Pomeron Physics and QCD 253
5.5 *L. Motyka:* Theory of Diffractive Meson Production . . 268
5.6 *G. Gustafson:* Unintegrated Parton Densities and Applications 278
5.7 *F. Schrempp and A. Utermann:* Instanton-Driven Saturation at Small x 289

6 New Physics at HERA

6.1 *K. O. Diener:* QCD Corrections to Single Top Quark and W Boson Production at HERA 305
6.2 *E. Gallo:* Search for Beyond the Standard Model Physics at HERA 315

7 Plans for HERA-II and Beyond

7.1 *E. C. Aschenauer:* Plans for the HERMES Experiment at HERA II 327
7.2 *M. Klein:* Physics with H1 at HERA II 330
7.3 *R. Yoshida:* Prospects and Status of ZEUS at HERA II . 347
7.4 *A. C. Caldwell:* A New Round of Experiments for HERA 352

List of Participants   **365**

1
Proton Structure

HIGH Q^2 STRUCTURE FUNCTIONS AND PARTON DISTRIBUTIONS

B. PORTHEAULT

Laboratoire de l'Accélérateur Linéaire
Univertisté Paris Sud,
F-91898 Orsay Cedex
E-mail: portheau@lal.in2p3.fr

This contribution reviews the main achievements in inclusive measurements made by the H1 and ZEUS collaborations during the first phase of HERA data taking. The QCD analysis of these data by both collaborations are described. The case for a common QCD analysis is briefly discussed, with an emphasis on the possible W mass extraction.

At leading order in the electroweak (EW) interaction, the double differential cross section of inclusive Deep Inelastic Scattering (DIS) can be expressed in terms of structure functions

$$\frac{d^2\sigma_{NC}^\pm}{dxdQ^2} = \frac{2\pi\alpha^2}{xQ^4}\left[Y_+\tilde{F}_2 - y^2\tilde{F}_L \mp Y_-x\tilde{F}_3\right], \quad (1)$$

for Neutral Currents (NC) where $Y_\pm = 1 \pm (1-y)^2$, and similarly for Charged Current (CC)

$$\frac{d^2\sigma_{CC}^\pm}{dxdQ^2} = \frac{G_F^2}{4\pi x}\left[\frac{M_W^2}{Q^2+M_W^2}\right]^2\left[Y_+F_2^{CC\pm} - y^2 F_L^{CC\pm} \mp Y_-xF_3^{CC\pm}\right], \quad (2)$$

where the structure functions exhibit a dependency upon the incoming lepton charge. The QCD factorisation theorem allows the separation of the long distance physics and the short distance physics, such that the structure functions can be expressed as convolutions of universal parton distributions (pdfs) and perturbatively computable kernels.

Rich physics can be extracted with the measurement of high Q^2 inclusive cross sections. On one hand the short distance physics can be tested. For example, the data can be used to test the structure of the EW interaction and to search for new physics. On the other hand, the long distance physics can be measured. Through suitable combinations of cross section or with the help of a prediction, the structure functions can be extracted out of Eqs. (1) and (2). The case of the F_L structure function is discussed in [1]. The QCD analyses (the so-called "QCD fits") aim at extracting the pdfs through the QCD evolution.

It is also possible to extract any parameter entering in the expression of the cross section. The best example of such an extraction is α_s, but in principle it also works for EW parameters.

In this contribution, inclusive cross section measurements and structure function extractions are reviewed. The QCD analysis done by the H1 and ZEUS collaborations are then detailed. Finally the case for a common QCD analysis is presented together with a strategy for an extraction of the W mass.

1. Inclusive measurement results

The inclusive differential cross sections as a function of Q^2 is shown in Fig. 1 (see[2,3,4,5,6,7,8,9,10]).

Important information can be extracted from this figure. At low-medium Q^2, where the NC cross section is the largest, the precision of the NC cross section is very high. At very large Q^2 where $\gamma - Z^0$ interference and Z^0 exchange start to play a non-negligible role, the e^+ and e^- cross sections start clearly to differ due to the opposite sign of the xF_3 contribution, which then can be extracted. For the CC cross section, the difference between the e^+ and e^- cross sections is clear, reflecting differences in the parton distributions probed and the different helicity factors. At $Q^2 \simeq M_W^2$ the CC and NC cross sections become of the same magnitude, which illustrates the deep relationship between the normalisation constants: $G_F^2/16\pi \sim 2\pi\alpha^2/M_W^4$. This is the HERA manifestation of EW unification.

1.1. Neutral Current

The F_2 structure function is shown in Fig. 2. It exhibits the QCD pattern of the scaling violations over several orders of magnitude in x and Q^2.

The typical accuracy is 2–3% in most of the phase space. At high x and Q^2, the potentially interesting region to look for exotics, one sees the statistically limited precision of the measurement compared to the fixed target experiments.

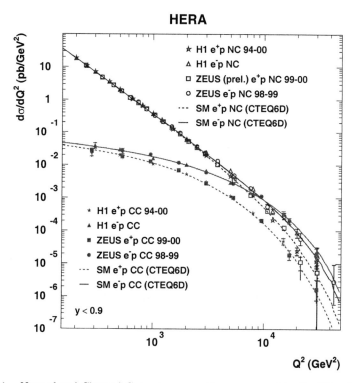

Figure 1. Neutral and Charged Current cross sections measured by the H1 and ZEUS collaborations.

The visible differences in the e^+ and e^- cross sections at high x allow the extraction of the xF_3 structure function. The results are shown in Fig. 3 in the left plot.

As $xF_3 = -a_e\chi_Z xF_3^{\gamma Z} + 2a_e v_e \chi_Z^2 xF_3^Z$ where $\chi_Z \propto Q^2/(Q^2 + M_Z^2)$, the propagator contribution and the small value of v_e have the effect that the xF_3^Z contribution to xF_3 is always below 3% in the measured range. This allows the extraction of $xF_3^{\gamma Z}$ as shown in the right plot of Fig. 3. As $xF_3^{\gamma Z} \propto 2xu_{val} + xd_{val}$ at leading order, this provides an access to the valences parton distributions. However there is still a large error of about 30% due mainly to the limited statistics of the e^- data sample available.

1.2. Charged Current

For the Charged Current process one can define a reduced cross section

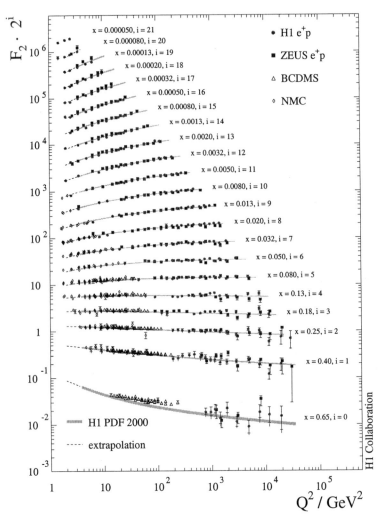

Figure 2. F_2 structure function measured by the H1 and ZEUS collaborations together with the H1PDF2000 fit. Note that the fixed target data shown are not included in the fit.

$$\tilde{\sigma}_{CC}^{\pm} = \frac{2\pi x}{G_F^2} \frac{(Q^2 + M_W^2)^2}{M_W^4} \frac{d^2 \sigma_{CC}^{\pm}}{dx dQ^2}. \quad (3)$$

The results of the H1 and ZEUS collaborations are shown in Fig. 4 for the e^+ and e^- data sets.

The good understanding of the hadronic response of the detectors has

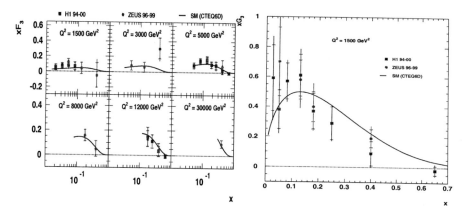

Figure 3. Left: xF_3 structure function. Right: $xF_3^{\gamma Z}$ structure function.

led to typical systematic errors of the order of 6% for the 99–00 data set. The total error is clearly dominated by statistics at the largest x and Q^2. At leading order the reduced cross sections read $\tilde{\sigma}_{CC}^+ = x\left[\bar{u} + \bar{c} + (1-y)^2(d+s)\right]$ and $\tilde{\sigma}_{CC}^- = x\left[u + c + (1-y)^2(\bar{d}+\bar{s})\right]$. This decomposition is useful to see the contribution of u–type quarks and d–type quarks to the total reduced cross section. One can see in Fig. 4 that at large x the e^+ cross section is dominated by the d quark distribution, and the e^- cross section is dominated by the u quark distribution. These data provide an important constraint at high x for the QCD analyses and are necessary for flavour separation of parton distributions.

Combining the e^+ and e^- data, the ZEUS collaboration performed an extraction of the structure function F_2^{CC} defined by $F_2^{CC} = F_2^{CC+} + F_2^{CC-}$. It is obtained with

$$F_2^{CC} = \frac{2}{Y_+}(\tilde{\sigma}_{CC}^+ + \tilde{\sigma}_{CC}^-) + \Delta(xF_3^{CC\pm}, F_L^{CC\pm}), \qquad (4)$$

where the correction $\Delta(xF_3^{CC\pm}, F_L^{CC\pm})$ is obtained with the ZEUS-S fit.

This reads $F_2^{CC} \propto u + d + s + \bar{u} + \bar{d} + \bar{s}$ which is similar to the NC expression but with an equal weight for each parton density. The result is shown in Fig. 5.

This result extends by two orders of magnitude the results of CCFR[13].

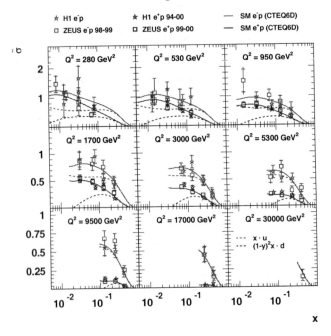

Figure 4. Charged Current reduced cross section for e^+ and e^- data.

2. QCD analysis and extraction of parton distributions

The precise data measured allow the fit of flavour separated parton distributions. All the fit details can be found in [2] and [11]. Rather than describing these details here, the main ideas and data sets used in both fits will be briefly reviewed. The emphasis will be placed only on these HERA parton distributions, and a few relevant technical details will be underlined. The possibility of a combined fit is briefly discussed.

2.1. The H1 and ZEUS QCD Fits

The ZEUS collaboration performed two fits. The first is the ZEUS–S fit, which uses the ZEUS 96–97 e^+ NC data with BCDMS[14], NMC[15], E665 proton F_2 data[16]. Deuterium data from E665 and NMC are also used with F_2^D/F_2^p results from NMC[17] and CCFR xF_3 iron data[18]. These data

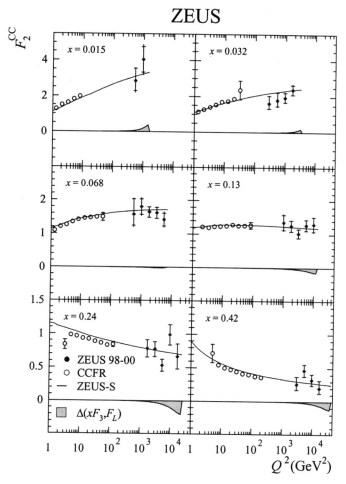

Figure 5. F_2^{CC} extraction with the ZEUS CC data and the ZEUS–S fit.

sets provide all the necessary constraints to extract flavor separated parton distributions. The ZEUS–O fit focuses on the ZEUS data and uses the NC and CC ZEUS e^+ and e^- data up to the 99 e^- data set. This is also the approach used in the H1PDF2000 fit which uses all the H1 HERA–I NC and CC data in addition to low Q^2 data from 96–97 [12]. For the use of only HERA data one has to solve technical problems due to the limitation of constraints. For the ZEUS–O fit this is achieved by arbitrarily fixing parameters to the values of the ZEUS–S fit. For the H1PDF2000 fit a novel ansatz of decomposition has been adopted, and only internal constraints between parameters have been used. Note that in any case assumptions have to be made which are the price to pay for the pdfs extraction using only HERA data.

Both fits handle the correlation of systematic errors in their parameters and errors estimations. Whereas ZEUS use the so-called offset method[19], H1 use the Pascaud-Zomer method[20]. The gist is that only a proper treatment of the correlated systematic errors allows the application of the "standard" $\Delta\chi^2 = 1$ statistical criteria for error estimate[21]. This is a major advantage with respect to the global analysis where the use of data sets providing only a total systematic error for each data point spoils the use of standard statistical tools. So the use of HERA data alone made possible a precise extraction with reliable error determination in the QCD fit.

The parametrisation of the input pdf at Q_0^2 is also a very delicate issue. The form generally adopted is $xf(x, Q_0^2) = Ax^B(1-x)^C P(x)$, where $P(x)$ may take several polynomial-like forms. It is not trivial to find a functional form for $P(x)$ such that the fit is flexible enough to ensure a good χ^2, whilst avoiding instabilities due to over-parametrisation. In the end, one has to keep in mind that any choice of parametrisation is more or less arbitrary, and that the distributions and their error depend on the parametric form chosen.

The results are shown in Fig. 6. The agreement is reasonable between the different fits given the different data sets and the different fitting schemes. From the H1PDF2000 the u-type and d-type quark distributions precisions are respectively 1% and 2% for $x = 10^{-3}$, 7% and 30% for $x = 0.65$. From the ZEUS–S fit the sea distribution precision is 5% between $x = 10^{-4}$ and 10^{-1}.

The extraction of parton densities from HERA data alone is a major achievement but the fits are at the limit of the technical fitting possibilities. This is why the use of all the combined H1 and ZEUS data could help to gain in flexibility, in particular to relax some of the assumptions made in the fitting ansatz. However one has to keep in mind that the combination would help, it may also trigger other problems as the elaboration of a fit is very delicate.

2.2. *Possibilities of a W mass extraction using a HERA fit.*

To confront the Standard Model with experimental data one needs to specify several parameters that enter in physical quantities. Several schemes

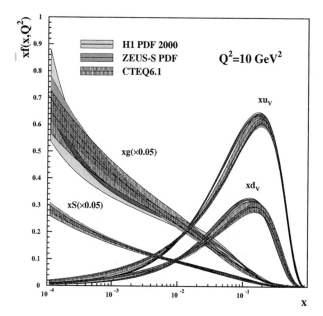

Figure 6. Parton distributions of the H1PDF2000 and ZEUS–S fits at 10 GeV^2 and comparison with the CTEQ6 fit.

are possible: the On Mass Shell (OMS) scheme uses masses as an input (α, M_W, M_Z, M_H) whereas the Modified OMS scheme uses the Fermi constant G_F instead of the W mass. The two schemes are related by the relation

$$G_F = \frac{\pi \alpha}{\sqrt{2}\left(1 - \frac{M_W^2}{M_Z^2}\right)} \times \frac{1}{1 - \Delta r(\alpha, M_W, M_Z, M_H, m_{top})} \quad (5)$$

where the radiative correction Δr is a function of the other parameters. The normalisation of the CC cross section depends on the scheme (besides the couplings of leptons to the Z^0 and its propagator normalisation). So several fitting strategies are possible. It is possible to fit M_W to the CC cross section as a "propagator mass", and this has been used several times by H1 and ZEUS, or it is possible to fit M_W in the OMS scheme as a propagator mass which also fixes the normalisation, assuming the SM is valid. This strategy was proposed in [22]. A breakthrough to reduce the uncertainty due to the proton structure is the use of a combined QCD–EW fit. This possibility is investigated by using the H1 and ZEUS data (up to the 99–00 ZEUS data set) in a QCD fit using the fitting scheme of the

... DF2000 fit. The principle is the following: a scan of W mass values is realized, and a full QCD fit is done for each W mass. The total χ^2 relative to its minimal value χ^2_{min} as a function of M_W is shown in Fig. 7. A one sigma experimental (statistical and systematic) error with a $\Delta\chi^2 = 1$ error criterion is used, which is possible due to the careful treatment of the correlated systematic errors.

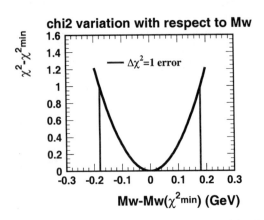

Figure 7. χ^2 shape obtained with a combined H1 and ZEUS QCD fit for variations of the W mass.

The error obtained is 190 MeV, which corresponds to $\delta M_W/M_W = 0.2\%$. The use of M_W as entering in the normalisation of the CC cross section allows this significant improvement with respect to previous "propagator mass" fits with a few percent precision. However note that the validity of this result is associated with the validity of the QCD analysis itself. In particular some theoretical uncertainty contribution to the total error may be present.

3. Summary and outlook

Many cornerstone results in DIS have been achieved in the HERA–I phase of data taking. The inclusive NC and CC cross sections have been measured with a good accuracy, although there is scope for considerable improvements using higher statistics and longitudinal lepton polarisation. These data have already been used in many QCD fits. However there is still an unexploited potential in a combined QCD analysis of the H1 and ZEUS data. This combination could settle many important physics issues such as the α_s value, or the determination of the gluon distribution. There is still the possibility of QCD–EW combined fits that could be used from now on to deliver the HERA physics message on parton distributions and Standard Model parameters.

Acknowledgments

The author wishes to thank his colleagues in the H1 and ZEUS collaborations for the measurements presented in this paper. He also thanks the organisers of the workshop for the invitation and for the unique hospitality at the Ringberg castle.

References

1. V. Shekelian, these proceedings
2. C. Adloff et al. [H1 Collaboration],
    Eur. Phys. J. C **13**, 609 (2000)
    [arXiv:hep-ex/9908059].
3. C. Adloff et al. [H1 Collaboration],
    Eur. Phys. J. C **19**, 269 (2001)
    [arXiv:hep-ex/0012052].
4. C. Adloff et al. [H1 Collaboration],
    arXiv:hep-ex/0304003.
5. J. Breitweg et al. [ZEUS Collaboration],
    Eur. Phys. J. C **12**, 411 (2000)
    [Erratum-ibid. C **27**, 305 (2003)]
    [arXiv:hep-ex/9907010].
6. S. Chekanov et al. [ZEUS Collaboration],
    Eur. Phys. J. C **21**, 443 (2001)
    [arXiv:hep-ex/0105090].
7. S. Chekanov et al. [ZEUS Collaboration],
    Phys. Lett. B **539**, 197 (2002)
    [Erratum-ibid. B **552**, 308 (2003)]
    [arXiv:hep-ex/0205091].
8. S. Chekanov et al. [ZEUS Collaboration],
    Eur. Phys. J. C **28**, 175 (2003)
    [arXiv:hep-ex/0208040].
9. S. Chekanov et al. [ZEUS Collaboration],
    arXiv:hep-ex/0307043.
10. A. Lopez-Duran-Viani [ZEUS Collaboration],
    Prepared for 9th International Workshop on Deep Inelastic Scattering (DIS 2001), Bologna, Italy, 27 Apr - 1 May 2001
11. S. Chekanov et al. [ZEUS Collaboration],
    Phys. Rev. D **67** (2003) 012007
    [arXiv:hep-ex/0208023].
12. C. Adloff et al. [H1 Collaboration],

Eur. Phys. J. C **21** (2001) 33
[arXiv:hep-ex/0012053].
13. U. K. Yang *et al.* [CCFR/NuTeV Collaboration], Phys. Rev. Lett. **86** (2001) 2742 [arXiv:hep-ex/0009041].
14. A. C. Benvenuti *et al.* [BCDMS Collaboration],
Phys. Lett. B **223**, 485 (1989).
15. M. Arneodo *et al.* [New Muon Collaboration],
Nucl. Phys. B **483** (1997) 3
[arXiv:hep-ph/9610231].
16. M. R. Adams *et al.* [E665 Collaboration],
Phys. Rev. D **54** (1996) 3006.
17. M. Arneodo *et al.* [New Muon Collaboration],
Nucl. Phys. B **487** (1997) 3
[arXiv:hep-ex/9611022].
18. W. C. Seligman *et al.*, Phys. Rev. Lett. **79** (1997) 1213.
19. M. Botje, Eur. Phys. J. C **14** (2000) 285 [arXiv:hep-ph/9912439].
20. C. Pascaud and F. Zomer, LAL 95-05
21. D. Stump *et al.*,
Phys. Rev. D **65** (2002) 014012
[arXiv:hep-ph/0101051].
22. H. Spiesberger,
J. Phys. G **28** (2002) 1155.

STRUCTURE FUNCTIONS AND THE TRANSITION REGION FROM PHOTOPRODUCTION TO DEEP INELASTIC SCATTERING

DORIAN KCIRA

University of Wisconsin
ZEUS/DESY
Notkestrasse 85
22607 Hamburg
GERMANY
E-mail: dorian.kcira@desy.de

Precise inclusive and dijet measurements are presented that extend the kinematic domain of HERA and cover the transition region from deep inelastic scattering to photoproduction. The rise of the structure function F_2 at low x persists but is shallower at lower Q^2. Dijet measurements can significantly constrain the parton distribution functions of the virtual photon.

1. Introduction

In deep inelastic ep scattering (DIS) at HERA, a photon of negative squared momentum (virtuality, Q^2) is exchanged. The highly virtual pointlike photon is used as a probe and precise determinations of proton structure are performed by inclusive measurements. Measurements of jet cross sections in DIS provide further information, are sensitive to the proton structure and allow precise tests of perturbative QCD (pQCD).

In photoproduction ($Q^2 \simeq 0$), the requirement of events with high transverse energy jets provides for a hard scale in the interaction and allows comparison to pQCD calculations. In photoproduction the measurements are sensitive to both the structure of the proton and the structure of the (quasi-real) photon.

Both the DIS and photoproduction regimes have been extensively studied at HERA and pQCD predictions have been successful in describing the data. The physical picture and the language used in the two regimes is not the same. Therefore, data and predictions in the region between them are particularly interesting.

Until recently, not much data had been gathered at HERA in the transition region between DIS and photoproduction mainly because of the experimental difficulty of identifying the scattered lepton at small scattering angles. Recently, the H1 and ZEUS Collaborations have performed measurements that "scan" this region using inclusive and dijet events. It must be noted that these are conceptually two different ways of crossing the transition region. In the case of inclusive measurements the transition is from a hard regime (DIS, with hard scale Q^2) to a soft regime (photoproduction). In the case of dijet measurements the requirement of high transverse energy jets allows application of pQCD throughout the transition from real ($Q^2 \simeq 0$) photons, which can have structure, to virtual pointlike photons. In this second transition two hard scales play a role: Q and E_T, the transverse energy scale of the jets. Depending on the interplay of these two hard scales, the photon may or may not display a structure. Both these types of measurements are presented in this paper.

2. Inclusive Measurements

Measurements of the proton structure function F_2 in the region of $Q^2 \gg 1$ GeV^2 have shown a good description of the data by perturbative QCD (pQCD) [1,2,3]. On the other hand the region $Q^2 < 1$ GeV^2 is described by models based on generalized vector meson dominance [4] and Regge theory [5]. Therefore, the region of Q^2 around 1 GeV^2 deserves particular attention.

In this section the latest HERA measurements in the transition region from low to high Q^2 are presented together with comparisons to perturbative and non-perturbative approaches. These measurements use different experimental techniques for accessing the low Q^2 data, which are reflected in the titles of the subsections.

2.1. *Initial State Radiation Analysis*

Events with QED initial state radiation (ISR) have been used to measure F_2. The measurement is based on the data taken during 1996 with the ZEUS detector (3.78 pb^{-1}) [6]. The emission of a hard photon from the initial-state positron leads to a reduction in the center-of-mass energy of the hard scattering particles. This means that lower values of Q^2 can be accessed.

The F_2 measurements are shown in Fig. 1 (left plot) as a function of x at different values of Q^2 together with the most recently published F_2 measurements [7,8] and the BPC/BPT [9] results. The data sets are in good

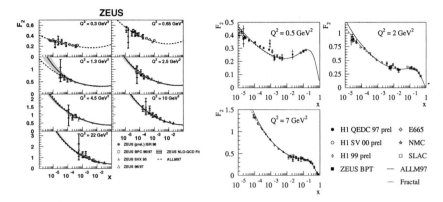

Figure 1. Left plot: F_2 versus x for fixed Q^2. The results of the ISR analysis are shown together with previously published ZEUS results. The yellow band shows the predictions from the ZEUS NLO QCD fit while the dashed curve those from ALLM97. Right plot: F_2 measurement from QEDC analysis (closed circles) compared with other HERA and fixed target measurements. The solid line is the prediction from the ALLM97 parametrization while the dashed line represents the fractal fit, plotted for $y > 0.0003$.

agreement in the region of overlap, completing the coverage of the x-Q^2 plane at low Q^2. The results are compared to the predictions of the ZEUS NLO QCD fit for $Q^2 \geq 1.3$ GeV^2 and to the predictions of ALLM97 [10] over the whole Q^2 range. Both predictions give a good description of the data.

2.2. QED Compton Analysis

The H1 Collaboration has performed measurements of F_2 using QED Compton scattering events[11]. The measurements use e^+p data with an integrated luminosity of 9.25 pb^{-1} collected during 1997.

The QED Compton process corresponds to low Q^2 and a large virtual electron mass. The final state topology is given by an azimuthal back-to-back configuration of the outgoing positron and photon detected under rather large scattering angles. Their transverse momenta balance such that very low values of Q^2 are experimentally accessible.

The F_2 values measured in QED Compton scattering are depicted in Fig. 1 (right plot) as a function of x at fixed Q^2 and compared to other HERA and fixed target data. The kinematic range is extended at very low Q^2 towards higher x values, complementing the standard inclusive and shifted vertex measurements. The results overlap with the region measured in fixed target experiments. Good agreement is observed.

2.3. Shifted Vertex Analysis

A special run was performed in 2000 at HERA in which the interaction vertex was shifted towards the proton beam direction by 70 cm. This increased the acceptance for positrons with smaller scattering angles. This again allows lower values of Q^2 to be accessed. The data presented here[12] correspond to an integrated luminosity of 0.6 pb^{-1}, a factor of four larger than the previous shifted vertex data samples at H1[13] and ZEUS [8]. The combination of larger statistics and the new Backward Silicon Tracker (BST) allowed the H1 Collaboration to make measurements with higher precision than previously.

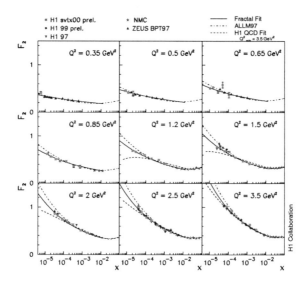

Figure 2. F_2 from H1 shifted vertex data (squares), H1 99 and 97 nominal vertex data (points and triangles), larger x data from BPC/BPT (triangles) and from NMC (stars).

In Fig. 2 the F_2 data are compared with phenomenological predictions obtained from fits to previous data. The best description is given by the fractal model parametrization [14] (solid curve) the parameters of which were determined by fitting the ZEUS BPC/BPT data and the published low Q^2 H1 data [1]. The ALLM97 parametrization (dashed-dotted curves) similarly used previous data which leads to an acceptable description of the present F_2 data with a slightly stronger rise with decreasing x for $Q^2 \geq 1$ GeV^2

than is observed in the H1 data. The prediction from the H1 NLO QCD fit [1] is also shown (dashed curves). This fit only included data with $Q^2 \geq 3.5$ GeV^2. Its backward extrapolation undershoots measurements at lower Q^2. At $Q^2 \leq 1$ GeV^2 the strong coupling constant $\alpha_S(Q^2)$ becomes large, making a comparison with NLO QCD meaningless.

In Fig. 3 (left plot) the photon-proton cross section, $\sigma_{\gamma^* p}$, which is proportional to F_2/Q^2, is plotted as a function of Q^2 at different values of $W^2 \simeq sy$. W is the invariant mass of the $\gamma^* p$ system. The new data fill the gap between previous measurements at lower and higher Q^2 and reach similar accuracy. The same three parametrizations are plotted as in Fig. 2. At $Q^2 \simeq 1$ GeV^2 the ALLM97 parametrization is too large at high W and the QCD fit fails, as expected. For $Q^2 \to 0$, the fractal model fit is too steep to describe the total photoproduction cross section (not shown here). However, this could be cured by considering mass effects in the fractal Ansatz [15].

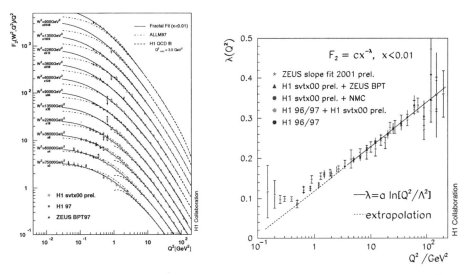

Figure 3. Left plot: measurements of the structure function F_2 represented as F_2/Q^2 compared to the Fractal fit, H1 QCD fit and the ALLLM97 parametrization. Right plot: determination of the exponent $\lambda(x, Q^2)$ from fits to F_2 as measured from H1 and ZEUS. The straight line represents a linear fit to the measurements for $Q^2 \geq 3.5$ GeV^2.

2.4. The Rise of F_2 at low x

To quantify the behaviour of the rise of F_2 towards low x at fixed Q^2 the H1 and ZEUS Collaborations have measured the derivative:

$$\left(\frac{\partial \ln F_2(x, Q^2)}{\partial \ln x}\right) \equiv -\lambda(Q^2)$$

Using the data presented in this paper, the measurement of $\lambda(Q^2)$ is extended to the transition region as plotted in Fig. 3 (right plot). For values $Q^2 < 1$ GeV^2 the derivative $\lambda(Q^2)$ is seen to significantly deviate from an extrapolation of the linear behaviour determined from the data in the deep inelastic scattering region ($Q^2 \geq 3.5$ GeV^2). From soft hadronic interaction models, it is expected that for $Q^2 \to 0$, $\lambda \to 0.08$. HERA data agree with these expectations.

3. Dijet Measurements

It has been established that the real photon ($Q^2 \simeq 0$) has a partonic structure [16]. In contrast, in hard collisions at high Q^2 the photon is considered to be pointlike. In leading order (LO) QCD two types of processes contribute to jet photoproduction: direct, in which the photon couples as a pointlike particle to quarks in the hard scatter; and resolved, in which the photon acts as a source of partons. Both processes lead to two jets in the final state. The x_γ^{obs} variable:

$$x_\gamma^{\text{obs}} = \frac{\sum_{\text{jets}}(E^{\text{jet}} - p_z^{\text{jet}})}{\sum_{\text{hadrons}}(E - p_z)},$$

which measures the fraction of the photon momentum participating in the production of the dijet system, is used to separate the two processes since resolved (direct) processes dominate at low (high) x_γ^{obs} values [17]. In the above definition the upper sum runs over the two jets with highest transverse energy and the lower sum runs over all final state hadrons in the laboratory frame.

Within fixed-order QCD, only pointlike photon interactions are expected to contribute to jet production in DIS. However, two scales play a role in the interaction: Q and the jet transverse energy, E_T^{jet}. QCD predicts that for high Q^2 ($Q^2 \gg (E_T^{\text{jet}})^2$) the photon will behave as a pointlike particle. For $Q^2 \ll (E_T^{\text{jet}})^2$ the photon may present a partonic structure, even for relatively large values of Q^2, which is resolved at a scale related to the transverse energy of the jets. Therefore, resolved processes may contribute significantly to the total cross section.

Dijet measurements in photoproduction, DIS and the transition region are presented in this section. Different LO and NLO pQCD calculations are compared with the data.

The H1 Collaboration has measured dijet cross sections in the $\gamma^* p$ center-of-mass frame using a sample of 16.3 pb^{-1} data collected in 1999[18]. The phase space is defined by: $2 < Q^2 < 80$ GeV^2, the inelasticity $0.1 < y < 0.85$, the transverse energy of the two leading jets $E_T^{\text{jet}1,2} > 5$ GeV, $\overline{E}_T = (E_T^{\text{jet}1} + E_T^{\text{jet}2})/2 > 6$ GeV and the pseudorapidity of the two leading jets $-2.5 < \eta^{\text{jet}1,2} < 0$.

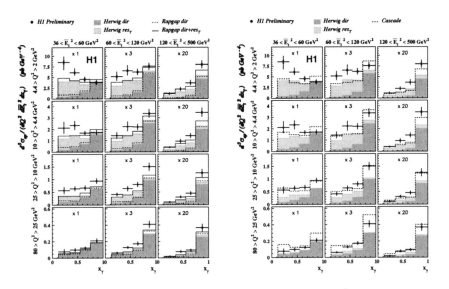

Figure 4. Triple differential cross section as a function of x_γ^{obs}, $\overline{E}_T^2$ and Q^2. The data are compared to HERWIG and RAPGAP MC predictions (left plot). Separately shown are the components of the cross section from direct (dir) and resolved γ^*. Predictions of the HERWIG and CASCADE MC are also compared to the data (right plot).

The triple differential dijet cross sections measured as a function of Q^2, $\overline{E}_T^2$, and x_γ^{obs} are shown in Fig. 4. The prediction of HERWIG [19] and RAPGAP [20] Monte Carlo (MC) simulations with the SaS 1D[21] parametrization of the virtual photon parton distribution function (PDF), as well as the pure direct contributions are compared to the data in Fig. 4 (left plot). In general HERWIG and RAPGAP tend to underestimate the measured cross sections. The direct contributions reasonably describe the data in the highest Q^2 bin, while a clear need for resolved processes is observed for $Q^2 \ll \overline{E}_T^2$. The measured cross sections are compared in Fig. 4 (right plot) to the predictions of the CASCADE [22] MC program, which uses the CCFM evolution scheme. This theoretical approach is based on non-k_T-ordered parton cascades and does not include the concept of virtual photon structure. CASCADE describes the data reasonably well but not perfectly.

The ZEUS Collaboration has studied the dependence of dijet production on Q^2 by measuring dijet cross sections in the range $0 \lesssim Q^2 < 2000$ GeV^2 using the 1996-1997 data sample[23], which corresponds to an integrated luminosity of 38.6 pb^{-1}. Inclusive dijet cross sections were measured for jets with transverse energy $E_T^{1,2} > 7.5, 6.5$ GeV and pseudorapidities in the photon-proton center-of-mass frame in the range $-3 < \eta^{jet} < 0$.

Fig. 5 presents the differential dijet cross section, $d\sigma/dQ^2$, for the direct-enhanced region ($x_\gamma^{obs} > 0.75$) and the resolved-enhanced region ($x_\gamma^{obs} < 0.75$) together with the total dijet cross section. The measurements are precise and cover a wide range of photon virtualities, including the transition region from photoproduction to DIS. As expected, the cross section for $x_\gamma^{obs} < 0.75$ falls more rapidly than that for $x_\gamma^{obs} > 0.75$. Even though the dijet cross section is dominated by interactions with $x_\gamma^{obs} > 0.75$ for $Q^2 \gtrsim 10$ GeV^2, there is still a contribution of approximately 24% from low-x_γ^{obs} events even for Q^2 as high as 500 GeV^2. The NLO QCD calculations of DISASTER++ [24] with $\mu_R^2 = Q^2 + (E_T^{jet})^2$ are compared to the measurements. The photoproduction measurement is compared with the NLO prediction from Frixione and Ridolfi where E_T^2 is used as the scale. The measured cross section for $x_\gamma^{obs} > 0.75$ is reasonably well described by the calculation but the measured cross section for $x_\gamma^{obs} < 0.75$ is significantly underestimated for high Q^2.

Fig. 6 shows the ratio $R = \frac{d\sigma}{dQ^2}(x_\gamma^{obs} < 0.75)/\frac{d\sigma}{dQ^2}(x_\gamma^{obs} > 0.75)$ as a function of Q^2 in three different regions of $\overline{E}_T^2$. The Q^2 dependence of the data is stronger at low $\overline{E}_T^2$ than at higher $\overline{E}_T^2$, showing that the low x_γ^{obs} component is suppressed at low Q^2 as $\overline{E}_T^2$ increases and vice-versa. Predictions from HERWIG assuming the SaS 1D and GRV[25] parametrizations of the photon PDFs are compared to the data in Fig. 6 (left plot). The SaS 1D parametrization contains the suppression of the virtual photon structure with increasing virtuality. The predictions fall with increasing Q^2 and reproduce the fall of the data relatively well. On the other hand the predictions using GRV which assumes no suppression of the virtual photon

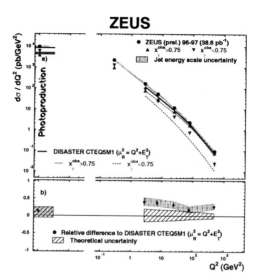

Figure 5. a) Measured dijet cross-sections $d\sigma/dQ^2$ for $x_\gamma^{\rm obs} > 0.75$ (upwards triangles) $d\sigma/dQ^2$ for $x_\gamma^{\rm obs} < 0.75$ (downwards triangles) and $d\sigma/dQ^2$ for the whole $x_\gamma^{\rm obs}$ region (black dots). The shaded band represents the uncertainty in the absolute energy scale of the jets. The NLO QCD calculations of DISASTER++ ($\mu_R^2 = Q^2 + (E_T^{\rm jet})^2$) and Frixione ($\mu_R^2 = (E_T^{\rm jet})^2$) are shown for each cross-section. b) Relative difference of the measured cross-section to the DISASTER++ and Frixione calculations. The hatched band shows the theoretical uncertainty of the calculations.

structure, are relatively constant at the level of the photoproduction ratio of the data. The NLO calculations are compared to the data in Figure 6 (right plot). For photoproduction, the calculations lie below the data for all $\overline{E_T}^2$. The NLO-prediction which used GRV for the photon PDFs is closer to the data than the one using AFG [26]. However, they underestimate the relative cross section at low $x_\gamma^{\rm obs}$. The DISASTER++ predictions show some suppression of the low $x_\gamma^{\rm obs}$ contribution with increasing Q^2, however they underestimate the proportion of low $x_\gamma^{\rm obs}$ relative to high $x_\gamma^{\rm obs}$.

4. Summary and Conclusions

New precise measurements of the structure function F_2 have been performed extending the kinematic domain of HERA and covering the transition region from DIS to photoproduction. The data are in agreement with the previous HERA data at higher and lower Q^2 and those from fixed targed experiments in the region of overlap.

Differential dijet cross sections have been measured in a large range of

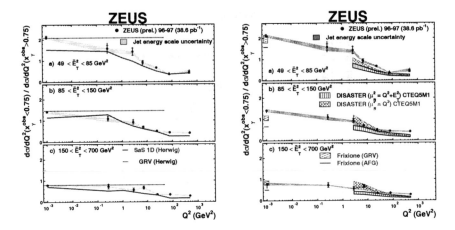

Figure 6. Measured ratio $R = \frac{d\sigma}{dQ^2}(x_\gamma^{obs} < 0.75)/\frac{d\sigma}{dQ^2}(x_\gamma^{obs} > 0.75)$ as a function of Q^2 in different regions of $\overline{E}_T^2$ (black dots). The LO calculations of HERWIG with using the SaS 1D and GRV photon pdfs (left plot) as well as the NLO predictions from DISASTER++ and Frixione-Ridolfi (right plot) are also shown.

photon virtualities, including the transition region from photoproduction to DIS. The data are reasonably well described by DGLAP LO models which introduce virtual photon structure, suppressed with increasing Q^2. The non-k_T-ordered CCFM-based approach also does a relatively good job in describing the measurements. These data have the potential to significantly constrain virtual photon PDFs and should be taken into account in future fits. The currently available NLO QCD calculations have large uncertainties at low Q^2, where the presence of the resolved photon contribution is expected. Improved higher-order or resummed calculations are needed. The NLO QCD predictions generally underestimate the cross section at low x_γ^{obs} relative to that of high x_γ^{obs}.

References

1. H1 Collaboration, C. Adloff et al., *Eur. Phys. J.* **C 21**, 33 (2001), [hep-ex/0012053].
2. H1 Collaboration, C. Adloff et al., *Eur. Phys. J.* **C 30**, 1 (2003).
3. ZEUS Collaboration, S. Chekanov et al., *Phys. Rev.* **D 67**, 012007 (2003).
4. J.J. Sakurai and D. Schildknecht, *Phys. Lett.* **B 40**, 121 (1972).
5. P.D.B. Collins, *An Introduction to Regge Theory and High Energy Scattering*, Cambridge University Press (1977).
6. ZEUS Collaboration, *Measurement of the proton structure functions F_2 and*

F_L *using initial state radiative events at HERA*, **EPS 2003 Conference**, Abstract 502 (2003).
7. ZEUS Collaboration, S. Chekanov et al., *Eur. Phys. J.* **C 21**, 443 (2001).
8. ZEUS Collaboration, J. Breitweg et al., *Eur. Phys. J.* **C 7**, 609 (1999).
9. ZEUS Collaboration, J. Breitweg et al., *Phys. Lett.* **B 487**, 53 (2000).
10. H. Abramowitz and A. Levy, *DESY 97-251* (1997).
11. H1 Collaboration, *Measurement of the Proton Structure Function F_2 at low Q^2 in QED Compton Scattering at HERA*, **EPS 2003 Conference**, Abstract 084 (2003).
12. H1 Collaboration, *Measurement of the Inclusive Deep Inelastic Scattering Cross Section at $Q^2 \simeq 1~GeV^2$ with the H1 Experiment*, **EPS 2003 Conference**, Abstract 082 (2003).
13. H1 Collaboration, C. Adloff et al., *Nucl. Phys.* **B 497**, 3 (1997).
14. T. Lastovicka, *Self-similar Properties of the Proton Structure at low x Eur. Phys. J.* **C** , to appear, [hep-ph/0203260] (2002).
15. T. Lastovicka, in "DIS 2002 - 10th Int. Workshop on Deep Inelastic Scattering", to appear in *Acta Phys. Polonica* **B**
16. PLUTO Collaboration, C. Berger et al., *Nucl. Phys.* **B 281**, 365 (1987).
17. ZEUS Collaboration, M. Derrick et al., *Phys. Lett.* **B 322**, 287 (1994).
18. H1 Collaboration, *Measurement of Dijet Cross-Sections at Low Q^2 at HERA*, **EPS 2003 Conference**, Abstract 085 (2003).
19. G. Marchesini et al., HERWIG: A MONTE CARLO EVENT GENERATOR FOR SIMULATING HADRON EMISSION REACTIONS WITH INTERFERING GLUONS, Comp. Phys. Comm. **67**, 465 (1992).
20. H. Jung, *Hard Diffractive Scattering in High Energy ep Collisions and the Monte Carlo Generator* RAPGAP, Comp. Phys. Comm. **86**, 147 (1995).
21. G.A. Schuler and T. Sjöstrand, *Parton Distributions of the Virtual Photon*, *Phys. Lett.* **B 376**, 193 (1996).
22. H. Jung and G. P. Salam, *Hadronic Final State Predictions from CCFM: The Hadron-Level Monte Carlo Generator* CASCADE, *Eur. Phys. J.* **C 19**, 351 (2001).
23. ZEUS Collaboration, *The Q^2 Dependence of Dijet Production at HERA*, **EPS 2003 Conference**, Abstract 585 (2003).
24. D. Graudenz, *DISASTER++ Version 1.0*, hep-ph/9710244, (1997).
25. M. Glück, E. Reya and A. Vogt, *Photonic Parton Distributions*, *Phys. Rev.* **D 46**, 1793 (1992).
26. P. Aurenche, J.P. Guillet and M. Fontannaz, *Parton Distributions in the Photon*, Z. Phys. **C 64**, 621 (1994).

DEEP-INELASTIC STRUCTURE FUNCTIONS AT NNLO AND BEYOND

S. MOCH

Deutsches Elektronensynchrotron DESY
Platanenallee 6, D-15738 Zeuthen, Germany

J.A.M. VERMASEREN AND A. VOGT

NIKHEF Theory Group
Kruislaan 409, 1098 SJ Amsterdam, The Netherlands

We have calculated the fermionic contributions to the flavour non-singlet structure functions in deep-inelastic scattering (DIS) at third order of massless perturbative QCD. We obtain complete results for the corresponding n_f-parts of the three-loop anomalous dimension and the three-loop coefficient functions for the structure functions F_2 and F_L. Our results agree with all partial and approximate results available in the literature. We discuss their implications for the threshold resummation at the next-to-next-to-leading logarithmic accuracy.

1. Introduction

The calculation of perturbative QCD corrections for deep-inelastic structure functions is an important task. The present and expected future experimental precision, for instance at HERA, calls for complete next-to-next-to-leading order (NNLO) predictions. These offer the possibility to determine the strong coupling constant α_s and to analyze the proton structure and its parton content with unprecedented precision. Knowledge of the latter is of particular importance for the analysis of hard scattering reactions at future LHC experiments.

At present, this level of accuracy is not yet possible, because the necessary anomalous dimensions governing the parton evolution at NNLO are not fully known. The two-loop coefficient functions of F_2, F_3 and F_L have been calculated some time ago [1,2,3,4,5], but for the corresponding three-loop anomalous dimensions $\gamma_{pp'}^{(2)}$, only a finite number of fixed Mellin moments [6,7,8] are available. As a first step towards the complete calculation, we have computed the fermionic three-loop contributions to the flavour non-singlet (NS) structure functions F_2 and F_L in unpolarized electromagnetic deep-inelastic scattering [9]. Already these results have immediate consequences for threshold resummation of soft gluons which we will discuss in the following.

2. Results

We are considering only the non-singlet structure functions and the method for the calculation of their moments closely follows ref. [6]. Hence there is not much need to explain the physics of the method here again. Thus we will discuss only the differences introduced by the fact that we now compute all moments simultaneously as a function of the moment number N. Since N is not a fixed constant, we cannot resort to the techniques of ref. [6], where the Mincer program [10,11] was used as the tool to solve the integrals. Instead, we will have to introduce new techniques. However, we can give N a positive integer value at any point of the derivations and calculations, after which the Mincer program can be invoked to verify that the results are correct. From a practical point of view this is the most powerful feature of the Mellin-space approach, as it greatly simplifies the checking of all programs.

Similar to the fixed-N computations of refs. [7,8], the diagrams are generated automatically with a special version of the diagram generator QGRAF [12]. For all the symbolic manipulations of the formulae we use the latest version of the program FORM [13]. The calculation is performed in dimensional regularization [14,15,16,17] with $D = 4 - 2\epsilon$. Hence the unrenormalized Mellin-space results will be functions of ϵ, N, and the values $\zeta_{3,...,5}$ of the Riemann ζ-function. The renormalization is carried out in the $\overline{\text{MS}}$-scheme [18,19] as described in ref. [6].

The major difference to the fixed-moment calculations is now that we obtain much longer results due to the presence of the parameter N in the answer. We have checked the correctness of each individual diagram for several values of N, by comparing with the results of a Mincer calculation. In addition we have compared the complete renormalized results with the results in the literature for the available values of N.

The perturbative expansion of the non-singlet coefficient functions and anomalous dimensions can be written as

$$C_{i,\mathrm{ns}}(\alpha_\mathrm{s}, N) = \sum_{n=0}^{\infty} \left(\frac{\alpha_\mathrm{s}}{4\pi}\right)^n c_{i,\mathrm{ns}}^{(n)}(N) \;, \qquad (1)$$

$$\gamma_{\mathrm{ns}}(\alpha_\mathrm{s}, N) = \sum_{n=0}^{\infty} \left(\frac{\alpha_\mathrm{s}}{4\pi}\right)^{n+1} \gamma_{\mathrm{ns}}^{(n)}(N) \qquad (2)$$

with $i = 2, L$ in eq. (1).

Our N-space results are expressed in terms of harmonic sums $S_{\vec{m}}(N)$. In the following all harmonic sums are understood to have the argument N, i.e., we employ the short-hand notation $S_{\vec{m}} \equiv S_{\vec{m}}(N)$. In addition we use operators $\mathbf{N}_\pm$ and $\mathbf{N}_{\pm \mathbf{i}}$ which shift the argument N of a given function by ± 1 or a larger integer i,

$$\mathbf{N}_\pm f(N) = f(N \pm 1) \;, \qquad \mathbf{N}_{\pm \mathbf{i}} f(N) = f(N \pm i) \;. \qquad (3)$$

We normalize the trivial leading-order (LO) coefficient function and recover, of course, the well-known result for the LO anomalous dimension [20,21]

$$c_{2,\mathrm{ns}}^{(0)}(N) = 1 \;, \qquad (4)$$

$$\gamma_{\mathrm{ns}}^{(0)}(N) = C_F \big(2(\mathbf{N}_- + \mathbf{N}_+) S_1 - 3\big) \;. \qquad (5)$$

At next-to-next-to-leading order we have computed all fermionic contributions to the splitting function $\gamma_{\text{ns}}^{(2)}$,

$$\gamma_{\text{ns}}^{(2)}(N) = 16\, C_A C_F n_f \left(\frac{3}{2}\zeta_3 - \frac{5}{4} + \frac{10}{9}S_{-3} - \frac{10}{9}S_3 + \frac{4}{3}S_{1,-2} - \frac{2}{3}S_{-4} \right.$$
$$+ 2S_{1,1} - \frac{25}{9}S_2 + \frac{257}{27}S_1 - \frac{2}{3}S_{-3,1} - \mathbf{N}_+\left[S_{2,1} - \frac{2}{3}S_{3,1} \right.$$
$$\left. - \frac{2}{3}S_4 \right] + (1-\mathbf{N}_+)\left[\frac{23}{18}S_3 - S_2 \right] - (\mathbf{N}_- + \mathbf{N}_+)\left[S_{1,1} \right.$$
$$+ \frac{1237}{216}S_1 + \frac{11}{18}S_3 - \frac{317}{108}S_2 + \frac{16}{9}S_{1,-2} - \frac{2}{3}S_{1,-2,1} - \frac{1}{3}S_{1,-3}$$
$$\left.\left. - \frac{1}{2}S_{1,3} - \frac{1}{2}S_{2,1} - \frac{1}{3}S_{2,-2} + S_1\zeta_3 + \frac{1}{2}S_{3,1} \right] \right) + 16\, C_F n_f^2 \left(\frac{17}{144} \right.$$
$$- \frac{13}{27}S_1 + \frac{2}{9}S_2 + (\mathbf{N}_- + \mathbf{N}_+)\left[\frac{2}{9}S_1 - \frac{11}{54}S_2 + \frac{1}{18}S_3 \right] \right)$$
$$+ 16\, C_F^2 n_f \left(\frac{23}{16} - \frac{3}{2}\zeta_3 + \frac{4}{3}S_{-3,1} - \frac{59}{36}S_2 + \frac{4}{3}S_{-4} - \frac{20}{9}S_{-3} \right.$$
$$+ \frac{20}{9}S_1 - \frac{8}{3}S_{1,-2} - \frac{8}{3}S_{1,1} - \frac{4}{3}S_{1,2} + \mathbf{N}_+\left[\frac{25}{9}S_3 - \frac{4}{3}S_{3,1} - \frac{1}{3}S_4 \right]$$
$$+ (1-\mathbf{N}_+)\left[\frac{67}{36}S_2 - \frac{4}{3}S_{2,1} + \frac{4}{3}S_3 \right] + (\mathbf{N}_- + \mathbf{N}_+)\left[S_1\zeta_3 \right.$$
$$- \frac{325}{144}S_1 - \frac{2}{3}S_{1,-3} + \frac{32}{9}S_{1,-2} - \frac{4}{3}S_{1,-2,1} + \frac{4}{3}S_{1,1} + \frac{16}{9}S_{1,2}$$
$$\left.\left. - \frac{4}{3}S_{1,3} + \frac{11}{18}S_2 - \frac{2}{3}S_{2,-2} + \frac{10}{9}S_{2,1} + \frac{1}{2}S_4 - \frac{2}{3}S_{2,2} - \frac{8}{9}S_3 \right] \right)$$
(6)

We havesubjected our results to a number of checks. First of all, we have calculated some lower even moments in an arbitrary covariant gauge with the Mincer program [10,11], keeping the gauge parameter ξ in the gluon propagator. All dependence on ξ does cancel in the final results. Secondly the n_f^2-contribution to $\gamma_{\text{ns}}^{(2)}$ is known from the work of Gracey [22] and we agree with his result. Furthermore the coefficients of $\ln^k N$, $k = 3,\ldots 5$, of $c_{2,\text{ns}}^{(3)}(N)$ agree with the prediction of the soft-gluon resummation [23]. Finally, we have checked the result of each individual diagram for several integer values of N by comparing with the results of a Mincer calculation. Thus, as the strongest check, our results reproduce the fixed even moments $N = 2,\ldots,14$ computed in refs. [6,7,8].

3. Threshold resummation

It is well known that perturbative QCD corrections to structure functions receive large logarithmic corrections, which originate from the emission of soft gluons. These corrections are relevant at large values of the scaling variable x (in Mellin space at large values of the Mellin moment N) and can be resummed to all orders in perturbation theory. It is interesting to investigate the implications of our three-loop results [9] for the threshold exponentiation [24,25,26] at next-to-next-leading logarithmic (NNLL) accuracy [27].

Here the quark coefficient function for F_2 can, up to terms which vanish for $N \to \infty$, be written as

$$C_2(\alpha_s, N) = (1 + a_s g_{01} + a_s^2 g_{02} + \ldots) \exp[G^N(Q^2)], \qquad (7)$$

where the resummation exponent G^N can be expanded as

$$G^N(Q^2) = L g_1(a_s L) + g_2(a_s L) + a_s g_3(a_s L) + \ldots \qquad (8)$$

with $a_s = \alpha_s/(4\pi)$ and $L = \ln N$. The functions g_l depend on universal coefficients $A_{i \leq l}$ and $B_{i \leq l-1}$ and process-dependent parameters $D_{i \leq l-1}^{\rm DIS}$ (see e.g. ref. [27] for the precise definitions of the functions $g_{1,2,3}$).

In a physical picture, the resummation of the perturbative expansion for C_2 rests upon the refactorization of C_2 (valid in the threshold region of phase space) into separate functions of the jet-like, soft, and off-shell quanta that contribute to its quantum corrections. Each of the functions organizes large corrections corresponding to a particular region of phase space [28].

To NNLL accuracy, the function g_3 involves the new coefficients A_3, B_2 and $D_2^{\rm DIS}$. These coefficients can be fixed by expanding eq. (7) in powers of α_s and comparing to the result of the full fixed-order calculation [9]. In the $\overline{\rm MS}$ scheme, the parameter A_3 is simply the coefficient of $\ln N$ in $\gamma_{\rm ns}^{(2)}(N)$ or, equivalently, of $1/(1-x)_+$ in $P_{\rm ns}^{(2)}(x)$. It reads

$$A_3 = (1178.8 \pm 11.5) \qquad (9)$$
$$+ C_A C_F n_f \left[-\frac{836}{27} + \frac{160}{9} \zeta_2 - \frac{112}{3} \zeta_3 \right] + C_F^2 n_f \left[-\frac{110}{3} + 32 \zeta_3 \right]$$
$$+ C_F n_f^2 \left[-\frac{16}{27} \right],$$

where $C_A = 3$, $C_F = 4/3$ and n_f is the number of light (massless) flavours. The estimate of the non-fermionic part [27] is based on the approximations of $P_{\rm ns}^{(2)}(x)$ constructed in ref. [29] using the first six even-integer moments [8] and its small-x limit [30]. The exact fermionic part has been obtained independently in refs. [9,31], while the n_f^2 contribution is already known from ref. [22].

The complete results for B_2 and D_2^{DIS} can be inferred from fermionic result of the three-loop coefficient function $c_{2,\text{ns}}^{(3)}$ in ref. [9], yielding

$$B_2 = C_F^2 \left[-\frac{3}{2} - 24\zeta_3 + 12\zeta_2 \right] + C_F C_A \left[-\frac{3155}{54} + 40\zeta_3 + \frac{44}{3}\zeta_2 \right] \quad (10)$$
$$+ C_F n_f \left[\frac{247}{27} - \frac{8}{3}\zeta_2 \right] ,$$

$$D_2^{\text{DIS}} = 0 . \quad (11)$$

As a matter of fact, the contribution to $c_{2,\text{ns}}^{(3)}$ involves only a linear combination, $\beta_0(B_2 + 2 D_2^{\text{DIS}})$, with β_0 being the one-loop coefficient of the QCD β-function. However, the different combination $B_2 + D_2^{\text{DIS}}$ has been determined in ref.[27] by comparing the expansion of eq. (7) to the two-loop coefficient function $c_{2,\text{ns}}^{(2)}$ of ref. [32]. Thus, B_2 and D_2^{DIS} can be disentangled. It is interesting to observe the vanishing of D_1^{DIS} and D_2^{DIS}, for which an all-order generalization has been proposed in ref. [33,34]. This is in contrast to the Drell-Yan process, where the functions D_l^{DY} are generally different from zero. For instance, D_2^{DY} has been derived in ref. [27].

Let us briefly illustrate numerically for large N the improvement due to the NNLL corrections for the soft gluon exponent G^N of deep-inelastic scattering. In fig. 1 on the left, we show the resummation exponent $G^N(Q^2)$ of eq. (8). Here, we choose $\mu_r^2 = \mu_f^2 = Q^2$, $n_f = 4$ and $\alpha_s = 0.2$, which corresponds to scales between about 25 and 50 GeV^2, depending on the precise value of $\alpha_s(M_Z^2)$. In fig. 1 on the right, we display the convolution with a schematic, but typical input evaluated with the so-called 'minimal-prescription' contour [35]. It is obvious from both figures that knowledge of the leading logarithmic (LL) and next-to-leading logarithmic (NLL) terms [24,25,26] alone, i.e., those enhanced by factors $\ln N(\alpha_s \ln N)^n$ and $(\alpha_s \ln N)^n$, is not sufficient for reliably determining the function G^N and its impact after convolution even for rather large values of N and x.

The NNLL corrections discussed here are rather small over a wide range. This indicates that the soft-gluon exponent $G^N(Q^2)$ stabilizes and that the soft-gluon effects on the $\overline{\text{MS}}$ quark coefficient function can be reliably estimated. Recall, however, that the NNLL corrections are large for the 'physical kernel' governing the scaling violations of the non-singlet structure function $F_{2,\text{ns}}$ [36].

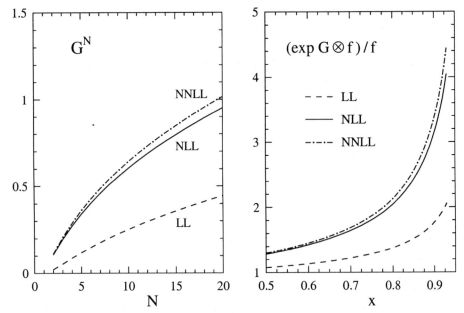

Figure 1. Left: The LL, NLL and NNLL approximations for the resummation exponent $G^N(Q^2)$ in eq. (8) at $\mu_r^2 = \mu_f^2 = Q^2$ for $\alpha_s(Q^2) = 0.2$ and four flavours. Right: These results convoluted with a typical input shape $xf = x^{1/2}(1-x)^3$.

References

1. W. L. van Neerven and E. B. Zijlstra, Phys. Lett. **B272**, 127 (1991).
2. E. B. Zijlstra and W. L. van Neerven, Phys. Lett. **B273**, 476 (1991).
3. E. B. Zijlstra and W. L. van Neerven, Phys. Lett. **B297**, 377 (1992).
4. E. B. Zijlstra and W. L. van Neerven, Nucl. Phys. **B383**, 525 (1992).
5. S. Moch and J. A. M. Vermaseren, Nucl. Phys. **B573**, 853 (2000), hep-ph/9912355.
6. S. A. Larin, T. van Ritbergen, and J. A. M. Vermaseren, Nucl. Phys. **B427**, 41 (1994).
7. S. A. Larin, P. Nogueira, T. van Ritbergen, and J. A. M. Vermaseren, Nucl. Phys. **B492**, 338 (1997), hep-ph/9605317.
8. A. Retey and J. A. M. Vermaseren, Nucl. Phys. **B604**, 281 (2001), hep-ph/0007294.
9. S. Moch, J. A. M. Vermaseren, and A. Vogt, Nucl. Phys. **B646**, 181 (2002), hep-ph/0209100.
10. S. G. Gorishnii, S. A. Larin, L. R. Surguladze, and F. V. Tkachev, Comput. Phys. Commun. **55**, 381 (1989).
11. S. A. Larin, F. V. Tkachev, and J. A. M. Vermaseren, NIKHEF-H-91-18.

12. P. Nogueira, J. Comput. Phys. **105**, 279 (1993).
13. J. A. M. Vermaseren, (2000), math-ph/0010025.
14. G. 't Hooft and M. Veltman, Nucl. Phys. **B44**, 189 (1972).
15. C. G. Bollini and J. J. Giambiagi, Nuovo Cim. **12B**, 20 (1972).
16. J. F. Ashmore, Lett. Nuovo Cim. **4**, 289 (1972).
17. G. M. Cicuta and E. Montaldi, Nuovo Cim. Lett. **4**, 329 (1972).
18. G. 't Hooft, Nucl. Phys. **B61**, 455 (1973).
19. W. A. Bardeen, A. J. Buras, D. W. Duke, and T. Muta, Phys. Rev. **D18**, 3998 (1978).
20. D. J. Gross and F. Wilczek, Phys. Rev. **D8**, 3633 (1973).
21. H. Georgi and H. D. Politzer, Phys. Rev. **D9**, 416 (1974).
22. J. A. Gracey, Phys. Lett. **B322**, 141 (1994), hep-ph/9401214.
23. A. Vogt, Phys. Lett. **B471**, 97 (1999), hep-ph/9910545.
24. G. Sterman, Nucl. Phys. **B281**, 310 (1987).
25. S. Catani and L. Trentadue, Nucl. Phys. **B327**, 323 (1989).
26. S. Catani and L. Trentadue, Nucl. Phys. **B353**, 183 (1991).
27. A. Vogt, Phys. Lett. **B497**, 228 (2001), hep-ph/0010146.
28. H. Contopanagos, E. Laenen, and G. Sterman, Nucl. Phys. **B484**, 303 (1997), hep-ph/9604313.
29. W. L. van Neerven and A. Vogt, Phys. Lett. **B490**, 111 (2000), hep-ph/0007362.
30. J. Blumlein and A. Vogt, Phys. Lett. **B370**, 149 (1996), hep-ph/9510410.
31. C. F. Berger, Phys. Rev. **D66**, 116002 (2002), hep-ph/0209107.
32. T. Matsuura, S. C. van der Marck, and W. L. van Neerven, Nucl. Phys. **B319**, 570 (1989).
33. S. Forte and G. Ridolfi, Nucl. Phys. **B650**, 229 (2003), hep-ph/0209154.
34. E. Gardi and R. G. Roberts, Nucl. Phys. **B653**, 227 (2003), hep-ph/0210429.
35. S. Catani, M. L. Mangano, P. Nason, and L. Trentadue, Nucl. Phys. **B478**, 273 (1996), hep-ph/9604351.
36. W. L. van Neerven and A. Vogt, Nucl. Phys. **B603**, 42 (2001), hep-ph/0103123.

GLUON DENSITY AND STRONG COUPLING FROM STRUCTURE FUNCTIONS AND MEASUREMENT OF F_L

VLADIMIR CHEKELIAN (SHEKELYAN)

Max-Planck-Institut für Physik
Werner-Heisenberg-Institut
Föhringer Ring 6, 80805 München, Germany
E-mail: shekeln@mppmu.mpg.de

In the first phase of data taking at the HERA *ep* collider various structure functions were measured. The precision of the measurements allows to make comprehensive analyses in the framework of QCD. Recent results on the gluon distribution function, the strong coupling constant $\alpha_s(M_Z^2)$ and the longitudinal structure function F_L are reviewed as well as studies of limits of applicability of the perturbative QCD approach.

1. Introduction

The study of the structure of hadrons has been a powerful means for establishing and testing of the theory of strong interactions and for the determination of the partonic structure of the nucleon. In Quantum Chromodynamics (QCD), structure functions are defined as a convolution of the universal parton momentum distributions inside the proton and coefficient functions, which contain information about the exchanged boson-parton interaction. At sufficiently large four-momentum transfer squared, Q^2, when the strong coupling α_s is small, a perturbative technique is applicable for QCD calculations of the coefficient and splitting functions. The latter represent the probability of a parton to emit another parton. Expressions for structure functions are determined by convolution integrals of appropriate sums over quarks of different flavours and the gluon, and they predict a logarithmic Q^2 dependence (evolution) of the structure functions. The perturbative QCD (pQCD) calculations are well established within the DGLAP [1] formalism to next-to-leading (NLO) order in the strong coupling and are close to completion to next-to-next-to-leading (NNLO) order [2].

Precision measurements of structure functions in a wide kinematic range allow comprehensive QCD analyses to determine the quark and gluon distributions inside the proton and the strong coupling constant $\alpha_s(M_Z^2)$. Such measurements allow also to explore pQCD application limits, for example at low x, where the parton density was found to rise steeply towards small x by the H1 and ZEUS experiments at HERA. Here, x is the fraction of proton momentum carried by the parton. In QCD this rise corresponds to an increase of the gluon density and is expected to slow down at highest energies (small x) due to gluon-gluon recombination, the so-called saturation effects.

Partons, i.e. quarks and gluons, enter differently into different structure functions. In deep inelastic scattering (DIS), the cross section of the neutral current (NC) process, mediated by a γ or Z^0-boson exchange, can be expressed in terms of three structure functions:

$$\frac{d^2\sigma^{e^\pm p \to e^\pm X}}{dxdQ^2} = \frac{2\pi\alpha^2}{Q^4 x}\left[Y_+ F_2(x,Q^2) - y^2 F_L(x,Q^2) \mp Y_- xF_3(x,Q^2)\right], \quad (1)$$

where $y = Q^2/xs$ is the inelasticity, s is the center of mass energy squared, α is the fine structure constant and $Y_\pm = 1\pm(1-y)^2$. The dominant contribution to the cross section is due to the proton structure function F_2, which is related in the framework of the quark-parton model (QPM) to the sum of the proton momentum fractions carried by the quarks and antiquarks in the proton weighted by the quark charges squared. The longitudinal structure function F_L, vanishing in the QPM, is directly sensitive to the gluon momentum distribution in the proton. The structure function xF_3 is related to valence quarks and is sizable only at large Q^2, with Q^2 comparable with the Z^0-boson mass squared.

In the first phase of data taking at the HERA ep collider (HERA I), the H1 and ZEUS experiments collected about 100 pb^{-1} of integrated luminosity each in the e^+p mode and about 15 pb^{-1} in the e^-p mode. These data provided information about structure functions, which are analysed in the framework of perturbative QCD. Further insight into the structure of the proton is obtained in semi-inclusive processes such as diffraction, heavy flavour and jet production, which are calculable in pQCD. This report concentrates on recent results on the gluon distribution function, the strong coupling constant $\alpha_s(M_Z^2)$ and checks for novel QCD effects.

2. QCD analyses of structure functions

The HERA I inclusive deep inelastic cross section measurements at low Q^2 and high Q^2 are reviewed in [3,4]. The determination of parton distribution functions (PDFs) in QCD analyses of these data, supplemented by data from fixed target experiments, is discussed in [4]. It is interesting to note that available measurements of the NC and CC (charged current) cross sections in $e^{\pm}p$ interactions at HERA allow PDF QCD fits [5,6], which lead to a quark flavour decomposition of parton densities based on HERA data only.

The improved quality of the recent QCD fits is achieved by more precise data and a new level of sophistication in the fitting techniques including a full treatment of correlated experimental systematic uncertainties. A new feature of the recent QCD analyses is also a pragmatic determination of PDF uncertainties and their physical predictions. One of the problems of the uncertainty estimation for fits with many data sets involved is related to a certain degree of inconsistency among them. Usually the one sigma error of a parameter in a fit is determined by variation of χ^2 by one unit from the minimum. However, the level of "inconsistency" should be reflected in the uncertainties. This can by achieved by the following modification of the above criterion: $\Delta\chi^2 = T^2$ [7,8,9], where T stands for a tolerance which is estimated from the level of (in)consistency of the data sets used in a particular QCD fit. For example, in the CTEQ6 fit the tolerance was taken to be 10 ($\Delta\chi^2 = 100$).

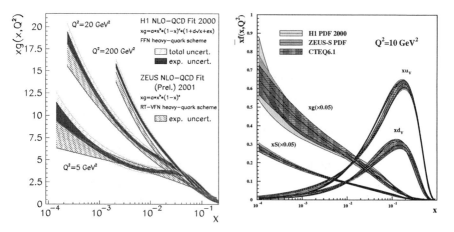

Figure 1. Gluon distributions for the dedicated xg&α_s fit of H1 and the ZEUS-S PDF fit for Q^2=5, 20, and 200 GeV^2 (left). Parton distributions for the H1 PDF 2000, ZEUS-S PDF and CTEQ6 fits for Q^2=10 GeV^2 (right). On the right plot the sea and gluon distributions are divided by a factor of 20.

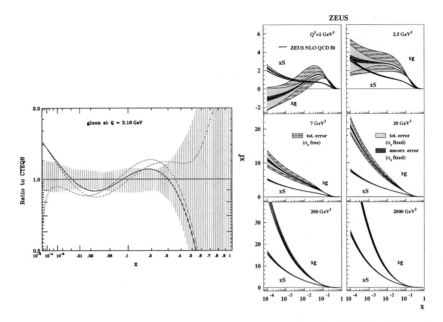

Figure 2. Relative gluon uncertainty as determined in the CTEQ6 fit (left). Gluon and sea distributions for the ZEUS-S PDF fit (right).

The H1 collaboration performed also a QCD fit [10] devoted especially to the determination of the gluon distribution and α_s ($xg\&\alpha_s$ fit). A novel flavour decomposition of F_2 used in this fit allows to reduce the number of parton distributions to just two combinations (apart from the gluon density): "valence-like" and "sea-like". Thus, for this fit there is no need to include deuteron data in the fit, and the number of data sets can be reduced to a minimum: the H1 cross section data covering low x and the BCDMS μp data covering the region of large x (see also section 2.2).

2.1. *Gluon distribution*

The gluon distributions for different Q^2 from the dedicated $xg\&\alpha_s$ fit [10] of H1, the H1 PDF 2000 [5] and the ZEUS-S PDF [6] fits are shown in Figure 1. The agreement of the different fits is reasonable, though there are clear differences in the shape of the gluon distribution beyond the quoted error bands. The CTEQ6 fit, shown in Figure 1 (right), bridges the H1 and ZEUS results. The relative uncertainty band for the gluon distribution obtained in the CTEQ6 fit is about 10% at $x < 0.3$, Figure 2 (left).

The differences between the H1 and ZEUS fits have many sources. The

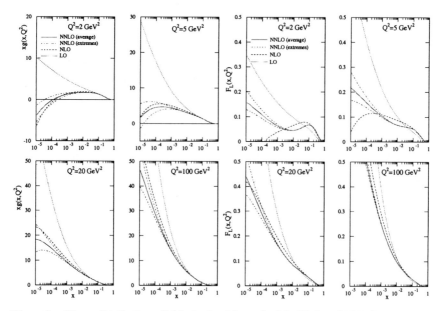

Figure 3. Gluon distributions (left) obtained from the LO, NLO and NNLO analyses of MRST and corresponding predictions for F_L (right) as a function of x for various values of Q^2.

assumptions and procedures are very different. This concerns for example the data sets used in the fits and the form of parameterisation of the parton densities at the starting scale for QCD evolution. Up to now there is no clear prescription how to take the last item properly into account in the uncertainty calculations. The HERA data themselves look very consistent, but it is not excluded that fits are sensitive to small features of the measurements.

The DGLAP evolution leads to a gluon distribution which rises towards small x at Q^2 above a few GeV^2, Figure 1. The gluon and sea distributions for the ZEUS-S PDF fit at different Q^2, including $Q^2=1$ GeV^2, are shown in Figure 2 (right). At the lowest Q^2 the rise of the sea density persists, while the gluon density flattens out at $Q^2=2.5$ GeV^2 and even becomes negative at $Q^2=1$ GeV^2. A simular feature is observed in the MRST NLO fit [8,11] shown in Figure 3 (left). While this is not necessarily a problem in itself, the gluon density from scaling violations is not an observable, it results in a distinct unphysical prediction for F_L shown in Figure 3 (right). Therefore, F_L measurements at HERA, discussed in section 3.3, are of a particular interest.

2.2. Strong coupling constant $\alpha_s(M_Z^2)$

Values for the strong coupling constant $\alpha_s(M_Z^2)$, determined in recent QCD analyses of DIS data, are summarised in Table 1. The results are rather consistent. However, the experimental uncertainties of $\alpha_s(M_Z^2)$ are very different due to different judgments on the $\Delta\chi^2$ criterion. The experimental errors are excellent (about 2%), if the canonical criterion $\Delta\chi^2 = 1$ is used in the fit.

The criterion $\Delta\chi^2 = 1$ can be applied only after careful consistency checks of the data sets used in the fit. For example in the dedicated $xg \& \alpha_s$ H1 fit only data from H1 and BCDMS μp are involved. Furthermore, for the BCDMS μp data a cut $y_\mu > 0.3$ is applied removing a region with large systematics uncertainties, where the deviations from the QCD fit (with full treatment of correlated systematics errors) exceed the systematic errors. Partial χ^2 of the H1 and BCDMS data points are shown as a function of $\alpha_s(M_Z^2)$ in Figure 4 (right). The χ^2 minima for both data sets are fully consistent and lie around $\alpha_s(M_Z^2) = 0.115$. The importance of the HERA data for the determination of α_s is illustrated in Figure 4 (left), where χ^2 for fits to "only H1" and "only BCDMS" data are shown. Without HERA data the gluon distribution from the fit to the BCDMS data becomes unrealistically flat at low x and drives $\alpha_s(M_Z^2)$ to about 0.110. The HERA data pin down the gluon density, and, as a result, $\alpha_s(M_Z^2)$ moves to 0.115 if both data sets are used in the fit.

Table 1. The strong coupling constant from NLO and NNLO QCD fits.

NLO QCD	$\Delta\chi^2$	$\alpha_s(M_Z^2)$
CTEQ6 [7]	100	$0.1165 \pm 0.0065(exp)$
ZEUS [6]	≈ 50	$0.1166 \pm 0.0008(uncor) \pm 0.0032(cor) \pm 0.0036(norm)$ $\pm 0.0018(model) \pm 0.004(theory)$
MRST01 [8]	20	$0.1190 \pm 0.002(exp) \pm 0.003(theory)$
H1 [10]	1	$0.1150 \pm 0.0017(exp)^{+\ 0.0009}_{-\ 0.0005}(model) \pm 0.005(theory)$
Alekhin [12]	1	$0.1171 \pm 0.0015(exp) \pm 0.0033(theory)$
NNLO QCD	$\Delta\chi^2$	$\alpha_s(M_Z^2)$
Alekhin [12]	1	$0.1143 \pm 0.0014(exp) \pm 0.0013(theory)$

Most of the $\alpha_s(M_Z^2)$ determinations in Table 1 are performed in the NLO QCD approach. A limitation of the precision of $\alpha_s(M_Z^2)$ is due to the associated theoretical error calculated by varying the renormalisation and

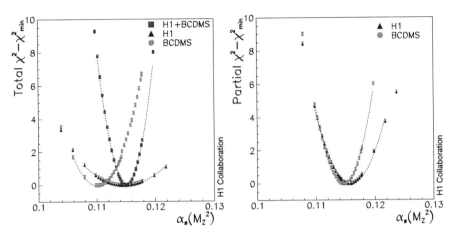

Figure 4. Total χ^2 in the $xg\&\alpha_s$ fits to H1 and BCDMS data separately and in the common fit to both data sets (left). Partial χ^2 contributions of the H1 and BCDMS data in the common fit (right).

factorisation scales. In the NNLO theory this uncertainty is reduced by a factor of 3, as one can see by comparing Alekhin's NNLO and NLO results, see also ref. [2].

3. The Low-x Regime

The region of low x accessed at HERA has been the subject of much debate (e.g. ref. [13]). The steep rise of F_2 discovered already in the first HERA measurements indicates a new partonic regime governed by a rising gluon density. This high parton density regime brings a possibility to reveal novel QCD effects related to e.g. unitarisation. The interpretation of experimental results in the lowest x domain should take into account the kinematical correlation between low x and low Q^2 at HERA. The latter domain brings its own complications related to the transition from the perturbative DIS regime to the photoproduction limit, where the perturbative approach is not working due to the large α_s and missing terms in theoretical calculations. The present experimental situation in inclusive measurements at low x obtained from detailed studies of derivatives of F_2 in $\ln x$ and $\ln Q^2$ and measurements of the longitudinal structure function F_L is discussed below.

3.1. *Rise of F_2 at low x*

In the double asymptotic limit, the DGLAP evolution equation can be solved and F_2 is expected to rise approximately as a power of x towards

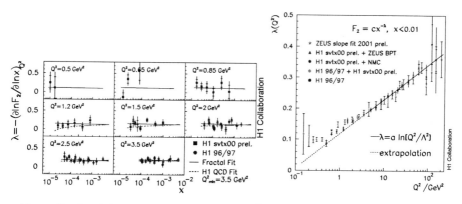

Figure 5. Local derivatives $\lambda = -(\partial \ln F_2(x,Q^2)/\partial \ln x)_{Q^2}$ (left) and fitted values of $\lambda(Q^2)$ (right).

low x. A power behaviour is also predicted in BFKL theory [14]. A damping of this rise would indicate the presence of novel QCD effects. A relevant observable for the investigation of the dynamics of this growth is the partial derivative of F_2 w.r.t. $\ln x$ at fixed Q^2

$$\lambda = -(\partial \ln F_2(x,Q^2)/\partial \ln x)_{Q^2} \ . \tag{2}$$

The high precision of the present F_2 data allowed H1 to measure this observable locally [15,16], as it is shown in Figure 5 (left) for the low Q^2 region. The measurements are consistent with no dependence of λ on x for $x < 0.01$. Thus, the monotonic rise of F_2 persists down to the lowest x measured at HERA, and no evidence for a change of this behaviour such as a damping of the growth is found.

The observed independence of the local derivatives in $\ln x$ at fixed Q^2 suggests that F_2 can be parameterised in a very simple form

$$F_2 = c(Q^2) x^{-\lambda(Q^2)} \ . \tag{3}$$

The results for $\lambda(Q^2)$ are shown in Figure 5 (right). The coefficient $c(Q^2) \approx 0.18$ and the parameterisation $\lambda(Q^2) = a \cdot \ln(Q^2/\Lambda^2)$ for $Q^2 \geq 2$ GeV^2 are in accord with pQCD predictions. In contrast, at $Q^2 \leq 1$ GeV^2 the behaviour is changing, and, in the low Q^2 limit, λ is approaching 0.08, which is expected from the energy dependence of soft hadronic interactions $\sigma_{tot} \sim s^{\alpha_P(0)-1} \approx s^{0.08}$.

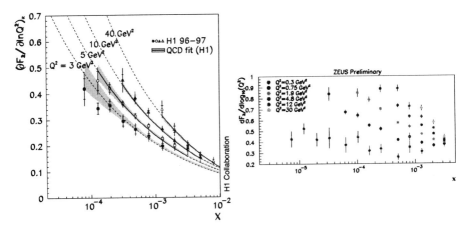

Figure 6. The derivative $(\partial F_2/\partial \ln Q^2)_x$ as function of x for different Q^2.

3.2. Scaling violations of F_2

Another important quantity in view of possible non-linear gluon interaction effects is the derivative

$$(\partial F_2(x, Q^2)/\partial \ln Q^2)_x \qquad (4)$$

describing scaling violations. Its behaviour in x is a direct reflection of the gluon density dynamics in the associated kinematic range.

The derivatives measured by H1 and ZEUS are shown in Figure 6 as a function of x for different Q^2. They show a continuous growth towards low x down to lowest $Q^2 = 0.3$ GeV^2, without an indication of a change in the dynamics. The derivatives are well described by the pQCD calculations for $Q^2 \geq 3$ GeV^2.

3.3. Longitudinal structure function $F_L(x, Q^2)$

Non-zero values of the structure function F_L appear in pQCD due to gluon radiation. Thus, F_L is the most appropriate quantity to test QCD to NLO and especially to examine the pathological effects of the negative gluon distribution discussed in section 2.1.

According to eq. 1, the F_L contribution to the inclusive cross section is significant only at high y. The conventional way to measure F_L is to explore the y dependence of the cross section at given x and Q^2 by changing the center of mass energy of the interaction. Such a measurement is not yet performed at HERA. The H1 collaboration developed other ways to

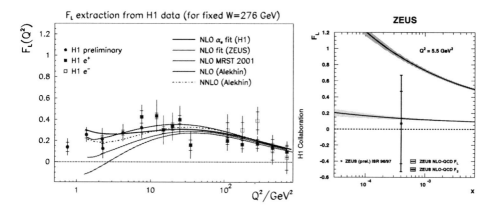

Figure 7. Summary of the F_L measurements by H1 at fixed photon-proton center of mass energy $W = 276$ GeV, $W \approx \sqrt{sy}$ (left). The F_L measurement by ZEUS using radiative events (right).

determine F_L from the measured turn over of the cross section at high y. This becomes possible due to the H1 measurements at very low scattered electron energy down to 3 GeV. The determinations of F_L are based on various methods to propagate F_2 measurements to the region corresponding to high y (low x).

A summary of the F_L measurements by H1 [17] is shown in Figure 7 (left). The results are significantly above zero everywhere, including the lowest Q^2. They are compared with pQCD predictions and different phenomenological models showing that already at the present level of precision the measurements can distinguish different calculations.

The ZEUS collaboration used DIS radiative events with detection of the photon from the initial state radiation (ISR) process in a zero angle photon tagger. ISR effectively reduces the energy of the electron participating in the interaction. The result [18] is shown in Figure 7 (right). Large errors of the measurement suggest that precise direct measurements of F_L at HERA can be performed only by reducing the beam energy and employing the highest y domain.

4. Conclusions

The inclusive DIS cross sections from HERA I cover a kinematic phase space extended towards low x and high Q^2 by two orders of magnitude compared with fixed target experiments. These data, studied in the framework of perturbative QCD in the form of NLO DGLAP analyses, provide an accurate determination of the gluon distribution function. The strong coupling constant $\alpha_s(M_Z^2)$ determined in such analyses has small experimental errors and is currently limited in precision by the accuracy of the NLO theory. No evidence for new QCD effects like saturation has been found in inclusive processes at low x.

The HERA II program, aiming for a 10-fold increase of statistics with a longitudinally polarised electron beam, with much improved H1 and ZEUS detectors, is just starting. A special run with lower proton beam energies is envisaged for direct measurements of F_L at low x.

References

1. Y. L. Dokshitzer, Sov. Phys. JETP **46** (1977) 641;
   V. N. Gribov and L. N. Lipatov, Sov. J. Nucl. Phys. **15** (1972) 438 and 675;
   G. Altarelli and G. Parisi, Nucl. Phys. B **126** (1977) 298.
2. S. Moch, Proc. of Ringberg Workshop "New Trends in HERA Physics 2003".
3. D. Kcira, Proc. of Ringberg Workshop "New Trends in HERA Physics 2003".
4. B. Partheault, Proc. of Ringberg Workshop "New Trends in HERA Physics 2003".
5. H1 Collab., C. Adloff et al., Eur. Phys. J. C **30** (2003) 1.
6. ZEUS Collab., S. Chekanov et al., Phys. Rev. **D67** (2003) 012007.
7. J. Pumplin, D. R. Stump, J. Huston, H. L. Lai, P. Nadolsky and W. K. Tung, JHEP **0207** (2002) 012.
8. A. D. Martin, R. G. Roberts, W. J. Stirling and R. S. Thorne, Eur. Phys. J. C **23** (2002) 73.
9. R. S. Thorne et al., J. Phys. G **28** (2002) 2717.
10. C. Adloff et al. [H1 Collaboration], Eur. Phys. J. C **21** (2001) 33.
11. A. D. Martin, R. G. Roberts, W. J. Stirling and R. S. Thorne, Phys. Lett. B **531** (2002) 216.
12. S. I. Alekhin, JHEP **0302** (2003) 015.
13. K. Golec-Biernat and M. Wusthoff, Phys. Rev. D **59** (1999) 014017;
    J. Bartels, K. Golec-Biernat and H. Kowalski, Phys. Rev. D **66** (2002) 014001.
14. E. A. Kuraev, L. N. Lipatov and V. S. Fadin, Sov. Phys. JETP **44** (1976)443;
    E. A. Kuraev, L. N. Lipatov and V. S. Fadin, Sov. Phys. JETP **45** (1977) 199;
    Y. Y. Balitsky and L. N. Lipatov, Sov. Journ. Nucl. Phys. **28** (1978) 822.
15. C. Adloff et al. [H1 Collaboration], Phys. Lett. B **520** (2001) 183.
16. H1 Collaboration, contributed paper to EPS03 (2003), Aachen, abstract 082.
17. H1 Collaboration, contributed paper to EPS03 (2003), Aachen, abstract 083.
18. ZEUS Collaboration, contrib. paper to EPS03(2003), Aachen, abstract 502.

POWER CORRECTIONS IN THE DEEP INELASTIC SCATTERING

S. ALEKHIN

Institute for High Energy Physics (IHEP),
Pobeda, 1, Protvino 142281 Russia
E-mail: alekhin@sirius.ihep.su

We outline the impact of the power corrections on the study of deep inelastic scattering and describe the determination of the power corrections from the existing deep inelastic scattering data.

One of the main ingredients of the high energy phenomenology of the hadrons interaction is factorization. It allows to greatly simplify theoretical calculations using the connection of different processes through convolution of the coefficient functions (i.e. parton cross sections) with the parton distributions functions (PDFs). However factorization is valid only for the hard interactions, i.e. in the limit of large masses, transverse momenta, etc. For the soft interactions it is broken due to parton correlations being non-negligible. The formalism for the description of the factorization breakdown is given by the operator-product expansion (OPE)[1]. In terms of this expansion the factorized part of the cross section corresponds to the contribution of the leading-twist (LT) operators and the factorization-breaking part – to the higher-twist (HT) ones. The coefficient functions for the LT terms and the dependence of the PDFs on the transverse momentum Q can be calculated in the pQCD-improved parton model. In accordance to these calculations the LT (twist-2) structure functions of deep inelastic scattering (DIS) demonstrate logarithmic Q-dependence. The Q-dependence of the HT terms is mainly defined by the factor of $1/Q^{\tau-2}$, where τ is the corresponding value of twist. Due to the $1/Q$ factors the HT terms rapidly fall with Q and in the limit of large Q the LT terms give the dominant contribution. However the coefficients at the power-like factors cannot be calculated using existing approaches and therefore the on-set scale of this asymptotic regime is a priori unclear. The importance of this problem was recognized just after first attempts to confront the pQCD predictions to the earliest available data since it was understood that the measured Q-dependence of the DIS structure functions can be equally well described both by the log-like and the power-like behavior[2]. With more precise measurements in the wide region of Q, which appeared later, we can better separate the LT and the HT terms, however the impact of the HT terms on different aspects of the DIS phenomenology is not understood completely. In this report we describe the modern status of this problem and outline possible ways to resolve it.

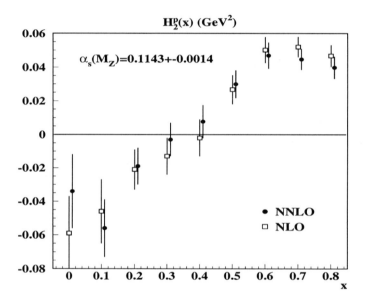

Figure 1. The x-dependence of twist-4 term in the proton structure function F_2 determined from the fit of Ref.[20] to the combined SLAC-BCDMS-NMC-H1-ZEUS data. The value of α_s given in the panel was obtained from the same fit.

For DIS the OPE up to twist-4 terms reads

$$F_i(x, Q^2) = F_i^{\text{LT,TMC}}(x, Q^2) + \frac{H_i(x, Q^2)}{Q^2} + O\left(1/Q^4\right), \quad (1)$$

where i marks different structure functions and $F^{\text{LT,TMC}}$ are the LT terms corrected for the target mass effects[3]. These target mass effects give contribution to the power-like terms of the OPE, but are regularly considered separately since they are just of kinematical origin and can be expressed as a convolution of the LT terms with the coefficient functions proportional to negative powers of Q (see[4]). The functions H give the dynamical HT contributions and correspond to the operators, which include correlations of the quark and gluon fields[5]. Calculation of matrix elements of these operators in field theory is very difficult[6] and only very limited results have been obtained so far[7,8]. The predictions of most of the models for the HT terms contain essential free parameters and cannot be used without a fit to the data. For this reason the main source of information about the HT comes from phenomenological studies. These studies are regularly based on the QCD extrapolation from the high-Q region to the low-Q one, where a discrepancy between the extrapolation and data is considered as a manifestation of the HT terms. This approach requires accurate treatment of all sources of theoretical uncertainties, which can have an impact on the extrapolation, and therefore to bias the results or even generate fake HT terms. Besides, since the the existing DIS data from any one single experiment do not cover the whole kinematic region necessary for the study of such kind, the combination of all available data is desirable. In this case the fake HT terms can be also generated due to incomparability of the separate data sets, especially if the kinematics of different experiments do not overlap.

The first firm evidence of the HT dynamical terms was found in[9]. In this analysis the twist-4 terms in the structure functions F_2 for proton and deuteron were determined from the next-to-leading order (NLO) QCD fit to the combined SLAC-BCDMS data with account of the HT terms parameterized in the factorized form:

$$F_2(x,Q^2) = F_2^{LT,TMC}(x,Q^2)\left[1 + \frac{h_2(x)}{Q^2}\right]. \quad (2)$$

This form is equivalent to (1) modulo the impact of the anomalous dimensions of LT and HT operators. The latter must exist as follows from the theoretical expectations[10] and were indeed found in the phenomenological analysis of[11]. However at $x \leq 0.75$, which are relevant for the analysis of[9], the results of fits based on (1) and (2) are comparable within the experimental errors[12]. This indicates that the logarithmic corrections to the Q-dependence of the HT terms, which are generated by the anomalous dimensions, are unimportant in this region of x. The interpretation of the HT terms observed in the NLO analysis of[9] for a long time was questionable due to the concern that they can be simulated by the missing higher-order QCD corrections. This concern is based on the observation that the Q-

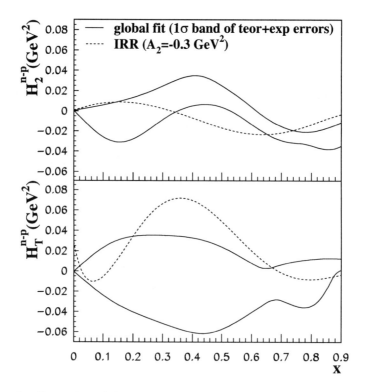

Figure 2. Comparison of the isospin asymmetry in the HT terms of the structure functions F_2 and $F_T = F_2 + F_L$ extracted from the global fit of Ref.[27] (solid curves) and calculated from the IRR model (dashes).

dependence of the terms proportional to the large powers of the strong coupling constant α_s can simulate power-like behavior in the limited range of Q (see[13]). However such a simulation is possible only for the specific choice of powers of α_s and Q and/or insufficiently precise data. In DIS the next-to-next-to-leading (NNLO) QCD corrections do not simulate the power-like behavior within the precision of existing data and as a result the change of the dynamical twist-4 terms from the NLO to the NNLO fit does not exceed the experimental errors[14] (similar stability was observed earlier in[15], although conclusion about the non-vanishing of the dynamical HTs was not drawn due to the fact that they were not separated from the target-mass effects). The same stability should be valid also for the higher-order corrections since the N³LO correction is important at very high and very low x only as follows from the analyses based on the resummation technique[16,17,18,19].

The x-dependence of the twist-4 term in the proton structure function F_2 extracted from the data with the assumption, that functions $H(x,Q)$ do not depend on Q, is given in Fig.1. This term is maximal at $x \sim 0.6$ and is determined in this region with the accuracy of $O(10\%)$ that allows to put a constraint on some matrix elements of the twist-4 operators[21].

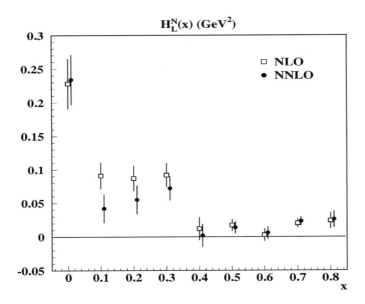

Figure 3. The HT terms in the nucleon structure function F_L determined from the global fit of Ref.[20].

The impact of the HT terms is important up to relatively large values of Q. E.g. at $x = 0.6$ the experimental error in F_2 averaged over global data set is less than the HT terms at $Q^2 \gtrsim 20$ GeV^2 only. In practice this means that inaccount of the HT terms can lead to bias of different quantities extracted from the DIS data. It is of particular importance for the extraction of α_s. Indeed, in the global fit with constraints $H = 0$ the value of $\alpha_s(M_Z) = 0.1215 + -0.0003$ (c.f. $\alpha_s(M_Z) = 0.1143 + -0.0014$ with the HT terms fitted)[22]. For the fit without account of the HT terms we observe the shift of the central value and an essential decrease of the error. Both effects are due to the value of α_s and are strongly anti-correlated with H_2. Essentially this means that for the global fit of existing data the error in α_s is determined by the uncertainty in the HT terms and if these terms could be reliably estimated (e.g. from lattice calculations) the error in α_s would be suppressed several times. The cut on data with $Q^2 \lesssim 20 \text{ GeV}^2$ aimed to decrease the uncertainty in α_s due to the HT terms is ineffective since in this case the error in α_s rises due to loss of data corresponding to the fastest rate of the QCD evolution. The promising way in this direction is to collect new statistically significant samples of data at HERA.

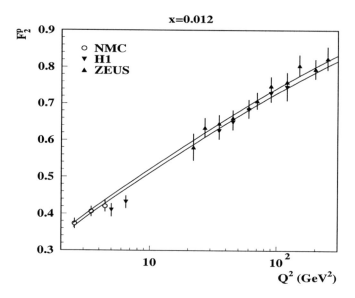

Figure 4. The 1σ band of the global fit (curves) compared to the data on F_2 used in the fit.

A good precision in the determination of H_2 allows a conclusive comparison to the theoretical models. One of the most predictive model of the HTs is based on the infrared renormalon (IRR) approach[23]. This model does not fix the general normalization of the HT terms, but puts a stringent constraint on their x-dependence. For the non-singlet terms the IRR model gives an especially simple predictions[24,25]

$$H_i^{\text{NS}} = A \cdot F_2^{\text{LT,NS}} \otimes C_i^{\text{IRR}}, \qquad (3)$$

where $\otimes$ denotes the Mellin convolution, C_i^{IRR} are the corresponding coefficient functions and A is the normalization factor. These predictions are in qualitative agreement to the data, however the subtle details are not reproduced. For example a study of the isospin asymmetry, which can be used for a test of the basic features of the IRR model, since C_i^{IRR} are flavor-independent[26], demonstrates significant disagreement between the HT terms extracted from the data and the IRR predictions (see Fig.2).

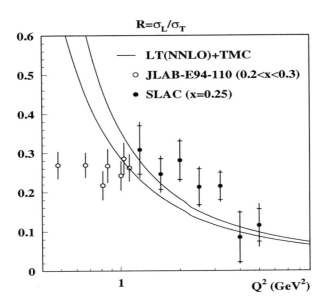

Figure 5. The JLAB and SLAC measurements of $R = \sigma_L/\sigma_T$ compared to 1σ band for $R^{\mathrm{LT,TMC}}$ extracted from the NNLO global fit of Ref.[20].

The HT term in F_L extracted from the global fit to the DIS data is given in Fig.3. The observed x-dependence of H_L is less regular than the one of H_2. It might be that due to F_L extracted from the global fit it is rather sensitive to disagreements in the data. This point is illustrated in Fig.4, which contains the comparison of the results of the global fit to the data on F_2 at $x \approx 0.012$. It is evident that due to the large gap between separate data sets any disagreement between the NMC data at low Q and/or the HERA ones at high Q can be compensated by the corresponding HT term in F_L, if one performs the fit to the data of the cross section in order to extract H_L. Therefore the spike at small x might be driven by such a disagreement and more data at small x and $Q^2 = 2 \div 30$ GeV^2 will be extremely useful for clarification of this point. An alternative explanation of this spike is the effect of missing higher-order QCD corrections that is partially supported by the relative perturbative instability of H_L, at least at $x \sim 0.2$. Indeed, the higher-order corrections at small x obtained using the k_T resummation technique are very large[28] and with the use of the approximate 3-loop coefficient function for F_L^{LT} from[15] based on the calculations of[28,29] one probably should observe a decrease of H_L extracted from the fit. However, one has to take into account that the uncertainty in the 3-loop coefficient function of[15] due to missing terms in the 3-loop corrections is very large and the impact of the exact 3-loop correction might turn insufficient to explain the observed spike. The spike might be generated by a new phenomena such as the parton saturation[30] and, eventually, the OPE up to twist-4 might be insufficient[31]. In agreement with the latter the non-vanishing twist-6 contribution to F_L was found in the analysis of[11]. This analysis was based on the data from the resonance region and therefore in order to obtain this result the duality hypothesis[32] was heavily used. However, the recent JLAB-E94-110 data[33] beyond the resonance region confirm this observation. In Fig.5 the JLAB measurements of $R = \sigma_L/\sigma_T$ are compared to the earlier SLAC data[34] and the LT term extracted from the global fit. Excess of the SLAC data over the LT term with account of the target mass correction at $Q^2 \sim 2$ GeV^2 is interpreted as a manifestation of the HT term in R and, correspondingly, in F_L[35]. At small Q the relation between data and the LT term is opposite and the most natural interpretation within OPE is that the effect of the positive twist-4 contribution is overwhelmed by the negative twist-6 contribution. If so, the structure of the HT terms is very non-trivial, at least for F_L. In summary, despite the predictions based on the HT terms given in Fig.3 are in good agreement with the independent measurements of F_L at small x and Q by the H1 collaboration (see[36] for the details), the status of the determination of H_L is rather unclear and for a better understanding we need efforts both from the theoretical and the experimental side.

The dynamical HT terms in the polarized structure functions are even less known than for the unpolarized ones, although the potential impact of the HT terms is very important in this case due to the typical relatively low Q of the data used in the analysis. The recent indication on the the HT terms in g_1[37] was not entirely confirmed in the latter study[38] and, evidently, a more careful examination is necessary.

The HT terms in the DIS of neutrinos are also poorly known because of the low precision of the available data (see recent review[39] for more details). The essential progress in this area can be achieved only with the very intense neutrino beams (neutrino factories)[40]. These facilities would allow to constraint the dynamical HT terms in the neutrino-nucleon structure function at the level of precision comparable to the case of the charged leptons' DIS. However, since the neutrino experiments provide a much wider set of constraints, this additional input would allow to significantly improve disentangling the matrix element of different operators, which describe the partons' correlation.

In conclusion, our understanding of the power corrections in the DIS structure functions is rather limited. The kinematical component of the power correction (target mass effects) can be calculated with a reasonable accuracy, but the dynamical HT terms are essentially unknown for the most of structure functions. Meanwhile further study of HT terms is necessary in order to outline the kinematical region, where the QCD factorization can be applied and the universality of PDFs can be proven. From a practical point of view it is necessary to provide unbiased extrapolations of PDFs to the high-energy region for the searches of the effects of new physics. Eventually, study of the dynamical HT terms give information about the regime intermediate between asymptotic freedom and confinement. All this makes the precise determination of the HT terms an important and promising topic.

The research was supported in part by the RFBR grant 03-02-17177 and by the National Science Foundation under Grant No. PHY99-07949.

References

1. K. G. Wilson, Phys. Rev. **179** (1969) 1499.
2. L. F. Abbott and R. M. Barnett, Annals Phys. **125**, 276 (1980);
   L. F. Abbott, W. B. Atwood and R. M. Barnett, Phys. Rev. **D22**, 582 (1980).
3. O. Nachtmann, Nucl. Phys. B **63** (1973) 237.
4. H. Georgi and H. D. Politzer, Phys. Rev. D **14** (1976) 1829.
5. E. V. Shuryak and A. I. Vainshtein, Nucl. Phys. B **199** (1982) 451;
   E. V. Shuryak and A. I. Vainshtein, Nucl. Phys. B **201** (1982) 141;
   R. L. Jaffe and M. Soldate, Phys. Rev. D **26** (1982) 49.
6. R. K. Ellis, W. Furmanski and R. Petronzio, Nucl. Phys. B **212** (1983) 29.
7. V. M. Braun and A. V. Kolesnichenko, Nucl. Phys. B **283** (1987) 723.
8. B. Dressler, M. Maul and C. Weiss, Nucl. Phys. B **578** (2000) 293 [arXiv:hep-ph/9906444].
9. M. Virchaux and A. Milsztajn, Phys. Lett. B **274** (1992) 221.
10. A. P. Bukhvostov, E. A. Kuraev and L. N. Lipatov, Yad. Fiz. **38** (1983) 439.
11. G. Ricco, S. Simula and M. Battaglieri, Nucl. Phys. B **555** (1999) 306 [arXiv:hep-ph/9901360].
12. S. I. Alekhin, Eur. Phys. J. C **12** (2000) 587 [arXiv:hep-ph/9902241].

13. Y. L. Dokshitzer, arXiv:hep-ph/9911299.
14. S. I. Alekhin, arXiv:hep-ph/0212370.
15. A. D. Martin, R. G. Roberts, W. J. Stirling and R. S. Thorne, Eur. Phys. J. C **18** (2000) 117 [arXiv:hep-ph/0007099].
16. G. Altarelli, R. D. Ball and S. Forte, Nucl. Phys. B **599** (2001) 383 [arXiv:hep-ph/0011270].
17. S. Schaefer, A. Schafer and M. Stratmann, Phys. Lett. B **514** (2001) 284 [arXiv:hep-ph/0105174].
18. E. Gardi and R. G. Roberts, Nucl. Phys. B **653** (2003) 227 [arXiv:hep-ph/0210429].
19. N. Bianchi, A. Fantoni and S. Liuti, Phys. Rev. D **69** (2004) 014505 [arXiv:hep-ph/0308057].
20. S. Alekhin, Phys. Rev. D **68** (2003) 014002 [arXiv:hep-ph/0211096].
21. S. Choi, T. Hatsuda, Y. Koike and S. H. Lee, Phys. Lett. B **312** (1993) 351 [arXiv:hep-ph/9303272].
22. S. I. Alekhin, JHEP **0302** (2003) 015 [arXiv:hep-ph/0211294].
23. M. Beneke, Phys. Rept. **317** (1999) 1 [arXiv:hep-ph/9807443].
24. E. Stein, M. Meyer-Hermann, L. Mankiewicz and A. Schafer, Phys. Lett. B **376** (1996) 177 [arXiv:hep-ph/9601356].
25. M. Dasgupta and B. R. Webber, Phys. Lett. B **382** (1996) 273 [arXiv:hep-ph/9604388].
26. S. J. Brodsky, arXiv:hep-ph/0006310.
27. S. I. Alekhin, S. A. Kulagin and S. Liuti, arXiv:hep-ph/0304210, to appear in PRD.
28. S. Catani and F. Hautmann, Nucl. Phys. B **427** (1994) 475 [arXiv:hep-ph/9405388].
29. S. A. Larin, T. van Ritbergen and J. A. M. Vermaseren, Nucl. Phys. B **427** (1994) 41;
S. A. Larin, P. Nogueira, T. van Ritbergen and J. A. M. Vermaseren, Nucl. Phys. B **492** (1997) 338 [arXiv:hep-ph/9605317];
A. Retey and J. A. M. Vermaseren, Nucl. Phys. B **604** (2001) 281 [arXiv:hep-ph/0007294].
30. L. V. Gribov, E. M. Levin and M. G. Ryskin, Phys. Rept. **100** (1983) 1.
31. S. Liuti, R. Ent, C. E. Keppel and I. Niculescu, Phys. Rev. Lett. **89** (2002) 162001 [arXiv:hep-ph/0111063].
32. E. D. Bloom and F. J. Gilman, Phys. Rev. D **4** (1971) 2901.
33. Yongguang Liang, Thesis, American Univ., Washington DC (2002).
34. L. W. Whitlow, S. Rock, A. Bodek, E. M. Riordan and S. Dasu, Phys. Lett. B **250** (1990) 193.
35. J. L. Miramontes, M. A. Miramontes and J. Sanchez Guillen, Phys. Rev. D **40** (1989) 2184.
36. S. Alekhin, arXiv:hep-ph/0311184.
37. E. Leader, A. V. Sidorov and D. B. Stamenov, Phys. Rev. D **67** (2003) 074017 [arXiv:hep-ph/0212085].
38. J. Bartelski and S. Tatur, arXiv:hep-ph/0307374.
39. A. L. Kataev, arXiv:hep-ph/0107247.
40. M. L. Mangano *et al.*, arXiv:hep-ph/0105155.

THE ART OF GLOBAL FITTING FOR PARTON DISTRIBUTIONS

JON PUMPLIN

Department of Physics and Astronomy,
Michigan State University
East Lansing MI 48824 USA

I review some aspects of the measurment of Parton Distribution Functions by QCD global fitting.

1. Introduction

High energy hadrons interact through their quark and gluon constituents. The interactions become weak at short distances thanks to the asymptotic freedom property of Quantum Chromodynamics, which allows perturbation theory to be applied to a rich variety of experiments. The nonperturbative nature of the proton for single hard interactions is thus characterized by Parton Distribution Functions (PDFs) $f_a(Q, x)$ of momentum scale Q and light-cone momentum fraction x. The evolution in Q is determined perturbatively by QCD renormalization group equations, so the non-perturbative physics can be characterized by functions of x alone at a fixed small Q_0. The ongoing project to extract those functions from experiment is the subject of this talk.

The goal of the CTEQ global fitting group (like that of our friendly rivals MRST) is to extract the universal non-perturbative parton densities of the proton—or potentially of the neutron or nucleus—from a large variety of hard-scattering experiments, based on three pillars:

- **Factorization** ⇒ Short distance and long distance are separable (see Fig. 1);
- **Asymptotic Freedom** ⇒ Hard scattering processes are perturbatively calculable;
- **DGLAP Evolution** ⇒ PDFs are characterized by functions of x at a fixed small Q_0, with the PDFs at all higher Q being determined from these by renormalization group evolution.

$$F_A^\lambda(x, \frac{m}{Q}, \frac{M}{Q}) = \sum_a f_A^a(x, \frac{m}{\mu}) \otimes \hat{F}_a^\lambda(x, \frac{Q}{\mu}, \frac{M}{Q}) + O((\frac{\Lambda}{Q})^2)$$

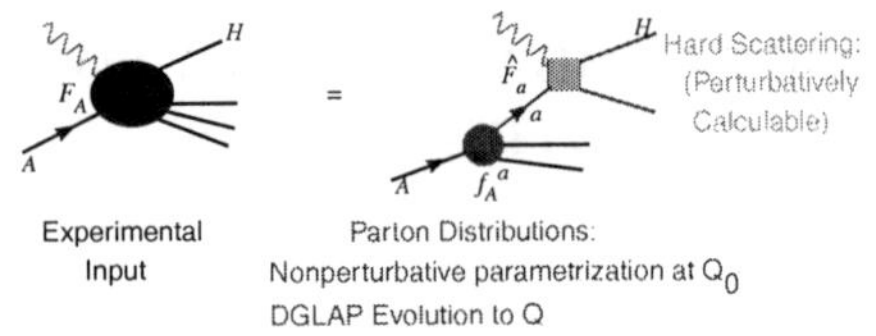

Figure 1. The factorization theorem

The goals of this work are to test the consistency of QCD theory with the collective body of experiments; to test the consistency of individual experiments with QCD and the others; to explore the range of uncertainties of the estimated PDFs; and to make state-of-the-art extracted PDFs widely available, since they are needed to interpret all experiments that use hadron beams or targets—HERA, RHIC, Tevatron, LHC, and non-accelerator.

The kinematic range of available data, after cuts to suppress effects at low Q and low W, are shown in Fig. 2. The data cover a wide range of scales. They are tied together by the DGLAP evolution equation and by the fact that the PDFs are universal. Consistency or inconsistency between different processes, and between different data points for the same process can be observed only by applying QCD to tie them together in a global fit. Future data from HERA, Tevatron run II (W, Z production), HERA II, and LHC will dramatically extend the range and accuracy of this global fit.

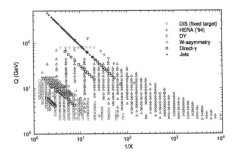

Figure 2. Kinematic region covered by data

2. Some details of the CTEQ6 Global analysis

Input from Experiment: ~ 2000 data points with $Q > 2\,\text{GeV}$, $W > 3.5\,\text{GeV}$ from e, μ, ν DIS; lepton pair production (DY); lepton asymmetry in W production; high p_T inclusive jets; $\alpha_s(M_Z)$ from LEP.

Input from Theory: NLO QCD evolution and hard scattering.

Parametrization at Q_0: Use the form $A_0\, x^{A_1}\,(1-x)^{A_2}\, e^{A_3 x}(1+e^{A_4 x})^{A_5}$ for $u_v = u - \bar{u}$, $d_v = d - \bar{d}$, $\bar{u} + \bar{d}$, and g.

Assumptions based on lack of information: $s = \bar{s} = 0.4\,(\bar{u}+\bar{d})/2$ at Q_0; no intrinsic b or c.

Procedure: Construct effective $\chi^2_{\text{global}} = \sum_{\text{expts}} \chi^2_n$, including published systematic error correlations. Minimize χ^2_{global} to obtain "Best Fit" PDFs.

Uncertainty estimates: Use the variation of χ^2_{global} in neighborhood of the minimum to estimate uncertainty limits as the region of parameter space where $\chi^2 < \chi^2(\text{BestFit}) + T^2$ with $T \approx 10$. This "Tolerance Factor" $T \sim 10$ is quite different from the traditional value 1 from Gaussian statistics, because of unknown systematic errors in theory and experiments. It can be estimated from the apparent inconsistencies between experiments when they are combined in the global fit.

A note on the details of parametrization at Q_0: For d_{val}, u_{val}, $\bar{u} + \bar{d}$ or g, we use the general form

$$xf(x, Q_0) = A_0\, x^{A_1}\,(1-x)^{A_2}\, e^{A_3 x}(1+e^{A_4 x})^{A_5}$$

This corresponds to

$$\frac{d}{dx}\ln(xf) = \frac{A_1}{x} - \frac{A_2}{1-x} + \frac{c_3 + c_4 x}{1 + c_5 x}$$

i.e., we add a 1:1 Padé approximation term to the singular terms of the time-honored $A_0\, x^{A_1}\,(1-x)^{A_2}$ parametrization, which is based on Regge theory and quark counting rules. A sufficiently flexible parametrization is important; but for convergence, there must not be too many "flat directions." For that reason, some of the parameters are frozen at arbitrary values for some flavors.

To measure a set of continuous PDF functions at Q_0 on the basis of a finite set of data points would appear to be an ill-posed mathematical problem. However, this difficulty is not so severe as might be expected since the actual predictions of interest that are based on the PDFs are discrete quantities. In particular, fine-scale structure in x in the PDFs at Q_0 tend to be smoothed out by evolution in Q. They correspond to flat directions in χ^2 space, so they are not accurately measured; but they have little effect on the applications of interest.

Some representative Best Fit parton distributions from the analysis are shown in Figs 3, 4. One sees that valence quarks dominate for $x \to 1$, and the gluon dominates for $x \to 0$, especially at large Q.

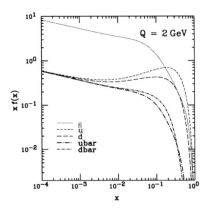

Figure 3. Parton distributions at $Q = 2$ GeV

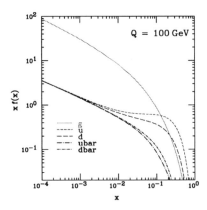

Figure 4. Parton distributions at $Q = 100$ GeV

3. Including Experimental Systematic Errors

The simplest definition of chi-squared,

$$\chi_0^2 = \sum_{i=1}^{N} \frac{(D_i - T_i)^2}{\sigma_i^2} \qquad \begin{cases} D_i = \text{data} \\ T_i = \text{theory} \\ \sigma_i = \text{``expt. error''} \end{cases}$$

works for random Gaussian errors,

$$D_i = T_i + \sigma_i r_i \quad \text{with} \quad P(r) = \frac{e^{-r^2/2}}{\sqrt{2\pi}}.$$

When correlated experimental systematic errors have been estimated, the errors become

$$D_i = T_i(a) + \alpha_i \, r_{\text{stat},i} + \sum_{k=1}^{K} r_k \, \beta_{ki},$$

where α_i is the 'standard deviation' of the random uncorrelated error, and β_{ki} is the 'standard deviation' of the kth (completely correlated!) systematic error on D_i.

In that case the parameters $\{r_k\}$ that control the corrections for systematic errors join the original PDF shape parameters $\{a_\lambda\}$ as parameters for which the global χ^2 is to be minimized. In effect, the other experiments in the global fit are used to determine the unknown systematic error coefficients. (This is an extension of a familiar situation in which the overall normalization of an experiment may be determined by fitting to others if it is not well measured directly.)

To take into account the systematic errors, we define

$$\chi'^2(a_\lambda, r_k) = \sum_{i=1}^{N} \frac{(D_i - \sum_k r_k \beta_{ki} - T_i)^2}{\alpha_i^2} + \sum_k r_k^2,$$

and minimize with respect to $\{r_k\}$. The result is

$$\widehat{r}_k = \sum_{k'} \left(A^{-1}\right)_{kk'} B_{k'}, \qquad \text{(systematic shift)}$$

where

$$A_{kk'} = \delta_{kk'} + \sum_{i=1}^{N} \frac{\beta_{ki} \beta_{k'i}}{\alpha_i^2}$$

$$B_k = \sum_{i=1}^{N} \frac{\beta_{ki} (D_i - T_i)}{\alpha_i^2}.$$

Since the dependence of χ^2 on the $\hat{r}_k$'s is quadratic, we are able to minimize with respect to those parameters explicitly for any set of PDF shape parameters $\{a_\lambda\}$. We then minimize the remaining $\chi^2(a)$ with respect to the PDF shape parameters $\{a_\lambda\}$. Then the $\{a_\lambda\}$ determine $f_i(x, Q_0^2)$. Meanwhile, $\{\hat{r}_k\}$ are the optimal "corrections" for systematic errors; i.e., systematic shifts to be applied to the data points to bring the data from different experiments into compatibility, within the framework of the theoretical model.

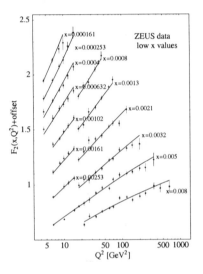

Figure 5. CTEQ6M fit to ZEUS data at low x

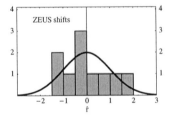

Figure 6. The 10 systematic error parameters for the ZEUS data. Curve is a Gaussian of width 1.

An example of how this works is shown in Fig. 5: The data points include the estimated corrections for systematic errors, i.e., the central values plotted have been shifted by an amount that is consistent with the estimated systematic errors, where the systematic error parameters are determined using other experiments via the global fit. The error bars are statistical errors only. Fig. 6 shows that the systematic error parameters come out of order 1 as they should. Fig. 7 shows that the residuals defined as (data-theory)/error also come out with the expected distribution. In contrast, Fig. 8 shows that if the systematic shifts are not included, the fit becomes bad.

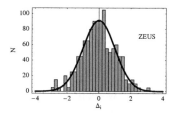

Figure 7. Residuals for the ZEUS data. Curve is a Gaussian of width 1.

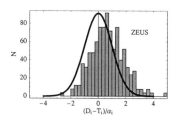

Figure 8. Residuals for the ZEUS data without including systematic shifts.

4. Uncertainties in PDFs

There are several **Sources of uncertainty:**

- Experimental errors included in χ^2
- Unknown experimental errors
- Parametrization dependence
- Higher-order corrections & Large Logarithms

- Power Law corrections ("higher twist")

There are some **Fundamental difficulties:**

- Good experiments run until systematic errors dominate, so the magnitude of remaining systematic errors involves guesswork.
- Systematic errors of the theory and their correlations are even harder to guess.
- Quasi–ill-posed problem: we must determine continuous functions from a discrete data set. (Because of the smoothing effect of DGLAP evolution, this is not as impossible as it sounds.)
- Some combinations of variables are unconstrained, e.g., $s - \bar{s}$ before NuTeV data.

There are several **Approaches to estimating the uncertainty.** In all of the uncertainty methods, we continue to use χ^2 as a measure of the quality of the fit; but vary weights assigned to the experiments to estimate the range of acceptable fits, rather than relying on the classical $\Delta\chi^2 = 1$.

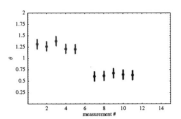

Figure 9. Hypothetical measurements of hypothetical parameter θ.

The essence of the Uncertainty Problem can be seen in the Fig. 9, which shows some hypothetical measurements of a single parameter θ. Suppose the quantity θ has been measured by two different experiments, or extracted using two different approximations to the True Theory. What would you quote as the Best Fit and the Uncertainty? The disagreements are of course not so obvious in the many-parameter global fit. However, the disagreements can be probed in one dimension by, for example, studying the variations of the Best Fit that result from assigning different weights to different experiments, or to different kinematic regions etc. Much of this is discussed in our papers; and more is work in progress.

5. Outlook

Parton Distribution Functions are a necessary infrastructure for precision Standard Model studies and New Physics searches at hadron colliders and at experiments using hadron targets. Some issues that were discussed, but not necessarily included in this writeup due to space limitations:

- PDFs of the proton are increasingly well measured.
- Useful tools are in place to estimate the uncertainty of PDFs and to propagate those uncertainties to physical predictions. There is adequate agreement between various methods for estimating the uncertainty:
  - "Hessian Method" based on the eigenvectors of the error matrix
  - "Lagrange Multiplier Method" based on finding the uncertainty on a predicted quantity by studying the variation of χ^2 as a function of that quantity
  - systematic reweighting of experiments (work in progress with John Collins)
  - random reweighting of experiments: a variant of the "well known" statistical bootstrap method
- The "Les Houches Accord" interface makes it easy to handle the large number of PDF solutions that are needed to characterize uncertainties. [hep-ph/0204316]
- Improvements in the treatment of heavy quark effects are in progress, and together with neutrino experiments they will allow improved flavor differentiation.
- Since PDFs summarize some fundamental nonperturbative physics of the proton, they should be considered a challenge to be computed! (Low moments of meson PDFs have indeed been calculated in lattice gauge theory.)
- Non-perturbative clues, e.g. for $s(x) - \bar{s}(x)$, may be helpful.
- HERA and Fermilab run II data will provide the next major experimental steps forward, followed by LHC.
- Theoretical improvements such as resummation to use direct photon and W transverse momentum data will be useful.
- In view of possible isospin breaking, and the importance of nuclear shadowing & anti-shadowing effects, HERA measurements on deuterons would be highly welcome.

I thank the organizers of Ringberg 2003 for an excellent workshop in an excellent setting. I thank my collaborators on this work, D. Stump, W.K. Tung, J. Huston, H. Lai, P. Nadolsky, F. Olness, S. Kuhlmann, J. Owens, S. Kretzer, and J. Collins for many valuable discussions.

2
Spin Physics

NEW RESULTS FROM HERMES

M. TYTGAT*

*University of Gent,
Dept. Subatomic and Radiation Physics,
Proeftuinstraat 86,
9000 Gent, Belgium
E-mail: michael@inwfsun1.UGent.be*

An overview is given of selected recent HERMES results obtained from measurements performed during the first running period of HERA. These topics include inclusive $g_1(x)$-measurements with a NLO QCD analysis, polarized quark distribution extraction, $b_1(x)$-measurement, double spin asymmetries in vector meson production, ρ^0-nuclear transparency and finally quark fragmentation in nuclei.

1. Introduction

The HERMES experiment uses the 27.5 GeV polarized HERA electron/positron beam to scatter deep-inelastically off a (un)polarized fixed gaseous nuclear target. Data taking began in 1995, during the first HERA running period and still continues with HERA-II. Although HERMES was originally designed to study the nucleon spin in polarized DIS, a wide range of physics topics has been addressed over the past years. Here only some of the recent results from HERA-I will be summarized. First results obtained during HERA-II are discussed elsewhere [1].

2. NLO QCD Analysis of the $g_1(x, Q^2)$ Structure Function

During HERA-I HERMES performed longitudinally polarized inclusive DIS measurements on ^{1}H, ^{2}H and ^{3}He. A compilation of the world data on the $g_1(x, Q^2)$-structure function including the HERMES results is displayed in figure 1. The HERMES data actually add most precision to g_1^d. Considering $g_1 = \frac{1}{2}\sum_q e_q^2 \Delta q$ in the naïve parton model with Δq being the polarized quark densities and neglecting the quark sea, one can qualitatively deduce

---
*On behalf of the HERMES Collaboration

from these inclusive data that Δu must be large and positive, while Δd should be negative.

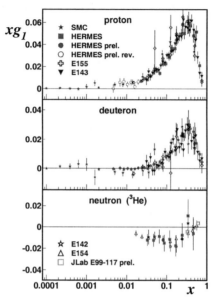

Figure 1. A compilation of recent world data on $xg_1(x, Q^2)$ for the proton, deuteron and neutron.

To obtain more quantitative information on the polarized quark densities a next-to-leading order QCD analysis of the $g_1(x, Q^2)$-data was carried out including the latest HERMES results. The $g_1(x, Q^2)$-structure function was parametrized as a Mellin convolution in terms of the polarized singlet density $\Delta\Sigma = \sum_{i=1}^{N_f}[\Delta q_i + \Delta\bar{q}_i]$, the polarized gluon density ΔG and the polarized non-singlet densities $\Delta q_i^{NS} = \Delta q_i + \Delta\bar{q}_i - 1/N_f \cdot \Delta\Sigma$. In the evolution equations the polarized splitting functions and Wilson coefficient functions are calculated in NLO in the $\overline{MS}$-scheme. Two independent methods were used to extract the parton densities. The first one is a numerical method with evolution in x-space and takes $\Delta\Sigma$, Δq_p^{NS}, Δq_n^{NS} and ΔG as distributions to be determined in the fit. The second method solves the evolution equations in Mellin-N space and uses Δu_v, Δd_v, $\Delta\bar{q}_s$ and ΔG as free distributions. The polarized parton densities at input scale $Q_0^2 = 4.0$ GeV^2 were parametrized in the form $x\Delta q_i(x, Q_0^2) = \eta_i A_i x^{a_i}(1-x)^{b_i}(1 + \gamma_i x + \rho_i x^{1/2})$, where the normalization constants A_i are chosen such that η_i represents the first moment of Δq_i.

Both methods produce compatible results and the final polarized parton densities as determined with the second method are displayed in figure 2 at input scale, where they are compared to previous analyses [2,3]. The shaded

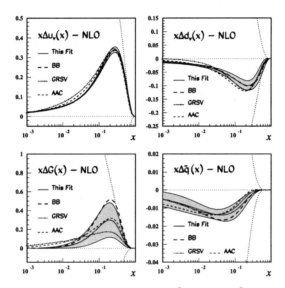

Figure 2. The polarized quark distributions at $Q_0^2 = 4.0$ GeV^2 compared to previous analyses. The dotted lines correspond to the positivity limits from unpolarized distributions.

areas denote the fully correlated 1σ statistical error bands and the dotted lines correspond to the positivity limits from unpolarized distributions. While Δu_v and Δd_v are well constrained by the current fit, ΔG and $\Delta \bar{q}$ remain poorly determined. For ΔG the theoretical uncertainty dominates the overall systematic error, such that not only an improved data set with a larger Q^2-range but also a NNLO calculation could improve the situation.

3. Flavor Decomposition of Polarized Quark Distributions

Another method to accomplish a flavor decomposition of the polarized quark densities is provided by flavor tagging in polarized semi-inclusive DIS [4] $e + p(d) \to e + h + X$. The measured photon-nucleon double spin asymmetry A_1^h for semi-inclusive events with a leading hadron of type h

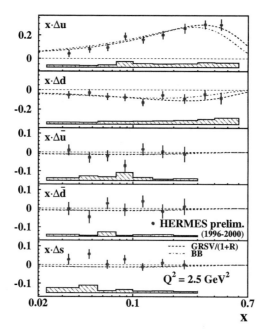

Figure 3. The polarized parton densities at $Q^2 = 2.5$ GeV^2.

can be written as

$$A_1^h(x,Q^2) \sim \sum_q \frac{e_q^2 q(x,Q^2) \int dz D_q^h(z,Q^2)}{\sum_{q'} e_{q'}^2 q'(x,Q^2) \int dz D_{q'}^h(z,Q^2)} \frac{\Delta q}{q}(x,Q^2) \quad (1)$$

$$\sim \sum_q P_q^h(x,Q^2) \frac{\Delta q}{q}(x,Q^2), \quad (2)$$

with D_q^h being the fragmentation function for a quark flavor q and Δq (q) being the (un)polarized quark density. The purity P_q^h represents the probability that the hadron h originates from an event in which a quark of flavor q was struck by the virtual photon. The purities are spin-independent quantities and were evaluated using the PEPSI/JETSET DIS Monte Carlo model [5]. The CTEQ5L parametrizations for unpolarized quark densities were used and the JETSET fragmentation parameters were tuned to reproduce the hadron multiplicities measured at HERMES. Equation 2 can be rewritten as a matrix equation $\vec{A} = P \cdot \vec{Q}$, where $\vec{A} = (A_{1p}, A_{1p}^{\pi^\pm}, A_{1d}, A_{1d}^{\pi^\pm}, A_{1d}^{K^\pm})$ contains the measured inclusive and semi-inclusive asymmetries on the proton and deuteron, P contains the purities for the proton and deuteron and $\vec{Q} = (\frac{\Delta u}{u}, \frac{\Delta d}{d}, \frac{\Delta \bar{u}}{\bar{u}}, \frac{\Delta \bar{d}}{\bar{d}}, \frac{\Delta s}{s}, \frac{\Delta \bar{s}}{\bar{s}})$ contains the quark and anti-quark polarizations to be determined. The matrix equation was solved by χ^2-minimization and the resulting polarized parton densities are depicted in figure 3 at the mean scale $<Q^2> = 2.5$ GeV^2. While Δu is positive in the measured x-range, Δd appears negative and only slightly dependent on x. Both distributions are consistent with previous measurements, but are now known with a much improved precision. The light sea quark polarization densities are compatible with zero. The densities in the figure are compared to two leading order QCD fits [2].

4. The $b_1(x, Q^2)$ Tensor Structure Function

When scattering off a spin-1 target an additional leading twist structure function $b_1(x, Q^2)$ enters the symmetric part of the hadronic tensor compared to the spin-1/2 case. This structure function measures the difference in the polarized quark distributions of a target with helicity $m = \pm 1$ and $m = 0$: $b_1(x, Q^2) = \frac{1}{2} \sum_i e_i^2 [2q_i^0(x, Q^2) - q_i^+(x, Q^2) - q_i^-(x, Q^2)]$. The cross section now depends on the vector and tensor polarization of the target $\sigma = \sigma_u[1 + P_b\, V\, A_\| + \frac{1}{2} T\, A_T]$, where $A_\| \sim g_1/F_1$ and $A_T \sim -2/3 \cdot b_1/F_1$ are the target vector and tensor asymmetry respectively, V and T the target vector and tensor polarization and P_b the beam polarization.

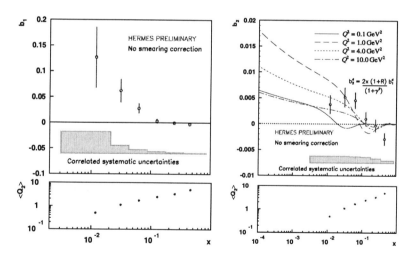

Figure 4. The measured $b_1(x, Q^2)$ and $b_2(x, Q^2)$ structure function on the deuteron.

The tensor asymmetry A_T was measured for the first time at HERMES on the deuteron target in the kinematic range $0.0021 < x < 0.85$ and

$0.1 < Q^2 < 20$ GeV^2. The resulting b_1 is shown in figure 4 together with b_2, which is related to the former via a Callan-Gross like relation. At small x both structure functions appear positive which can be interpreted as being due to the quadrupole deformation of the deuteron. The b_2-data is in qualitative agreement with the double scattering model calculations from Ref. 6.

5. Double-Spin Asymmetries in Vector Meson Production

The longitudinally polarized beam and target allows HERMES to investigate double-spin asymmetries in vector meson production [7]. Averaged over the kinematic acceptance the asymmetries for exclusive ρ^0-production were found to be $0.23 \pm 0.14 \pm 0.02$ on the proton and $-0.040 \pm 0.076 \pm 0.013$ on the deuteron for average kinematics $< W > = 4.9$ GeV, $< Q^2 > = 1.8$ GeV^2 and $< -t' > = 0.15$ GeV^2. The same asymmetries extracted for the ϕ-meson were compatible with zero. Figure 5 displays the ρ^0 asymmetries for both targets as function of Bjorken x and Q^2. Using the HERMES data

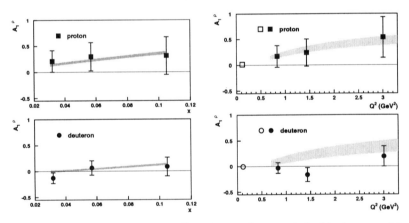

Figure 5. Double-spin asymmetry in ρ^0-production from a polarized hydrogen and deuterium target as function of Bjorken x and Q^2.

on A_1^N and the simple relation $A_1^\rho = 2A_1^N/(1 + (A_1^N)^2)$ between the vector meson and inclusive DIS asymmetry as derived in Ref. 8, one obtains the shaded bands in figure 5 for the x-dependence of the asymmetries. Both the proton and deuteron asymmetries are consistent with this prediction. The Q^2-dependences are compared to a Regge model calculation from Ref. 9. In this model the parameters of the Reggeons contributing to ρ^0-production were extracted from fits to g_1^N and F_2^N. While the proton measurement seems to agree with the prediction, the deuteron data lie below the calculation. The datapoints at the lowest Q^2-value are both compatible with zero and were obtained with a special trigger for events in which the beam particle scatters at too low angles to be detected.

In Ref. 8 the non-zero asymmetry result for ρ^0-production arises from an interference between natural and unnatural parity exchange amplitudes. Since the Pomeron has natural parity, it is suggested that at HERMES energies the transverse ρ^0-production amplitude on the proton receives significant contributions from Reggeon or di-quark exchange. This not only agrees with the previous observation that at HERMES the longitudinal production amplitude on the proton is dominated by quark exchange [10], but also is not in contradiction with the vanishing asymmetry found at higher energies at SMC [11], where ρ^0-production is dominated by gluon exchange and also the inclusive asymmetry is much smaller. The zero asymmetry result for ϕ-production, which is dominated by Pomeron exchange even at HERMES energies, can also be understood in the same way.

6. Color Transparency in ρ^0-Production

A fundamental prediction of QCD is the existence of the color transparency (CT) phenomenon, i.e. a hadron which is produced at sufficiently high Q^2 will traverse a nuclear medium with vanishing initial and final state interactions. In the search for CT one commonly uses the nuclear transparency $T = \sigma_A/(A\sigma_p)$, which is the ratio of the nuclear cross section per nucleon to the one on the proton. The distance which the $q\bar{q}$-fluctuation of the virtual photon in ρ^0-production can propagate is the coherence length $l_c = 2\nu/(Q^2 + M_{q\bar{q}}^2)$. This latter quantity is important for CT-studies, especially for incoherent ρ^0-production, where at l_c-values smaller than the nuclear size T increases with Q^2 due to the coherence length effect previously observed at HERMES [12]. For coherent ρ^0-production the drop of the nuclear form factor at high Q^2 (small l_c) causes a decreasing behavior of T with Q^2. In order to get a clean signature of a CT-effect, one has to disentangle the coherence length effects from those of CT, which can be done by studying T as function of Q^2 for fixed l_c [13]. Figure 6 shows the measured T for both coherent and incoherent exclusive ρ^0-production on nitrogen as function of Q^2 for small intervals of l_c [14]. The data were fitted with a common Q^2-slope equal to 0.074 ± 0.023 GeV^{-2}, which agrees with the theoretical prediction of about 0.058 GeV^{-2}. This positive slope of the

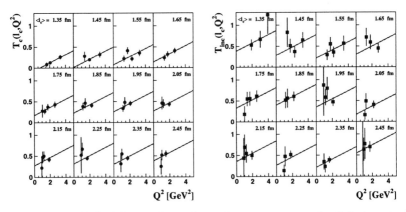

Figure 6. The left (right) plot shows the nuclear transparency for (in)coherent ρ^0-production on nitrogen as function of Q^2 for several coherence length intervals.

Q^2-dependence represents a signature of color transparency.

7. Quark Fragmentation in Nuclei

In deep-inelastic scattering off a nucleus, the nuclear medium acts as an ensemble of targets for the quark struck by the virtual photon and for the produced hadrons and will therefore influence the hadronization process. Due to multiple interactions with the surrounding medium and induced gluon radiation the struck quark will lose energy. A final hadron formed inside the nucleus can also interact via the hadronic interaction cross section with the nuclear medium. Both effects will induce a softening of the leading hadron spectra compared to the free nucleon case. The hadronization process in nuclear media is usually described either in phenomenological string models including final state interactions of the produced hadron or in more QCD-inspired models by e.g. parton energy loss or gluon bremsstrahlung. A clean tool to study nuclear effects in quark propagation and hadronization is semi-inclusive DIS, where one usually examines the multiplicity ratio for a hadron h on a nucleus A compared to the deuteron,

$$R_M^h(z,\nu,p_t^2,Q^2) = \frac{(N_h(z,\nu,p_t^2,Q^2)/N_e(\nu,Q^2))|_A}{(N_h(z,\nu,p_t^2,Q^2)/N_e(\nu,Q^2))|_D}, \qquad (3)$$

with N_h the semi-inclusive yield normalized to the inclusive yield N_e.

The multiplicity ratios for π, K, p and $\bar{p}$ measured by HERMES [15] on ^{84}Kr are displayed in figure 7 as function of ν and z. Contributions due to target fragmentation were suppressed by requiring $z > 0.2$ and large

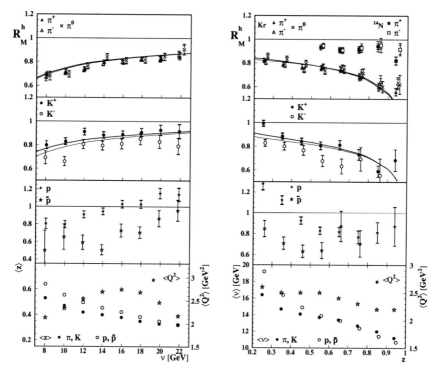

Figure 7. Multiplicity ratios for π, K, p and $\bar{p}$ for ^{84}Kr as function of ν and z. The z-dependence of the π multiplicity on ^{14}N is also shown.

acceptance effects in z were avoided by taking $\nu > 7$ GeV. The plot also includes the π-multiplicity z-dependence on ^{14}N [16]. The multiplicity ratios for π^+, π^- and π^0 on krypton appear to be similar, which is also true for the nitrogen $\pi^\pm$-data. Whereas the multiplicity ratio for K^- is compatible with that for pions, the K^+-multiplicity ratio is significantly larger and an even larger difference is observed between p and $\bar{p}$. One way to understand these differences is to interpret them in terms of different hadron formation times for mesons and baryons in combination with different hadron-nucleon interaction cross sections. The curves in the plots represent the result of a model calculation [19] based on a rescaling of the quark fragmentation due to the nuclear medium. In Ref. 17 the differences are attributed to a mixing of quark and gluon fragmentation functions in nuclei due to multiple parton scattering, such that quark and anti-quark fragmentation functions receive different medium modifications. The π-data on both targets exhibit the same trend, however as expected, the attenuation in krypton appears much stronger than in nitrogen. The mass dependence of the nuclear attenuation $1 - R_M^h$ appears closer to a $A^{2/3}$-dependence as predicted in Ref. 17 than to a $A^{1/3}$-dependence suggested by models based on nuclear absorption only.

Another nuclear effect is the Cronin effect [18], which concerns the broadening of the hadron p_t-distribution on a nuclear target compared to the proton due to multiple scattering of the propagating quark and hadron. This phenomenon was also observed by HERMES [15] as an enhancement of the charged hadron multiplicity ratio on ^{14}N and ^{84}Kr at high p_t^2. It is similar to the effect seen in proton-nucleus and nucleus-nucleus collisions and may help to interpret the new heavy-ion data from SPS and RHIC.

Acknowledgments

The author would like to acknowledge the support of the FWO-Flanders, Belgium, for this work.

References

1. E.C. Aschenauer, Plans of the HERMES experiment at HERA-II, same proceedings.
2. J. Blümlein and H. Böttcher, Nucl. Phys. **B636**, 225 (2002); M. Glück *et al.*, Phys. Rev. **D63**, 094005 (2001).
3. Y. Goto *et al.*, Phys. Rev. **D62**, 034017 (2000).
4. A. Airapetian *et al.*, Phys. Rev. Lett (in press), hep-ex/0307064
5. L. Mankiewicz *et al.*, Phys. Comm. **71**, 305 (1992); T. Sjöstrand *et al.*, Comp. Phys. Comm. **135**, 238 (2001).
6. K. Bora and R.L. Jaffe, Phys. Rev. **D57**, 6906 (1998).
7. A. Airapetian *et al.*, Phys. Lett. **B513**, 301 (2001);Eur. Phys. J. **C29**, 171 (2003).
8. H. Fraas, Nucl. Phys. **B113**, 532 (1976).
9. N.I. Kochelev *et al.*, Phys. Rev. **D65**, 097504 (2002); Phys. Rev. **D67**, 074014 (2003)
10. A. Airapetian *et al.*, Eur. Phys. J. **C17**, 389 (2000).
11. A. Tripet for the SMC Collaboration, Nucl. Phys. **B79**, 529 (1999).
12. K. Ackerstaff *et al.*, Phys. Rev. Lett. **82**, 3025 (1999).
13. B.Z. Kopeliovich *et al.*, Phys. Rev. **C65**, 035201 (2002).
14. A. Airapetian *et al.*, Phys. Rev. Lett. **90**, 052501 (2003).
15. A. Airapetian *et al.*, Phys. Lett. B (in press); hep-ex/0307023.
16. A. Airapetian *et al.*, Eur. Phys. J. **C20**, 479 (2001).
17. X.N. Wang and X. Guo, Nucl. Phys. **A696**, 788 (2001); E. Wang and X.N. Wang, Phys. Rev. Lett. **89**, 162301 (2002)
18. J.W. Cronin *et al.*, Phys. Rev. **D11**, 3105 (1975).
19. A. Accardi, V. Muccifora, H.J. Pirner, Nucl. Phys. **A720**, 131 (2003).

THEORETICAL ASPECTS OF SPIN PHYSICS

DANIËL BOER

Department of Physics and Astronomy, Vrije Universiteit Amsterdam
De Boelelaan 1081, NL-1081 HV Amsterdam, The Netherlands

A summary is given of how spin enters in collinearly factorizing processes. Next, theoretical aspects of polarization in processes beyond collinear factorization are discussed in more detail, with special focus on recent developments concerning the color gauge invariant definitions of transverse momentum dependent distribution and fragmentation functions, such as the Sivers and Collins effect functions. This has particular relevance for azimuthal single spin asymmetries, which currently receive much theoretical and experimental attention.

1. Introduction

The goal of QCD spin physics is to understand the spin structure of hadrons in terms of quark and gluon properties. For this purpose one studies polarization effects in high energy collisions, where one or more large energy scales may allow a factorized description. This means that cross sections factorize into quantities that describe the soft, nonperturbative physics and those that describe the short distance physics, which is calculable.

2. Spin in collinearly factorizing processes

The polarized structure functions g_1 and g_2 of Deep Inelastic Scattering (DIS) of polarized electrons off polarized protons (or other spin-1/2 hadrons), $\vec{e}\vec{p} \to e' X$, appear in the parametrization of the hadronic part of the cross section, i.e., in the antisymmetric part of the hadron tensor

$$W_A^{\mu\nu} = \frac{i\epsilon^{\mu\nu\rho\sigma}q_\rho}{P \cdot q}\left[S_\sigma g_1(x_B, Q^2) + \left(S_\sigma - \frac{S \cdot q}{P \cdot q}P_\sigma\right)g_2(x_B, Q^2)\right], \quad (1)$$

with hadron momentum P and spin vector S, photon momentum q, $x_B = Q^2/2P \cdot q$ and $Q^2 = -q^2$. The definition of structure functions is independent of the constituents of the hadron. However, the operator product expansion or the pQCD improved parton model allows one to go to the quark-gluon level, such that the structure functions are expressed in terms of parton distribution functions (see Fig. 1). The two-quark

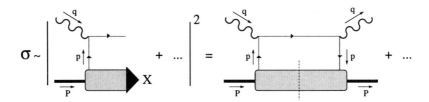

Figure 1. The $\gamma^* p$ cross section can be expanded in terms of parton correlators.

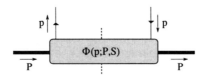

Figure 2. The two-quark correlation function Φ, which depends on the hadron momentum P, quark momentum $p = xP$ and hadron spin vector S.

correlation function $\Phi(p; P, S)$ (Fig. 2), or $\Phi(x)$ in short, is defined as

$$\Phi(x) = \int \frac{d\lambda}{2\pi} e^{i\lambda x} \langle P, S | \overline{\psi}(0) \mathcal{L}[0, \lambda] \psi(\lambda) | P, S \rangle, \tag{2}$$

where the path-ordered exponential (also simply called 'link')

$$\mathcal{L}[0, \lambda] = \mathcal{P} \exp\left(-ig \int_0^\lambda d\eta\, A^+(\eta n_-)\right), \tag{3}$$

is not inserted in an ad hoc way to make $\Phi(x)$ color gauge invariant, but can actually be *derived* [1] (n_- in Eq. (3) is a lightlike direction). This $\Phi(x)$ is parametrized in terms of parton distribution functions. For longitudinal spin or helicity the (leading twist) parton distributions are $\Delta q, \Delta \bar{q}, \Delta g$ and for transverse spin they are $\delta q, \delta \bar{q}$ ($\delta g = 0$ due to helicity conservation):

$$\text{Tr}\left[\Phi(x)\gamma^+\right] \sim q(x),$$
$$\text{Tr}\left[\Phi(x)\gamma^+\gamma_5\right] \sim \lambda \Delta q(x),$$
$$\text{Tr}\left[\Phi(x)\gamma_T^i \gamma^+\gamma_5\right] \sim S_T^i \delta q(x).$$

From inclusive DIS, or more specifically, from the measurement of the structure function $g_1(x)$, one has obtained experimental information on $\Delta q(x) + \Delta \bar{q}(x)$, and implicitly on $\Delta g(x)$ via evolution. More information about $\Delta \bar{q}$ and Δg will be obtained from polarized pp collisions at

RHIC (BNL) and from (semi-)inclusive DIS data of COMPASS (CERN), HERMES (DESY) and JLAB. In contrast, transversity (δq) is completely unknown (no data). It cannot be measured in inclusive DIS, where it is heavily suppressed. The reason is that it must be probed together with another helicity flip. There are two types of collinearly factorizing processes that serve this purpose:

- Processes with two transversely polarized hadrons, e.g. $p^\uparrow p^\uparrow \to \ell \bar{\ell} X$, $p^\uparrow p^\uparrow \to \text{jet } X$, $e\, p^\uparrow \to \Lambda^\uparrow X$ or $p p^\uparrow \to \Lambda^\uparrow X$
- Processes sensitive to the two-hadron interference fragmentation functions [2,3,4,5], such as $e\, p^\uparrow$ or $p p^\uparrow \to (\pi^+ \pi^-) X$, where the angular distribution of final state hadron pairs is expected to be correlated with the transverse spin direction

This last option exploits the fact that the direction of produced hadrons can be correlated with the polarization of one or more particles in the collision. This is not merely a theoretical idea, but also has been seen in experiments, namely in single spin asymmetries in hadron and lepton pair production. Large single spin (left-right) asymmetries have been observed in $p p^\uparrow \to \pi X$ [6,7,8], where the pions prefer to go left or right of the plane spanned by the beam direction and the transverse spin, depending on whether the transverse spin is up or down and depending on the charge of the pions. Similar types of asymmetry have been observed in $p p \to \Lambda^\uparrow X$ [9] and $\nu_\mu p \to \mu \Lambda^\uparrow X$ [10]. It is expected that the underlying mechanisms of these different asymmetries are related, but it is also fair to say that single transverse spin asymmetries are not really understood, i.e., it is not yet clear how to explain them on the quark-gluon level. The suggested mechanisms can be roughly labeled as: semi-classical models; k_T-dependent distributions; and, higher twist. Motivated by recent developments, the next section will mainly be about k_T-dependent distributions.

First some short comments on the helicity dependence of transversity. A transverse spin state is an off-diagonal state in the helicity basis, which means that amplitudes with proton helicity + interfere with those of helicity −, see Fig. 3. For the transversity function $\delta q(x)$ the helicity flip of the proton states, is accompanied by helicity flip of the quark states, due to helicity conservation. In case one does not have helicity flip of the quark states, then one can satisfy helicity conservation by having an additional ±1 helicity gluon, at the cost of a suppression by one power of a large energy scale of the process. It is a twist-3 quark-gluon correlation inside a transversely polarized proton (Fig. 3). Neither δq, nor g_T ($= g_1 + g_2$) lead

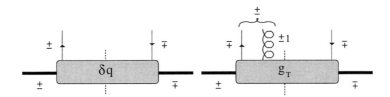

Figure 3. The helicity dependence of transversity δq and the twist-3 function g_T.

to single transverse spin asymmetries in collinearly factorizing processes.

3. Beyond collinear, leading twist factorization

In order to describe single transverse spin asymmetries within a factorized approach, several ideas have been put forward, summarized in Fig. 4. Qiu

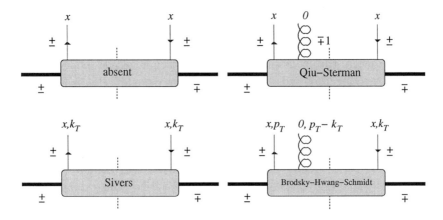

Figure 4. Pictorial representations of the Qiu-Sterman effect, the Sivers effect and the contribution considered by Brodsky, Hwang and Schmidt to generate nonzero SSA.

and Sterman ('91) [11] showed that the contribution where the gluon in the above-mentioned twist-3 quark-gluon correlation has vanishing momentum does give rise to a (suppressed) single spin asymmetry (SSA). Around the same time Sivers ('90) [12] suggested to consider quark momenta that are not completely collinear to the parent hadron's momentum. In that case one does not need helicity flip on the quark side to satisfy helicity conservation and an unsuppressed SSA could occur. However, Collins ('93) [13] demonstrated that this "Sivers effect" must be zero due to time reversal invariance. This demonstration turned out to be incorrect, as became clear after Brodsky, Hwang and Schmidt ('02) [14] obtained an unsuppressed SSA from a k_T-dependent quark-gluon correlator (see Fig. 4) that *is* allowed by time reversal invariance. Belitsky, Ji and Yuan ('02) [15] showed that this particular correlator is a part of the proper gauge invariant definition of the Sivers function. After taking into account all numbers of gluons in this correlator, one obtains a path-ordered exponential in the off-lightcone, non-local operator matrix element that defines the Sivers function:

$$f_{1T}^\perp \propto \langle P, S_T | \overline{\psi}(0) \, \mathcal{L}[0, \xi] \, \gamma^+ \, \psi(\xi) | P, S_T \rangle, \qquad (4)$$

where ξ has (apart from an n_- component) a transverse component ξ_T. Collins ('02) [16] realized that the fact that the gauge invariant definition of the Sivers function in DIS contains a future pointing Wilson line (l.h.s. picture in Fig. 5), whereas in Drell-Yan (DY) it is past pointing (r.h.s. picture in Fig. 5), implies $(f_{1T}^\perp)_{\text{DIS}} = -(f_{1T}^\perp)_{\text{DY}}$. This calculable process

Figure 5. The links in DIS (l.h.s.) and DY run in opposite directions along the lightcone towards lightcone infinity, where an excursion in the transverse direction is taken.

dependence is an indication that the factorization is in terms of intrinsically nonlocal matrix elements, which are sensitive to certain aspects of the process as a whole. This does leave the still open question: what about more complicated processes?

After k_T integration both links reduce to the same link, namely the one we already encountered in $\Phi(x)$ (cf. Eq. (3)). On this latter quantity time reversal does pose the constraint that Collins initially derived for the Sivers function [13], leading to the conclusion that no SSA can arise in *fully* inclusive DIS. This fact was already known at the level of structure functions: Christ and Lee ('66) [17] concluded that for the one-photon exchange approximation in inclusive DIS, only time-reversal violation can lead to a $\sin \phi_S^e$ SSA in $e\, p^\uparrow \to e'\, X$.

Another way to represent the Sivers function and the three other leading k_T-dependent (and often-called 'T-odd') functions is given in Figs. 6–9, where they are depicted as differences of probabilities.

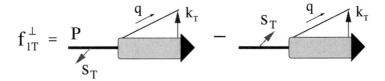

Figure 6. The Sivers effect distribution function. The proton (P) is transversely polarized in direction S_T and the quark (q) has a transverse momentum k_T, such that the probability is proportional to $S_T \times k_T$.

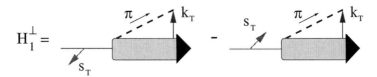

Figure 7. The Collins effect fragmentation function. Here the fragmenting quark is transversely polarized in direction s_T and the outgoing hadron (e.g. a pion) has k_T.

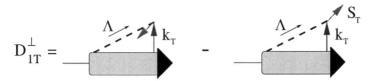

Figure 8. The fragmentation function $D_{1T}^{\perp}$ [18]. Now the outgoing, transversely polarized hadron (here a Λ hyperon) has a transverse momentum k_T.

4. Azimuthal single spin asymmetries

Apart from the left-right asymmetries, *azimuthal* spin asymmetries have been observed. In semi-inclusive DIS, $ep \to e'\pi X$ (SIDIS), the HERMES Collaboration [20] has measured a nonzero $\sin\phi$ asymmetry in $e\vec{p}$ scattering (A_{UL}) (for the definition of ϕ see Fig. 10). Also, preliminary data has been released (at this workshop) by HERMES on $ep^{\uparrow}$ scattering (A_{UT}), suggesting that both Sivers and Collins effects are nonzero. In addition, the CLAS Collaboration (Jefferson Lab) has observed [21] a nonzero $\sin\phi$ in $\vec{e}p$ scattering (A_{LU}). These DIS data are at low Q^2 ($\langle Q^2 \rangle \sim 1-3$ GeV^2), so the interpretation of the asymmetries is not a straightforward matter. But they do demonstrate nontrivial spin effects, possibly related to the asymmetries of the pp experiments.

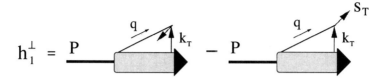

Figure 9. The distribution function $h_1^\perp$ [19] describes transversely polarized quarks with nonzero transverse momentum inside an unpolarized hadron.

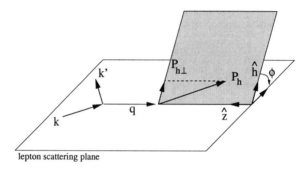

Figure 10. Kinematics of the semi-inclusive DIS process. The angle $\phi \equiv \phi_\pi^e$ is of the transverse momentum $\boldsymbol{P}_{h\perp}$ of the pion w.r.t. the lepton scattering plane, around the photon direction $\boldsymbol{q}$.

4.1. Sivers and Qiu-Sterman effects

The Sivers effect leads to a nonzero A_{UT} (and also to A_{UL}, when the longitudinal spin is taken along the beam direction instead of the photon direction) and has the following characteristic angular dependence [19]:

$$A_{UT} \propto \sin(\phi_\pi^e - \phi_S^e) f_{1T}^{\perp[+]} D_1, \qquad (5)$$

where the superscript [+] indicates that a future pointing Wilson line appears in the Sivers function in this process. It is important to note that this Sivers effect asymmetry does not depend on the lepton scattering plane orientation, because at the parton level it arises from unpolarized quark-photon scattering. Note also that there is no suppression by $1/Q$ (except in A_{UL}).

Qiu and Sterman originally showed that the twist-3 matrix element

$$T_F^{(V)}(x,x)|_{A^+=0} \propto \langle \overline{\psi}(0) \, \Gamma_\alpha \int d\eta \, F^{+\alpha}(\eta n_-) \, \gamma^+ \, \psi(\lambda n_-) \rangle \qquad (6)$$

can lead to a SSA in prompt photon production [11] (Γ_α is an S_T-dependent

Lorentz structure). But it can also lead to a nonzero A_{UT} in SIDIS

$$A_{UT} \propto \sin(\phi_S^e) \frac{T_F^{(V)}(x,x) D_1}{Q}, \qquad (7)$$

where the expression applies *after* integrating over the transverse momentum of the pion. Note that this is not in conflict with the absence of a $\sin(\phi_S^e)$ asymmetry in inclusive DIS [17], where all final state hadrons are integrated out fully.

Recently, it was demonstrated that there is a direct relation between the Sivers and Qiu-Sterman effects [22]:

$$f_{1T}^{\perp(1)[+]}(x) = \frac{g}{2M\vec{S}_T^2} T_F^{(V)}(x,x). \qquad (8)$$

This expression contains a *weighted* Sivers function:

$$f_{1T}^{\perp(1)}(x) = \int d^2\bm{p}_T \, \frac{\bm{p}_T^2}{2M^2} \, f_{1T}^\perp(x, \bm{p}_T^2). \qquad (9)$$

So we conclude that the Sivers and Qiu-Sterman effects are not really different mechanisms after all.

4.2. *Collins effect*

The Collins effect is the only mechanism (within the formalism considered) that can lead to asymmetries A_{UT}, A_{UL} and A_{LU}. For A_{UT} it leads to

$$A_{UT} \propto \sin(\phi_S^e + \phi_\pi^e) \, |\bm{S}_T| \, \delta q \, H_1^{\perp [-]}, \qquad (10)$$

which does depend on the orientation w.r.t. the lepton scattering plane, because at the parton level it is transversely polarized quark scattering off the virtual photon. The asymmetry A_{LU} from the Collins effect is $1/Q$ suppressed [23], but can also be generated perturbatively at $\mathcal{O}(\alpha_s^2)$ [24,25] (only relevant if $|\bm{P}_\perp^\pi|^2 \sim Q^2$, whereas here we consider $|\bm{P}_\perp^\pi|^2 \ll Q^2$).

In Eq. (10) we have indicated the direction of the link in the definition of the Collins function in SIDIS. However, for fragmentation functions the implications of the link structure are not yet clear. On the basis of symmetry restrictions alone one finds schematically [22]

$$(H_1^\perp)_{\text{SIDIS}} \equiv A + B \quad \Rightarrow \quad (H_1^\perp)_{e^+e^-} = A - B.$$

On the other hand, a model calculation by Metz [26] shows that $B = 0$. If this turns out to be true in general, it would simplify the comparison of Collins effect asymmetries from different processes. Clearly, this (calculable) process dependence must be studied further.

Similar considerations apply to the process that may perhaps be of interest to the H1 and ZEUS experiments, namely $\smash{\stackrel{(\to)}{\ell}} p \to \ell' \Lambda^\uparrow X$ [27],

$$P_N \propto K_1 \sin(\phi_\Lambda^\ell - \phi_S^\ell) f_1 D_{1T}^{\perp[-]} + K_3 \sin(\phi_\Lambda^\ell + \phi_S^\ell) h_1^{\perp[+]} H_1. \qquad (11)$$

Here we would like to emphasize that all these asymmetry expressions apply to current fragmentation only.

4.3. Scale dependence

Sivers and Collins effect asymmetries are interesting observables, but are complicated from a theoretical viewpoint. The dependence on the hard scale Q is highly non-trivial. Collinear factorization does not apply, since it is a multiscale process: $M, |\boldsymbol{P}_\perp^\pi|$ and Q with $|\boldsymbol{P}_\perp^\pi|^2 \ll Q^2$. If one considers the differential cross section for this not-fully-inclusive process, $d\sigma/d^2\boldsymbol{P}_\perp^\pi$, beyond tree level, then one finds that soft gluon corrections do not cancel, but rather exponentiate into Sudakov factors [28]. These factors lead to a lowering and broadening (in transverse momentum) of the asymmetry with increasing Q. This decrease can be substantial, but one can define specific weighted asymmetries that are unaffected [29] (apart from logarithmic corrections).

For the azimuthal spin asymmetries one finds [29] that in general, higher harmonics in the azimuthal angle ϕ_π^e decrease faster with Q^2. This is different from the azimuthal asymmetries generated perturbatively at higher orders in α_s, where for instance the ratio $\langle\cos\phi\rangle/\langle\cos 2\phi\rangle$ does not depend on Sudakov factors [30].

5. Conclusions

Striking single spin asymmetries have been observed in experiment (left-right asymmetries and $\sin\phi$ azimuthal asymmetries), but these are still not understood. By using collinear factorization at leading twist, one will not be able to describe these asymmetries, even if one includes higher order perturbative QCD corrections.

Some insights about possible mechanisms for single spin asymmetries are that: the Sivers effect is allowed by time reversal invariance; in SIDIS and Drell-Yan it is opposite in sign; $\boldsymbol{k}_T$-dependent functions may lead to unsuppressed asymmetries; and, the Qiu-Sterman and Sivers effects are directly related. Issues that require further study are: the calculable process dependence of $\boldsymbol{k}_T$-dependent functions (especially of fragmentation functions); the possible connection between the Sivers effect and orbital angular momentum [31]; and, the Q^2 dependence of azimuthal spin asymmetries.

Acknowledgments

I thank the organizers of this interesting workshop for their kind invitation. Some results presented here were obtained in collaboration with Piet Mulders and Fetze Pijlman. The research of D.B. has been made possible by financial support from the Royal Netherlands Academy of Arts and Sciences.

References

1. A.V. Efremov and A.V. Radyushkin, *Theor. Math. Phys.* **44**, 774 (1981).
2. X. Ji, *Phys. Rev.* **D49**, 114 (1994).
3. J.C. Collins, S.F. Heppelmann, G.A. Ladinsky, *Nucl. Phys.* **B420**, 565 (1994).
4. R.L. Jaffe, X. Jin and J. Tang, *Phys. Rev. Lett.* **80**, 1166 (1998).
5. A. Bianconi et al., *Phys. Rev.* **D62**, 034008 (2000).
6. FNAL E704 Collaboration, D.L. Adams et al., *Phys. Lett.* **B261**, 201 (1991).
7. K. Krueger et al., *Phys. Lett.* **B459**, 412 (1999).
8. STAR Collaboration, J. Adams et al., hep-ex/0310058.
9. G. Bunce et al., *Phys. Rev. Lett.* **36**, 1113 (1976).
10. NOMAD Collaboration, P. Astier et al., *Nucl. Phys.* **B588**, 3 (2000).
11. J. Qiu and G. Sterman, *Phys. Rev. Lett.* **67**, 2264 (1991).
12. D. Sivers, *Phys. Rev.* **D41**, 83 (1990); *Phys. Rev.* **D43**, 261 (1991).
13. J.C. Collins, *Nucl. Phys.* **B396**, 161 (1993).
14. S.J. Brodsky, D.S. Hwang and I. Schmidt, *Phys. Lett.* **B530**, 99 (2002).
15. A.V. Belitsky, X. Ji and F. Yuan, *Nucl. Phys.* **B656**, 165 (2003).
16. J.C. Collins, *Phys. Lett.* **B536**, 43 (2002).
17. N. Christ and T.D. Lee, *Phys. Rev.* **143**, 1310 (1966).
18. P.J. Mulders and R.D. Tangerman, *Nucl. Phys.* **B461**, 197 (1996).
19. D. Boer and P.J. Mulders, *Phys. Rev.* **D57**, 5780 (1998).
20. HERMES Collab., A. Airapetian et al., *Phys. Rev. Lett.* **84**, 4047 (2000).
21. CLAS Collaboration, H. Avakian et al., hep-ex/0301005.
22. D. Boer, P.J. Mulders and F. Pijlman, *Nucl. Phys.* **B667**, 201 (2003).
23. J. Levelt and P.J. Mulders, *Phys. Lett.* **B338**, 357 (1994).
24. K. Hagiwara, K. Hikasa and N. Kai, *Phys. Rev.* **D27**, 84 (1983).
25. M. Ahmed and T. Gehrmann, *Phys. Lett.* **B465**, 297 (1999).
26. A. Metz, *Phys. Lett.* **B549**, 139 (2002).
27. D. Boer, R. Jakob and P.J. Mulders, *Nucl. Phys.* **B564**, 471 (2000).
28. J.C. Collins and D. Soper, *Nucl. Phys.* **B193**, 381 (1981).
29. D. Boer, *Nucl. Phys.* **B603**, 195 (2001).
30. P. Nadolsky, D.R. Stump and C.-P. Yuan, *Phys. Lett.* **B515**, 175 (2001).
31. M. Burkardt and D.S. Hwang, hep-ph/0309072.

EFFECTS OF E± POLARIZATION ON FINAL STATES AT HERA

U. STÖSSLEIN

DESY Hamburg
Notkestrasse 85, 22607 Hamburg, Germany
E-mail: uta.stoesslein@desy.de

At HERA II, longitudinally polarized electrons and positrons collide with unpolarized protons at centre-of-mass-energies of 318 GeV and at high luminosities. This offers unique opportunities for testing the Standard Model and probing new physics areas. A brief overview is given of the possibilities offered by the use of beam charge and polarization dependent final states at HERA, in particular as regards electroweak structure functions, generalized parton distributions and Λ-baryon production.

1. Polarization and Luminosity

The need for luminosity greater than the design value as well as for longitudinally polarized beams at HERA has been anticipated by a series of physics workshops [1]. After tapping the potential of the HERA I phase, in 2000/2001 a major upgrade [2] was performed to the two interaction regions (H1 and ZEUS) to improve focussing at the interaction points and to thus obtain higher luminosity values. Each interaction region was rebuilt and complemented by a pair of spin rotators, allowing the transverse polarization generated via synchrotron radiation in HERA-e to be converted to longitudinal polarization. With one spin rotator pair, which has been working since 1995 at the HERA-e fixed target experiment HERMES, typical polarization values of up to 60% were achieved. In March 2003, successful operation with three spin rotator sets was achieved and the longitudinal polarization of the positrons reached values of up to 54% in colliding mode [2]. The HERA-e polarimeter continuously measure both the transverse and the longitudinal beam polarization. These devices are being upgraded to allow polarization measurements of a relative systematic uncertainty of 1% with negligible statistical errors [3], as is required for precision measurements. Absolute luminosities of $2.7 \cdot 10^{31}$cm^{-2}s^{-1} have been reached, suggesting that luminosities of about $5 \cdot 10^{31} \mathrm{cm}^{-2} \mathrm{s}^{-1}$ are achievable at full beam currents and design specific luminosity. This should allow the collection of about 200 pb^{-1} of data per year [2].

This talk discusses briefly the effects on the final states of scattering polarized e$^\pm$ off unpolarized protons at HERA. A collider programme of nucleon and photon spin physics and on electron-ion collisions may be realized after the present HERA II phase [4].

2. Beam Polarization Effects in Electroweak Interactions

High luminosities and high polarization values allow inclusive deep inelastic scattering (DIS) measurements to be extended to large photon virtualities Q^2 where the event rates are small. Basic quantities of interest are the proton structure functions which are effective descriptions of the nonperturbative part of the inclusive ep-scattering process. Utilizing the lepton charge and helicity dependence of the DIS charged current (CC) and neutral current (NC) cross sections, more precise and new information on the proton dynamics can be obtained, or alternately, using the knowledge of proton structure functions, electroweak parameters can be tested.

An impressive, 'textbook' test of the electroweak theory can be per-

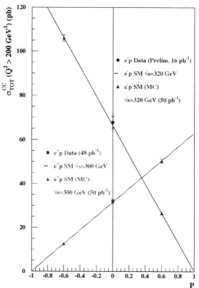

Figure 1. Measured unpolarized (HERA I data) and simulated polarized e$^\pm$p CC cross sections as a function of lepton polarization assuming an integrated luminosity of 50 pb^{-1} for each of the polarized data sets. Figure taken from [5].

formed by measuring of the polarized $e^{\pm}p$ CC cross sections as shown in Fig. 1 [5]. The expected linear dependence of the CC cross section on the beam polarization, i.e. the handedness of the lepton, is a direct consequence of the Standard Model, in which the massive W^+ and W^--particles, the carriers of the weak force, only act on left-handed particles and right-handed antiparticles. Any deviation from a straight line would point to physics beyond the Standard Model. Presently, the H1 and ZEUS experiments are taking data with right-handed positrons and a first analysis of the CC cross section can be expected soon. However, to get the complete picture depicted in Fig. 1, four beam charge and helicity combinations have to be measured with high polarization values and integrated luminosities of about 50 pb^{-1} per sample.

Competitive searches for new physics can be performed by investigating parity violation in NC DIS, where the cross section depends on both beam polarization and charge, see e.g. [6] for a systematic consideration of cross sections and their combinations. The interference of neutral electromagnetic (γ) and weak (Z) currents leads to additional vector and axial-vector contributions to the $e^{\pm}p$ cross section. These are parameterized in terms of two new structure functions, G_2 and xG_3 (not to be confused with the well known polarized structure functions). Both contain the quark couplings to the Z-boson. Since the strength of the γZ interference itself is about $10^{-4} \cdot \text{GeV}^2/Q^2$, high Q^2 values are preferable if these structure functions are to be measured. Measuring the asymmetries introduced by electroweak effects has the advantage that common systematic uncertainties, like acceptances and global inefficiencies, cancel.

Varying the beam polarization for a fixed lepton charge delivers a parity-violating asymmetry ($\propto G_2/F_2$), studied first at SLAC in 1978 in a high statistics ed-scattering experiment at a Q^2 of 1.6 GeV^2 [7]. Measured at large x_{Bj}, the parity-violating asymmetry is dominated by the ratio of d to u-valence quarks. Simulated HERA data covering the range of Q^2 from 1000 until 30000 GeV^2 unveil the feasibility of a G_2 measurement at $x_{\text{Bj}} > 0.01$ using the excellent knowledge of the electromagnetic structure function F_2 already obtained from the HERA I data. High polarization values, of about 50%, and high luminosity values of about 200 pb^{-1} per sample are needed [8]. Variation of both lepton polarization and charge, first performed in μC-scattering in 1982 at CERN [9], allow the determination of a beam conjugation asymmetry which is predominantly sensitive to the interference structure function xG_3. At HERA this delivers new information on the behavior of valence quark distributions in the sea quark range, $x_{\text{Bj}} > 0.01$.

High polarization values of at least 50% and integrated luminosities of about 150 pb^{-1} per sample can significantly improve the tests of electroweak parameters, for example the determination of the aforementioned vector and axial-vector couplings of the Z-boson to the light quarks [10]. The best sensitivity to the mass of the W-boson and the value of $\sin^2\theta_W$ can be achieved using the polarization dependent neutral to charged current cross section ratio for electrons; slightly less precise results are obtained using positrons [11].

3. DVCS Cross Section and Asymmetries

With the advent of HERA, a new class of large rapidity gap events was discovered in DIS [12] at small $x_{\rm Bj} < 0.01$. These diffractive events were not expected in the quark-parton model (QPM). An intuitive picture of these events was offered by the color dipole model [13], a nearly forgotten picture at this time. With increased luminosity, in 1999 another class of diffractive events was discovered at HERA [14,15,16] in which there is a large rapidity gap between the recoiling proton and a real photon in the final state, see Fig. 2a. These deeply virtual, or off-forward, Compton scattering (DVCS) events at small momentum transfer t to the proton can also be described by the color dipole model; see [16,17] for comparisons with recent ZEUS and H1 data. Furthermore, the observation of DVCS triggered the further development and use of another theoretical concept for diffractive and exclusive processes, generalized parton distribution (GPD) functions. GPDs were revived in 1996 when the spin physics community realized the unique potential they have for unravelling the spin structure of the nucleon [18]. Only recently, the exciting potential of GPDs in the study of three dimensional hadron structure has been recognized [19]. In the case that the initial and final state differ only in their transverse momenta, GPDs encode simultane-

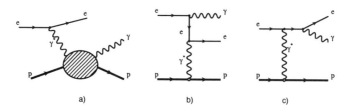

Figure 2. a) DVCS graph, b) BH with photon from initial state lepton and c) with photon from final state lepton. Figure taken from [15].

ously information on the longitudinal momentum, determined by x_{Bj}, and transverse position of partons in the infinite momentum frame, described by an impact parameter $b \sim 1/\sqrt{-t}$. Nowadays, GPD based model calculations deliver fair descriptions of the measured, t-integrated DVCS cross section and its kinematic dependence [16,17], and are theoretically rather well understood, see [20] for a review. More, and more precise, measurements are required to pin down the various theoretical model uncertainties.

The diffractive electroproduction of a real photon, depicted in Fig. 2a, interferes with the Bethe-Heitler (BH) process shown in Figs. 2b,c. Both deliver the same final states, but these can be distinguished experimentally by the different weights they have in different kinematic regions. Furthermore, the BH process is a pure QED process with a calculable cross section. This kind of background is well controlled and can be subtracted allowing DVCS cross section measurements [15,16]. Moreover, with the advent of HERA II, the complex and rich angular structure of the DVCS cross section can be employed to access more observables which may help to unfold GPDs.

In general, the amplitude-squared of the real photon electroproduction cross section receives contributions from pure DVCS (Fig. 2a), from pure BH (Figs. 2b,c) and from their interference (with a sign governed by the lepton charge): $|\mathcal{T}|^2 \propto |\mathcal{T}_{DVCS}|^2 + |\mathcal{T}_{BH}|^2 + \mathcal{I}$. Each term can be expanded in a Fourier series in the azimuthal angle ϕ between the electron scattering plane and the real photon reaction plane [21]. This can be used to filter out the relevant information about the Compton process. The terms are given by

$$|\mathcal{T}_{BH}|^2 \propto \left[c_0^{BH} + \sum_{n=1}^{2} c_n^{BH} \cos(n\phi) + s_1^{BH} \sin(\phi) \right] \quad (1)$$

$$|\mathcal{T}_{DVCS}|^2 \propto \left[c_0^{DVCS} + \sum_{n=1}^{2} [c_n^{DVCS} \cos(n\phi) + s_n^{DVCS} \sin(n\phi)] \right] \quad (2)$$

$$\mathcal{I} \propto -\text{sign}(e) \left[c_0^{\mathcal{I}} + \sum_{n=1}^{3} [c_n^{\mathcal{I}} \cos(n\phi) + s_n^{\mathcal{I}} \sin(n\phi)] \right] \quad (3)$$

According to the detailed formulae in [22], ϕ dependent contributions in the scattering of polarized leptons with helicity λ off an unpolarized proton can be expected to arise from:

i) pure BH (Eq. 1) in addition to the known kinematical ϕ dependence of the BH propagator giving

(a) $\cos\phi$ and $\cos 2\phi$ terms (leading twist-2)

(b) no $\sin\phi$ term (appears only for transversely polarized proton)
   (c) no beam helicity dependence (appears only for longitudinally or transversely polarized protons)

ii) pure DVCS (Eq. 2) giving

   (a) suppressed $\cos\phi$ and $\lambda\sin\phi$ terms (twist-3)
   (b) α_S power suppressed $\cos 2\phi$ term (leading twist-2, related to gluon transversity; $\sin 2\phi$ will appear only for longitudinally or transversely polarized protons)

iii) DVCS-BH Interference (Eq. 3) in addition to the known kinematical ϕ dependence of the BH propagator and the dependence on the sign of the lepton charge giving

   (a) $\cos\phi$ and $\lambda\sin\phi$ terms (leading twist-2)
   (b) suppressed $\cos 2\phi$ and $\lambda\sin 2\phi$ terms (twist-3)
   (c) α_S power suppressed $\cos 3\phi$ and $\sin 3\phi$ terms (leading twist-2, related to gluon transversity).

The Fourier coefficients c_n and s_n in Eqs. 2-3 are directly related to the amplitudes of the γ^*p Compton process which can be parameterized by the so-called Compton Form Factors (CFF) [22]. CFFs for their part are convolutions of QCD coefficient functions and GPDs denoted usually by H, $\tilde{H}$, E, $\tilde{E}$. CFFs, and hence the GPDs, appear in quadratic combinations in the DVCS cross section, but in linear combinations in the interference term (Eq. 3). Particularly for HERA kinematics, measurements of the dominant interference Fourier coefficients $c_1^{\mathcal{I}}$ and $s_1^{\mathcal{I}}$ gives access to, respectively, the real and the imaginary part of the CFF $\mathcal{H}$ and thus to the leading twist-2 GPD H, since at small x_{Bj} possible contributions from the GPDs $\tilde{H}$ and E can be safely neglected [23]. In principle, a complete separation of the four leading twist-2 GPDs would require in addition data taken with transversely and longitudinally polarized protons.

The real photon production observable containing information on the $s_1^{\mathcal{I}}$ term is the single beam spin asymmetry (SSA), measured at Q^2 of about 2 GeV^2 by the fixed target experiments CLAS and HERMES for the first time [24]. A measurement of the SSA requires high lepton beam and virtual photon polarizations (high y). A simulation [23] for the HERA-positron beam reveals significant negative asymmetry values, shown in Fig. 3 (left). This asymmetry is positive for an electron beam. Experimentally, the ϕ dependence of the difference between the number of events with positive beam helicity and negative beam helicity has to be formed. Using an un-

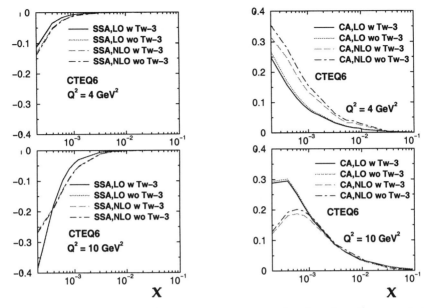

Figure 3. DVCS asymmetries for HERA kinematics as a function of x_{Bj} for two typical values of Q^2 and $t_{max} = -0.5$ GeV^2, modeled in LO and NLO QCD with (w) and without (wo) twist-3 contributions (Tw-3). Left: t integrated single beam spin asymmetry (SSA) for a positron beam. Right: t integrated beam charge asymmetry (CA). Figures taken from [23].

polarized beam sample to get rid of the $\sin\phi$ amplitude contributions, and positron and electron beam data, the beam charge asymmetry sensitive to the $c_1^{\mathcal{I}}$ term can be formed for real photon production events. Only HERMES has been able to measure the CA so far [25], employing HERAs unique feature of delivering positron and electron beams. The CA simulation [23] for HERA-collider kinematics shows, in Fig. 3 (right), encouragingly large asymmetry values. A further measurement of the $c_1^{\mathcal{I}}$ term can be performed via an azimuthal angle asymmetry that is predicted to have a size similarly to that of the CA [23]. However, this measurement would require a detector with excellent ϕ resolution and control of twist-3 contributions whereas the twist-3 effects for the CA and SSA have been estimated to be negligible [23].

In the second phase of HERA, it is anticipated that the size of the data samples available for the DVCS cross section measurements will be about 10 times larger than those used so far. The measurement of the DVCS associated asymmetries will remain challenging with the present H1 and

ZEUS detectors, since the recoiling proton will not generally be observed, making the determination of ϕ and t difficult [26]. The very forward proton spectrometer (VFPS) [27] recently installed at H1 may help to further reduce the proton-dissociative background in the DVCS sample and to deliver a first measurement of the t-dependence of the DVCS cross section at centre-of-mass energies of the γ^*p-system W of about 20 GeV.

4. Beam Spin Transfer to Λ-Baryons

Studying Λ ($\bar{\Lambda}$) baryon production in unpolarized and polarized DIS processes, allows the exploitation of the 'self-analyzing' decay of the Λ. The Λ-polarization can be measured by studying the angular distribution of the $\Lambda \to p\pi$ decay (in the Λ helicity rest frame). The polarization of the fragmenting parton is determined by the elementary Standard Model interactions and the initial parton's spin state. Neglecting weak interactions, there are four general DIS observables for involving Λ's that employ the various possible combinations of (un)polarized leptons and nucleons, see e.g. [28].

Particularly interesting for the upgraded HERA is to consider the spin transfer from a longitudinally polarized charged lepton (helicity S) to the Λ while the proton is unpolarized (helicity 0). In this DIS process, the scattered quark will be polarized and its spin will be transferred to the baryon produced in the fragmentation of this quark. In the QPM, the longitudinal spin transfer to the outgoing Λ is given by [28]:

$$P_{S,0} = \frac{y(2-y)}{1+(1-y)^2} \frac{\sum_q e_q^2\, q(x)\, \Delta D_{\Lambda/q}(z)}{\sum_q e_q^2\, q(x)\, D_{\Lambda/q}(z)}. \qquad (4)$$

Here, $P_{S,0}$ is the polarization of the hyperon Λ which is measurable in semi-inclusive DIS, and q and D are the usual unpolarized quark distribution and fragmentation functions. The fragmentation of a longitudinally polarized parton into a longitudinally polarized Λ is described by $\Delta D_{\Lambda/q}(z)$. This polarized fragmentation function may be further related, via the so-called Gribov-Lipatov relation [29], to the polarized quark distribution function of the Λ [30].

From Eq. 4, it follows that a Λ-spin transfer measurement requires high beam polarizations, high virtual photon polarizations (high y values) and a broad range in z, the hadron momentum fraction in the lab frame. For Λ's produced in the current fragmentation region in the DIS process, i.e. originating from the struck quark, the spin of the Λ is entirely due to the strange quark within the naive QPM. On the other hand, e.g. using SU(3)$_f$, the quark distributions of the Λ can be related to those in the proton [31], or, if both the proton and the Λ spin structures are known, $SU(3)_f$ symmetry may be tested. In recent years, many theoretical models have been proposed for the longitudinal polarization of Λ baryons in DIS, addressing the question of the relationships between the spin structure of the proton and of the other baryons [32,33]. Recent DIS results on longitudinal Λ-polarization have come from HERMES [34] in NC charged lepton-nucleon scattering and from NOMAD in ν_e charged current interactions [35], but all these data are for low W values, which complicates their interpretation. Here, HERA II data could give a significant input, due to the much larger W ranges accessible, although detailed simulations have yet to be performed. The production of Λ (Λ̄) baryons in polarized charged current $e^\pm p$ DIS could provide information on flavor separated polarized quark fragmentation functions, in a manner analogous to the scattering of a neutrino beam on a hadronic target.

5. Conclusion and Outlook

HERA II, with its high luminosity and with longitudinally polarized electrons and positrons, opens new horizons in the study of electroweak theory and parton dynamics, both in inclusive DIS and via the selection of particular final states. The experimental prospects have been illustrated in this talk by discussing the most promising channels: electroweak structure functions and charged current interactions, real photon electroproduction which extends the framework of DIS to the off-forward region of the virtual Compton process and the study of Λ spin structure via its longitudinal polarization.

Beyond the measurements mentioned, there are further subjects awaiting investigation. One may think of e.g. an even more detailed mapping of the GPDs which can be done by studying the complex angular dependence of the cross section for the electroproduction of lepton pairs off an unpolarized or polarized nucleon target [36]. Unfortunately, the cross section is very low thus requiring very high luminosities, but first studies may start at HERA II. Another interesting test of NLO versus twist-3 effects at HERA II may be performed by measuring a single-beam spin asymmetry in semi-inclusive pion production, as was done recently by CLAS at low Q^2, hence making interpretation difficult [37]. Single-beam spin and charge asymmetries may also be observed in the diffractive electroproduction of a $\pi^+\pi^-$ pair. These observables are expected to be sensitive to Pomeron-Odderon interference [38] and could thus give first evidence for the Odderon, which has so far escaped detection.

HERA II has just started. More detailed investigations are required of the huge variety of channels available, and unexpected results may well appear.

Acknowledgments

I would like to thank Markus Diehl, Andreas Freund and Henri Kowalski for fruitful discussion and Tim Greenshaw for carefully reading the manuscript.

References

1. B. Wiik, *Proceedings of the Workshop Future Physics at HERA 1995/96*, http://www.desy.de/ heraws96 (1996).
2. G. Hoffstätter, M. Vogt, F. Willeke, *ICFA Beam Dyn. Newslett.* **30**, 7 (2003).
3. J. Böhme for the POL2000 group, *Acta Phys. Polon.* **B33**, 3949 (2002).
4. H. Abramowicz et al., MPI-PhE/2003/06 (2003);
   T. Alexopoulos et al., DESY/03-194 (2003).
5. A. Mehta, talk given at *10th International Workshop on Deep Inelastic Scattering (DIS2002), Cracow, Poland, 30 Apr - 4 May 2002* ; *Acta Phys. Polon.* **B33**, 3937 (2002).
6. M. Klein, T. Riemann, *Z. Phys.* **C24**, 151 (1984).
7. C. Y. Prescott et al., *Phys. Lett.* **B84**, 524 (1979).
8. M. Klein, *Proceedings of 9th International Workshop on Deep Inelastic Scattering (DIS 2001), Bologna, Italy, 27 Apr - 1 May 2001*, 409 (2002)
9. BCDMS, A. Argento et al., *Phys. Lett.* **B120**, 245 (1983).
10. R. Cashmore et al., *Proceedings of the Workshop Future Physics at HERA 1995/96*, 163 (1996).
11. J. Blümlein, M. Klein, T. Riemann, *Proceedings, HERA Workshop, Hamburg 1987, vol. 2*, 687 (1987).
12. ZEUS Collaboration, M. Derrick et al., *Phys. Lett.* **B315**, 481 (1993).
    H1 Collaboration, T. Ahmed et al., *Nucl. Phys.* **B429**, 477 (1994).
13. J.D. Bjorken, J.B. Kogut, D.E. Soper, *Phys. Rev.* **D3**, 1382 (1970).
14. P.R.B. Saull for the ZEUS collaboration, *Proc. of the International Europhysics Conference on High-Energy Physics (EPS-HEP 99), Tampere, Finland, 15-21 July 1999*, 420, (2000), hep-ex/0003030 (2000).
15. H1 Collaboration, C. Adloff et al., Phys. Lett. B **517** (2001) 47, hep-ex/0107005 (2001).
16. ZEUS Collaboration, S. Chekanov et al., Phys. Lett. B **573** (2003) 46, hep-ex/0305028 (2003).
17. L. Favart for the H1 Collaboration, hep-ex/0312013 (2003).
18. X.-D. Ji, *Phys. Rev. Lett.* **78**, 610 (1997), hep-ph/9603249 (1996).
19. M. Burkardt, *Phys. Rev.* **D62**, 071503 (2000) [Erratum-ibid. **D66**, 119903 (2002)], hep-ph/0005108 (2000).

20. M. Diehl, *Phys. Rept.* **388**, 41 (2003), hep-ph/0307382 (2003).
21. M. Diehl, T. Gousset, B. Pire and J. P. Ralston, *Phys. Lett.* **B411**, 193 (1997), hep-ph/9706344 (1997).
22. A. V. Belitsky, D. Muller, A. Kirchner, *Nucl. Phys.* **B629**, 323 (2002), hep-ph/0112108 (2001).
23. A. Freund, *Phys. Rev.* **D68**, 096006 (2003), hep-ph/0306012 (2003).
24. HERMES Collaboration, A. Airapetian et al., *Phys. Rev. Lett.* **87**, 182001 (2001), hep-ex/0106068 (2001).
    CLAS Collaboration, S. Stepanyan et al., *Phys. Rev. Lett.* **87**, 182002 (2001), hep-ex/0107043 (2001).
25. F. Ellinghaus for the HERMES Collaboration, *Nucl. Phys.* **A711** 171 (2002), hep-ex/0207029 (2002).
26. R. Stamen, talk given at *HERA-III Workshop 2002, Munich, Germany, 18 - 20 Dec 2002*, http://wwwhera-b.mppmu.mpg.de/hera-3/hera3/index.html.
27. L. Favart et al., PRC-01/00 (2000) http://web.iihe.ac.be/h1/vfps/documents.html.
28. M. Anselmino, hep-ph/0302008 (2003).
29. V.N. Gribov, L.N. Lipatov, *Phys.Lett.* **B37**, 78 (1971); *Sov. J. Nucl. Phys.* **15**, 675 (1972).
30. B.Q. Ma, I. Schmidt, J. Soffer and J.J. Yang, *Eur. Phys. J.* **C16**, 657 (2000), hep-ph/0001259 (2000).
31. B.Q. Ma, I. Schmidt, J. Soffer and J.J. Yang, *Phys. Rev.* **D65**, 034004 (2002), hep-ph/0110029 (2001).
32. J. Soffer, *Nucl. Phys. Proc. Suppl.* **105**, 140 (2002) hep-ph/0111054 (2001).
33. J.R. Ellis, A. Kotzinian, D.V. Naumov, *Eur. Phys. J.* **C25**, 603 (2002) hep-ph/0204206 (2002).
34. HERMES Collaboration, A. Airapetian et al., *Phys.Rev.* **D64**, 112005 (2001) 112005, hep-ex/9911017 (1999).
    S. Belostotski, O. Grebenyuk, Yu. Naryshkin for the HERMES Collaboration, *Acta Phys. Polon.* **B33**, 3785 (2002).
35. D.V. Naumov for the NOMAD Collaboration, *Acta Phys. Polon.* **B33**, 3791 (2002).
36. A.V. Belitsky and D. Müller, *Phys. Rev.* **D68**, 116005 (2003), hep-ph/0307369 (2003).
37. CLAS Collaboration, H. Avakian et al., hep-ex/0301005 (2003).
38. P. Hagler et al., *Nucl. Phys. Proc. Suppl.* **121**, 155 (2003), hep-ph/0209242 (2002).

3
Production of Hadrons, Jets and Photons

PHOTOPRODUCTION OF JETS AND PROMPT PHOTONS AT HERA[*]

J. CVACH

*Centre for Particle Physics, Institute of Physics AS CR
Na Slovance 2,
Prague 8, CZ-182 21, Czech Republic
E-mail: cvach@fzu.cz*

Recent measurements of jet and prompt photon production at HERA are reviewed. The results are compared to NLO QCD calculations which describe the data well except in extreme regions of phase space. Studies of the inclusive single jet cross section provide very accurate value of α_s.

1. Introduction

At the positron–proton HERA collider positrons[a] of energy 27.6 GeV collide with protons with energy 820 (until 1997) or 920 GeV. The positrons are the source of quasi-real photons which interact with the protons. We usually call their interactions photoproduction at HERA if the photon virtuality $Q^2 < 1$ GeV^2. Scattered positrons at higher virtualities are detected in the main detector and lead to processes which are called deep inelastic scattering. Photoproduction represents the dominant hard scattering process and provides, therefore, the most accurate results in a large kinematic range. In the lowest order of the perturbative series expansion in α_s, a quasi-real photon interacts with a parton from the proton directly (Fig. 1a) or as resolved (Fig. 1b) via one of its partons generated in a hadronic fluctuation. Due to the QCD radiation and fragmentation processes, the partons appear as collimated jets of particles accompanied in general by the proton remnant and, in the case of resolved processes, by the photon

---

[*]This work is supported by the Ministry of Education of the Czech Republic under the projects INGO-LA116/2000 and LN00A006.
[a]We use positron for the HERA lepton beam where electrons are used as well. As the electron data sample has significantly smaller luminosity we mention electrons explicitly where it is relevant.

remnant and by particles resulting from a possible remnant-remnant interaction, also called the underlying event. The cross sections for direct and

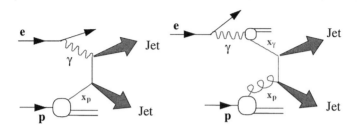

Figure 1. Example LO QCD diagrams for jet-jet photoproduction in direct (a) and resolved (b) photon interactions.

resolved photon interactions are given to LO by:

$$\sigma_{\rm dir}^{ep \to ejetjet(\gamma)X} = \sum_j f_{i/p} \otimes \hat{\sigma}_{ej \to {\rm jetjet}(\gamma)}$$

and for the resolved photon interaction by:

$$\sigma_{\rm res}^{ep \to ejetjet(\gamma)X} = \sum_{i,j} f_{\gamma/e} \otimes f_{i/\gamma} \otimes f_{j/p} \otimes \hat{\sigma}_{ij \to {\rm jetjet}(\gamma)},$$

where $f_{k/a}$ is the parton density function (PDF) for parton k in the particle $a = \gamma, p$ and $f_{\gamma/e}$ is the photon flux in the positron given by the equivalent photon approximation, $\hat{\sigma}_{ej}$ is the cross section for positron–quark scattering using the exact matrix element (which is also valid in the NLO case) and $\hat{\sigma}_{ij}$ is the cross section for parton–parton scattering. PDFs depend on the factorization scale μ_f and the cross sections $\hat{\sigma}$ on the renormalization scale μ_r via α_s. Here both scales are set to be equal.

The new HERA results on jet photoproduction are based on luminosities of several tens of pb^{-1}. They allow comparisons to the NLO calculations up to higher jet E_T and in a more differential way. Studying high E_T jets has the advantage that the hadronisation of the parton from the hard scattering influences the jet less. Also, the impact of soft processes from the proton–photon remnant interaction are less important. Having achieved good agreement with the NLO calculations, data are used to check different parametrization of the PDFs of the photon, which are mostly known from F_2^γ measurements at LEP. HERA data are unique due to the high sensitivity to the gluon density in the photon via the dominant boson–gluon fusion process. The sensitivity to the PDFs is high in the region where the difference between the rapidities of both jets is small. In contrast, the region with large jet rapidity difference is sensitive to the form of the matrix element.

We will first present results on single jet inclusive cross section measurements which provide a very accurate value of α_s. Then we review the results on dijet inclusive cross sections. From the four-momenta of both jets, the fraction of the photon momentum carried by the interacting parton x_γ^{jet} can be estimated. This quantity is used to separate direct and resolved events and to investigate the dynamics of these classes separately. Finally, new data on direct photon production are compared to NLO QCD calculations.

Since most of the results use a similar experimental procedure, we shall summarize here their main features:

(a) $Q^2 < 1$ GeV^2, $0.1 < y < 0.9$ (H1), $0.2 < y < 0.85$ (ZEUS).
(b) Jets were obtained using the inclusive k_t algorithm[1].
(c) Data were corrected for detector effects using two of three MC event generators, PYTHIA[2] and PHOJET[3], which both use LUND strings for fragmentation and HERWIG[4] with its cluster fragmentation scheme. Soft processes are modeled in PYTHIA as Multiple Interactions between the proton and photon remnants, PHOJET uses the Dual Parton Model[5] and HERWIG adds an adjustable fraction of the soft underlying events.
(d) The dominant systematic errors come from two sources: uncertainty in the knowledge of the energy scale in the hadron calorimeter causes errors of 20% at low E_T which drop below 10% at high E_T and model dependence of the detector corrections which causes errors below 10%.

For the comparison of data with NLO predictions the following calculations are used:

(a) NLO weighted parton MC of Frixione[6] using GRV-HO[7] for the photon and CTEQ5M[8] for the proton PDF as the standard parametrization. To test the dependence of the calculation on the PDF parametrization, other choices were used such as MRST99 and CVTEQ5HJ for the proton and AFG-HO for the photon.
(b) The program of Klasen, Kleinwort and Kramer[9] using the phase space (PS) slicing method to separate the infrared and mass singular

phase space region.

(c) The program of Fontannaz, Guillet and Heinrich[10] for isolated prompt photon production combines the PS slicing and subtraction methods. It uses AFG for the photon and MRST2 for the proton PDFs.

(d) As the calculations are done for partons, the main uncertainty comes from the hadronisation correction and decreases from 20% for low E_T to less than 10% at high E_T. The renormalization and factorization scale uncertainty is below 10%.

Experimental data are displayed in figures with the inner error bars denoting the statistical error. Systematic errors without the uncertainty due to the jet energy scale are added in quadrature with the statistical error and shown as the outer error bars. The uncertainty due to the jet energy scale is shown separately as a shaded band.

The dominant partonic subprocesses responsible for jet production at HERA are $\gamma g_p \to q\bar{q}$ and $q_\gamma g_p \to qg$. The kinematics of these two-to-two subprocesses are such that the majority of jets in the region of jet pseudorapidity $\eta^{\text{jet}} < 0$ are predicted to originate from outgoing quarks while the fraction of gluon-initiated jets increases as η^{jet} increases[11]. Therefore, the experimental investigation of jet properties as a function of jet pseudorapidities allows the study of the characteristics of quark- and gluon-initiated jets. To distinguish between both types of jets, jet shapes and mean subjet multiplicities using the k_t-cluster algorithm were used as an analyser. The jets become broader and the mean subjet multiplicity increases for final-state gluon-initiated jets. The use of simultaneous cuts on the jet shapes and subjet multiplicity allowed the definition of enriched samples of quark- ("thin") and gluon- ("thick") initiated jets.

The distribution of the angle between the jet-jet axis and the beam direction in the dijet centre-of-mass system $\cos\theta^*$ is sensitive to the spin J of the exchanged particle $d\sigma/d\cos\theta^* \sim (1 - |\cos\theta^*|)^{-2J}$ as $|\cos\theta^*| \to 1$. In the case of direct photon interactions the contributing processes in LO QCD are $\gamma q(\bar{q}) \to gq(\bar{q})$ and $\gamma g \to q\bar{q}$ which involve quark exchange in the s, t and u channels. In the case of resolved photon interactions, the contributing processes $qg \to qg$, $qq' \to qq'$, $gq \to gq, \ldots$ involve gluon exchange. The study of the angular distribution of dijet events thus provides a handle for the investigation of the underlying parton dynamics[12].

2. Single jet production and α_s

The new measurements from H1[13] cover an extended range of jet transverse energies between 5 and 75 GeV. For $E_T^{\rm jet} \geq 21$ GeV, the hadronisation corrections to the NLO QCD calculation only slightly improve the agreement with data, whereas for $5 < E_T^{\rm jet} < 21$ GeV good agreement can only be obtained by applying them. The NLO predictions are in agreement with the $d\sigma/d\eta^{\rm jet}$ by measurements in different $E_T^{\rm jet}$ bins except in the lowest range $5 < E_T^{\rm jet} < 12$ GeV (see Fig. 2). The data seem to indicate a rise of the cross section with increasing $\eta^{\rm jet}$ which is faster than in the theoretical predictions. Inadequacy of the photon PDFs in this kinematic range or the

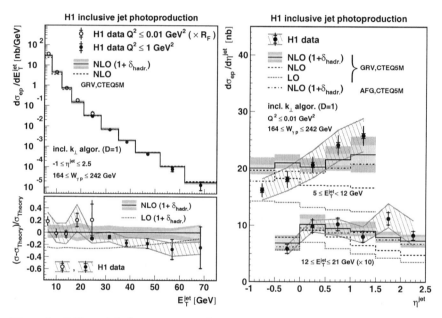

Figure 2. Differential e^+p cross section for inclusive jet production as a function of (a) $E_T^{\rm jet}$ integrated over $-1 < \eta^{\rm jet} < 2.5$. At the bottom of the figure, the relative difference between data or LO QCD and the NLO calculations, including hadronisation correction, are shown. (b) $\eta^{\rm jet}$ integrated over the two lowest $E_T^{\rm jet}$ ranges. The data are compared with LO and NLO QCD predictions obtained by using GRV or AFG photon PDFs and CTEQ5M proton PDFs.

absence of higher order corrections beyond NLO may be responsible.

ZEUS measured[14] the inclusive jet cross section as a function of $E_T^{\rm jet}$ in the range between 17 and 95 GeV. No significant deviation with respect to QCD NLO calculation was observed up to the highest scale studied. The measured cross section $d\sigma/dE_T^{\rm jet}$ was used to determine $\alpha_s(M_Z)$ using the method presented previously[15]. The $\alpha_s(M_Z)$ dependence of the predicted

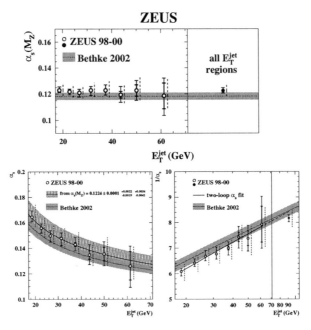

Figure 3. (top) The $\alpha_s(M_Z)$ values determined from the QCD fit to the measured $d\sigma/dE_T^{\rm jet}$ values in different $E_T^{\rm jet}$ regions (open circles) and the combined value. (lower left) The $\alpha_s(E_T)$ values determined from the QCD fit of the measured $d\sigma/dE_T^{\rm jet}$ values as a function of $E_T^{\rm jet}$ (open circles). The solid line represents the prediction of the renormalization group equation; the light shaded area displays its uncertainty. (lower right) The $1/\alpha_s(E_T)$ values as a function of $E_T^{\rm jet}$ (open circles). The solid line represents the result of the two-loop α_s fit to the measured values, the dashed line the extrapolation of the fit to $E_T^{\rm jet} = M_Z$. The point plotted at $E_T^{\rm jet} = M_Z$ represents the inverse of the combined value from (top).

$d\sigma/dE_T^{\rm jet}$ in each bin of $E_T^{\rm jet}$ was parametrised according to

$$[d\sigma/dE_T^{\rm jet}(\alpha_s(M_Z))]_i = C_1^i \alpha_s(M_Z) + C_2^i \alpha_s^2(M_Z)$$

where C_1^i and C_2^i are constants, by using the NLO QCD calculations corrected for hadronisation effects. Finally, a value of $\alpha_s(M_Z)$ was determined in each bin of the measured cross section as well as from all data by a χ^2 fit. The value of $\alpha_s(M_Z)$ is

$$\alpha_s(M_Z) = 0.1224 \pm 0.0001({\rm stat.})^{+0.0022}_{-0.0019}({\rm exp.})^{+0.0054}_{-0.0042}({\rm th.}).$$

This value of $\alpha_s(M_Z)$ is consistent with the current world average[16] of 0.1183 ± 0.0027. The largest contribution to the experimental uncertainty comes from the jet energy scale and amounts to $\pm 1.5\%$. The QCD prediction for the energy scale dependence of α_s was tested by determining α_s from $d\sigma/dE_T^{\text{jet}}$ at different E_T^{jet} values. The measured $\alpha_s(E_T)$ values are shown in Fig. 3 and are in good agreement with the predicted running of the strong coupling constant over a large range of E_T^{jet}.

The cross section $d\sigma/d\eta^{\text{jet}}$ is presented in Fig. 4a for samples of "thick" and "thin" jets separated according to the method explained in Section 1. The measured cross sections for the two samples of jets exhibit different behaviour: the η^{jet} distribution for "thick" jets peaks at $\eta^{\text{jet}} \sim 2$, whereas the distribution from "thin" jets peaks at $\eta^{\text{jet}} \sim 0.7$. The same selection methods has been applied to the jets of hadrons in the Monte Carlo samples (PYTHIA, HERWIG and PYTHIA MI). The MC predictions provide a good description of the shape of the "thin"-jet data distribution. From

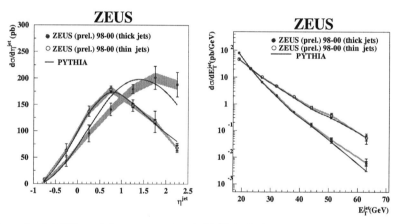

Figure 4. (a) Differential ep cross section $d\sigma/d\eta^{\text{jet}}$ for inclusive jet production integrated over $E_T^{\text{jet}} > 17$ GeV for "thick" and "thin" jets. (b) Differential $d\sigma/dE_T^{\text{jet}}$ cross section integrated over $-1 < \eta^{\text{jet}} < 2.5$. The PYTHIA prediction normalised to the total cross section is shown for comparison.

the calculation of PYTHIA the sample of "thick" jets is predicted to contain 52% (16%) of events resulting from $gq(gg)$ subprocesses in the final state. The sample of "thin" jets contains 65% (33%) of events from $qq(qg)$ subprocesses. The cross section $d\sigma/dE_T^{\text{jet}}$ is shown separately for "thick" and "thin" jet samples. The "thin" jet sample exhibits a harder spectrum than the "thick"-jet sample.

3. Dijet production and photon structure

Requiring at least two jets per event makes possible the determination of the distribution of fractional parton momentum $x_p(x_\gamma)$ in the proton (photon) and of the centre-of-mass scattering angle $\cos\theta^*$ of the partons using two-to-two particle scattering kinematics via the expressions:

$$x_p^{\mathrm{obs}} = \sum_{i=1}^{2} E_T^{\mathrm{jet}i}\exp(\eta^{\mathrm{jet}i})/2E_p, \qquad x_\gamma^{\mathrm{obs}} = \sum_{i=1}^{2} E_T^{\mathrm{jet}i}\exp(-\eta^{\mathrm{jet}i})/2E_\gamma,$$

$$\cos\theta^* = \tanh(\eta^{\mathrm{jet}1} - \eta^{\mathrm{jet}2})/2.$$

Both experiments have published new cross sections from photoproduction. The H1 experiment[17] concentrated on double differential x_p and x_γ measurements. The data at the highest $x_p \sim 0.6$ are sensitive to the gluon density of the proton, which contributes almost 40% of the cross section. The $d\sigma/dx_\gamma$ cross sections are presented for different regions of the variables x_p and $E_{T,\mathrm{max}}^{\mathrm{jet}}(= E_T^{\mathrm{jet}1}$, the transverse energy of the highest E_T jet). The data (Fig. 5 left) are compared to NLO calculations with two different

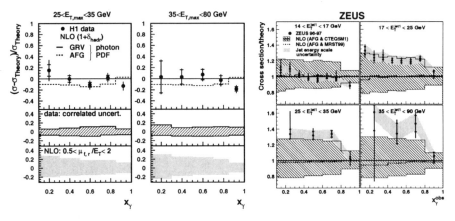

Figure 5. (left) The x_γ dependence of the relative difference of the measured H1 dijet cross sections to the NLO prediction with hadronisation corrections. The two figures show the relative difference for two $E_{T,\mathrm{max}}$ bins. The error due to the uncertainty in the calorimeter energy scale is shown in the middle. The band in the lower plot shows the renormalization and factorization scale uncertainties. (right) Ratio of cross sections from the ZEUS experiment to the NLO prediction using AFG-HO and CTEQ5M1 as the photon and proton PDFs, with the scale set to $E_T/2$. The theoretical uncertainty is shown as a hatched band.

parameterizations of the photon structure. The prediction describes the data well. The conclusion of the analysis is that the data do not require significant changes in the parameterizations of the PDFs but do provide useful further constraints.

ZEUS published[18] dijet cross sections as a function of E_T^{jet1} in different regions of the pseudorapidity of the two jets for $x_\gamma^{\text{obs}} > 0.75$ and $x_\gamma^{\text{obs}} < 0.75$. In general, the description of the data by NLO predictions is reasonable except when both jets are produced in the range $1 < \eta^{\text{jet}} < 2.4$. Here the NLO prediction lies below the data at low transverse energy. Similarly to H1, ZEUS tests the current parametrization of the photon PDFs in different bins of E_T^{jet1} (see Fig. 5 right). The predictions using the GRV-HO parametrization agree with the data in the region of lowest transverse energy, but are below the data for the higher E_T^{jet1} bins. The predictions using AGF-HO are similar in shape to those using GRV-HO but are (10-15)% lower.

To improve the understanding of the cross section in different regions of transverse energy, the sensitivity to the value of the transverse energy cut on the second jet has been studied. Starting at a minimum of 11 GeV, the cut on E_T^{jet2} was raised in both data and theory for the region $25 < E_T^{\text{jet1}} < 35$ GeV. The data fall less steeply with the increasing transverse energy than do the NLO QCD predictions and show sensitivity to the parton densities of the photon. Neither the AFG-HO nor the GRV-HO parametrization, convoluted with the NLO matrix element, fully describe the data. The data and NLO calculation cross each other at $E_T^{\text{jet2}} \sim 15$ GeV which is the cut used in the H1 analysis. This may explain why in the interval $25 < E_T^{\text{jet1}} < 35$ GeV shown in Fig. 5 left, H1 observes good agreement with the NLO QCD calculation with both photon PDFs, as H1 uses a higher cut $E_T^{\text{jet2}} > 15$ GeV than ZEUS.

ZEUS defined dijet data samples of "thick-thick", "thin-thin" and "thick-thin" events according to the criteria mentioned in Section 1. The differential dijet cross section as a function of $\cos\theta^*$ was measured[12] in the range of $|\cos\theta^*| < 0.8$ for dijet invariant mass $M^{\text{jj}} > 52$ GeV. The region of phase space in the $(M^{\text{jj}}, \cos\theta^*)$ plane was chosen in order to minimise the biases introduced by selecting jets with $E_T^{\text{jet1}} > 17$ GeV and $E_T^{\text{jet2}} > 14$ GeV. Fig. 6a shows the measured dijet cross section as a function of $\cos\theta^*_{\text{thick}}$. Calculations using PYTHIA and HERWIG give a good description of the shape of the measured cross section. The measured and predicted cross sections were normalised so as to have a value of unity at $\cos\theta^*_{\text{thick}} = 0.1$. The prediction of HERWIG for the partonic content is 41% of qg, 6% of gg and 53% of qq subprocesses. The observed asymmetry is adequately reproduced by the calculation and is understood in terms of the dominance

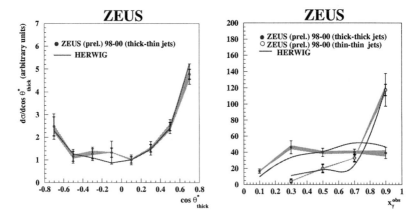

Figure 6. Differential ep cross section (a) $d\sigma/d\cos\theta^*_{\text{thick}}$ for dijet production. The cross section is for dijet events with "thick–thin" selection. The HERWIG prediction is normalised to data at $\cos\theta^*_{\text{thick}} = 0.1$. (b) $d\sigma/dx^{\text{obs}}_\gamma$ for dijet production with "thick–thick" (black dots) and "thin–thin" (open circles) jet selection. The HERWIG prediction is normalised to the total measured cross section. Cross sections in (a) and (b) are integrated over $E^{\text{jet1}}_T > 17$ GeV, $E^{\text{jet2}}_T > 14$ GeV and $-1 < \eta^{\text{jet}} < 2.5$.

of the resolved subprocess $q_\gamma g_p \to qg$. The $\cos\theta^*_{\text{thick}}$ distribution is asymmetric due to the different dominant diagrams in the different kinematic regions.

As the virtuality of the photon increases, the photon structure gradually changes. Whereas for quasi-real photons, the PDF approach is an indispensable theoretical tool, for $Q^2 \gg \Lambda^2_{\text{QCD}}$ the data can be analysed using perturbative calculations of the direct process. However, the resolved photon concept is still useful phenomenologically provided the photon virtuality remains smaller than a measure of the hardness of the process, e.g. E^{jet}_T. In the new analysis[19] from H1 of electroproduction of dijets at $2 < Q^2 < 80 \text{ GeV}^2$, triple differential dijet cross sections were analysed within several theoretical approaches, differing in the way higher order QCD effects were taken into account. The NLO direct calculation of JETVIP[20] is shown in Fig. 7 by a broken line and is in marked disagreement with the data in the region of small Q^2, small E^{jet}_T and small x_γ. JETVIP is the only NLO program which calculates also the NLO resolved photon contribution. The sum of both cross sections direct + resolved, is shown by the full line in Fig. 7. The agreement between data and calculations improves, but is still not satisfactory. The conclusion is that the data show clear evidence of the need for $\alpha\alpha^3_s$ and higher order terms in the theory. The best description of

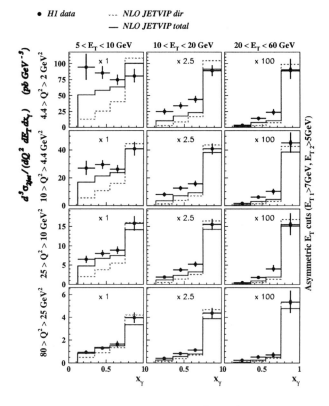

Figure 7. Triple differential dijet cross section $d^3\sigma/dQ^2 dE_T dx_\gamma$ for the H1 data depicted by points compared to the prediction of JETVIP NLO calculations corrected for hadronisation effects. The dashed line shows the NLO direct cross section, the full line the sum of NLO direct and NLO resolved contributions.

the data (not shown here) is achieved with the LO HERWIG generator with the inclusion of longitudinally polarised resolved photon and initial and final state parton showers as an effective approximation to higher order QCD processes.

4. Prompt photon production

Prompt photons are produced in direct and resolved interactions when a hard photon emerges instead of a gluon jet. Since the properties of hard photons can be more accurately measured than those of a jet, prompt photons are expected to allow cleaner tests of parton dynamics. Two obstacles

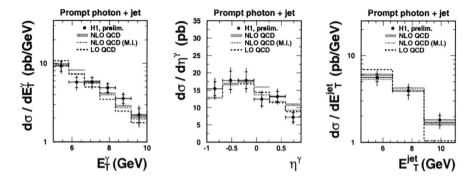

Figure 8. Prompt photon differential cross sections as a function of E_T^γ, η^γ and η^{jet}. The data are compared to LO (dashed) and NLO calculations. The error bands show the effect of a variation of the renormalization and factorization scales in the NLO calculations from $0.5 E_T^\gamma$ to $2 E_T^\gamma$. Dotted line shows the NLO result corrected for multiple interaction effects using PYTHIA.

make these advantages less clear: prompt photons are born in an environment of copious production of photons from π^0 decays and their production cross section is a factor of α/α_s smaller than that of gluon jets. H1 has presented new results[21] on the photoproduction of isolated prompt photons, both inclusively and in association with jets. The cross sections are given for transverse photon energies and pseudorapidities of $5 < E_T^\gamma < 10$ GeV, $-1 < \eta^\gamma < 0.9$ and for jets in the range $E_T^{\mathrm{jet}} > 4.5$ GeV, $-1 < \eta^{\mathrm{jet}} < 2.3$. The cross sections for a prompt photon with a jet are presented in Fig. 8 and compared to NLO calculations[10]. The NLO corrections are substantial and lead to a good description of the data. Taking into account the multiple interaction effects (PYTHIA) improves the data description in particular at $\eta^\gamma > 0$ and $x_\gamma < 0.5$. The NLO prediction is not corrected for hadronisation effect. The H1 cross sections were also compared to the measurements[22] of the ZEUS Collaboration. The results are consistent but the H1 data are somewhat lower at small η^γ where the ZEUS data appear to exceed the NLO calculation.

5. Conclusions

The H1 and ZEUS Collaborations have provided accurate measurements of jet and prompt photon cross sections at the lowest photon virtualities. Data were compared to several NLO calculations which agree very well at large values of the hard scale (usually E_T^{jet}). The QCD analysis of inclusive jet $d\sigma/dE_T^{\rm jet}$ cross sections leads to accurate value of α_s which are in good agreement with the world average. At the edges of the phase space of the current measurements (low $E_T^{\rm jet}$, large $\eta^{\rm jet}$), especially when cross sections are presented in a multi-differential way, differences between data and NLO calculations become significant. Data provide constraints on the photon PDF.

Acknowledgments

I am indebted to all members of the H1 and ZEUS collaborations who contributed to the results. Especially, I would like to thank to my colleagues J. Chýla, T. Greenshaw, G. Grindhammer, T. Kluge, R. Lemrani, P. Van Michelen, M. R. Sutton, J. Terrón and A. Valkárová who helped me with the preparation of this review. I would like to thank the organizers of the Ringberg workshop for preparing a stimulating meeting.

References

1. S. D. Ellis and D. E. Soper, *Rev. Phys.* **D48**, 3160 (1993).
2. T. Sjöstrand, *Comput. Phys. Commun.* **82**, 14 (1994).
3. R. Engel, *Z. Phys.* **C66**, 203 (1995).
4. G. Marchesini et al., *Comput. Phys. Commun.* **67**, 465 (1992).
5. A. Capella et al., *Phys. Rept.* **236**, 225 (1994).
6. S. Frixione, *Nucl. Phys.* **B507**, 295 (1997).
7. M. Glück, E. Reya and A. Vogt, *Phys. Rev.* **D45**, 9863 (1992).
8. H. L. Lai et al., [CTEQ Collaboration], *Eur. Phys. J.* **C12**, 375 (2000).
9. M. Klasen, T. Kleinwort and G. Kramer *Eur. Phys. J. Direct* **C1**, 1 (1998).
10. M. Fontannaz, J. P. Guillet and G. Heinrich, *Eur. Phys. J.* **C21**, 303 (2001).
11. J. Breitweg et al., *Eur. Phys. J.* **C2**, 61 (1998).
12. ZEUS Collaboration, Substructure dependence of jet cross section at HERA, Abstract 518, IECHEP Aachen, Germany, 2003.
13. C. Adloff et al., DESY 02-225, hep-ex/0302034.
14. S. Chekanov et al., *Phys. Lett.* **B560**, 7 (2003).
15. J. Breitweg et al., *Phys. Lett.* **B507**, 70 (2001); S. Chekanov et al., *Phys. Lett.* **B547**, 164 (2002).
16. S. Bethke, *Nucl. Phys. Proc. Suppl.* **121**, 4 (2003).
17. C. Adloff et al., *Eur. Phys. J.* **C25**, 13 (2002).
18. S. Chekanov et al., *Eur. Phys. J.* **C23**, 615 (2002).
19. A. Aktas et al., H1 Collaboration, Measurement and QCD analysis of dijet production at low Q^2 at HERA, to be published in *Eur. Phys. J.*
20. B. Pötter, *Comp. Phys. Commun.* **133**, 105 (2001).
21. H1 Collaboration, Measurement of Prompt Photon Production in γp Interactions, Abstract 093, IECHEP Aachen, Germany, 2003.
22. J. Breitweg et al., *Phys. Lett.* **B472**, 175 (2000).

PHOTOPRODUCTION OF ISOLATED PHOTONS, SINGLE HADRONS AND JETS AT NLO

GUDRUN HEINRICH

II. Institut für theoretische Physik, Universität Hamburg,
Luruper Chaussee 149, 22761 Hamburg, Germany

The photoproduction of large-p_T charged hadrons and of prompt photons is discussed, for the inclusive case and with an associated jet, using predictions from the NLO partonic Monte Carlo program EPHOX. Comparisons to recent HERA data are also shown.

1. Introduction

At HERA, the photoproduction of jets has been measured with high statistics over the past years[1,2,3]. Data on single charged hadron[4] and prompt photon[5,6,7] production are also available and have been compared to theoretical predictions[8–15], where in the prompt photon case the statistics is of course lower as the cross sections are small.

In photoproduction reactions, an almost real photon, emitted at small angle from the electron, interacts with a parton from the proton. The photon can either participate directly in the hard scattering ("direct photon") or act as a "resolved photon", in which case a parton stemming from the photon takes part in the hard interaction. Therefore photoproduction reactions offer a unique opportunity to constrain the parton distributions in the photon, especially the gluon distribution $g^\gamma(x)$. The latter is rather poorly constrained by LEP $\gamma^*\gamma$ data as it enters in this reaction only at next-to-leading order (NLO), whereas in ep photoproduction it enters already at leading order.

On the other hand, photoproduction reactions at HERA also could serve to constrain the gluon distribution in the *proton*. To determine the latter to a better accuracy is of particular interest in view of the LHC with its large gluon luminosity, $gg \to H$ being the dominant production mode for a light Higgs boson. As the photon represents a rather "clean" initial state, photoproduction reactions can be used to probe the proton in a way which is complementary to other experiments.

In what follows, the kinematic regions where gluon initiated subprocesses play a significant role will be identified and it will be argued that their relative contribution to the cross section can be enhanced by appropriate cuts, for the proton as well as for the photon case. For a more detailed study, the reader is referred to[16].

Comparisons to HERA data, especially to very recent (preliminary) H1 data on prompt photon plus jet production, will also be shown.

2. Theoretical aspects of charged hadron and prompt photon production

We will concentrate here on the two reactions $\gamma p \to h^\pm (+\text{jet}) + X$ and $\gamma p \to \gamma (+\text{jet}) + X$. Identifying a jet in addition to the prompt photon or hadron allows for a more detailed study of the underlying parton dynamics, and in particular for the definition of observables x_{obs}, x_{LL} which are suitable to study the parton distribution functions.

Comparing the reaction of charged hadron production versus prompt photon production, one is tempted to say that prompt photon production is more advantageous: Due to photon isolation, the dependence of the theoretical prediction on the non-perturbative fragmentation functions is negligible. Further, as will be shown in the following, the NLO prediction for the prompt photon cross section is not very sensitive to scale changes, which cannot be said for the $h^\pm (+\text{jet})$ cross section[9]. On the other hand, the $\gamma + \text{jet}$ cross section is orders of magnitude smaller than the $h^\pm + \text{jet}$ cross section, and the identification of the prompt photon events among the huge background from the decay of neutral pions is experimentally not an easy task.

Photon isolation

Prompt photons can originate from two mechanisms: Either they stem directly from the hard interaction, or they are produced by the fragmentation of a hard parton. In order to distinguish them from the background stemming from the decay of light mesons, an *isolation criterion* has to be imposed: The amount of hadronic transverse energy E_T^{had}, deposited inside a cone with aperture R centered around the photon direction in the rapidity and azimuthal angle plane, must be smaller than some value $E_T^{\max}$:

$$(\eta - \eta^\gamma)^2 + (\phi - \phi^\gamma)^2 \leq R^2$$
$$E_T^{\text{had}} \leq E_T^{\max} . \quad (1)$$

For the numerical results we will follow the HERA conventions $E_T^{\max} = 0.1\,p_T^\gamma$ and $R = 1$.

Apart from reducing the background from secondary photons, isolation also substantially reduces the fragmentation component, such that the total cross section depends very little on the photon fragmentation functions.

Factorisation and scale dependence

The cross section for prompt photon production can be written symbolically as

$$d\sigma^{ep\to\gamma X}(P_e, P_p, P_\gamma) = \sum_{a,b}\int dx_e\,dx_p\,F_{a/e}(x_e, M)F_{b/p}(x_p, M)\{d\hat\sigma^{\text{dir}} + d\hat\sigma^{\text{frag}}\}$$

$$d\hat\sigma^{\text{dir}} = d\hat\sigma^{ab\to\gamma X}(x_a, x_b, P_\gamma, \mu, M, M_F)$$

$$d\hat\sigma^{\text{frag}} = \sum_c \int dz\, D_{\gamma/c}(z, M_F) d\hat\sigma^{ab\to c X}(x_a, x_b, P_\gamma/z, \mu, M, M_F).$$

In the case of the production of a single hadron, the partonic cross section $d\hat\sigma^{\text{dir}}$ is of course zero, and the fragmentation functions $D_{\gamma/c}(z, M_F)$ have to be replaced by the ones for hadrons, $D_{h/c}(z, M_F)$.

In photoproduction, $F_{a/e}(x_e, M)$ is a convolution

$$F_{a/e}(x_e, M) = \int dx^\gamma \int dy\, \delta(x^\gamma y - x_e)\, f_{\gamma/e}(y)\, F_{a/\gamma}(x^\gamma, M) \qquad (2)$$

where the spectrum of quasi-real photons emitted from the electron is described by the Weizsäcker-Williams approximation

$$f_{\gamma/e}(y) = \frac{\alpha}{2\pi}\left\{\frac{1+(1-y)^2}{y}\text{Log}\frac{Q^2_{\max}(1-y)}{m_e^2 y^2} - \frac{2(1-y)}{y}\right\}.$$

The function $F_{a/\gamma}(x^\gamma, M)$ in eq. (2) denotes the parton distribution function for a parton of type "a" in the resolved photon. In the case of a direct initial photon, one has $F_{a/\gamma}(x^\gamma, M) = \delta_{a\gamma}\delta(1 - x^\gamma)$. Therefore one can try to switch on/off the resolved photon by suppressing/enhancing large x^γ.

As the perturbative series in α_s for the partonic cross section is truncated, the theoretical prediction is scale dependent, depending on the renormalisation scale μ as well as on the initial/final state factorisation scales M/M_F. While the leading order cross section depends very strongly on the scales, the NLO cross section is already much more stable, as can be seen for example from Fig. 1 for the prompt photon inclusive cross section.

Observables x^γ, x^p

In order to reconstruct the longitudinal momentum fraction of the parton stemming from the proton respectively the photon from measured quantities, one can define the observables

$$x^\gamma_{obs} = \frac{p_T\, e^{-\eta} + E_T^{jet}\, e^{-\eta^{jet}}}{2 E \gamma} \quad (3)$$

$$x^p_{obs} = \frac{p_T\, e^{\eta} + E_T^{jet}\, e^{\eta^{jet}}}{2 E p}$$

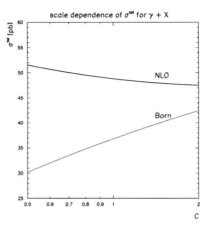

Figure 1. The scales have been set to $\mu = M = M_F = C p_T^\gamma$ with C varied between 0.5 and 2.

where p_T is the momentum of the prompt γ or the hadron, and η its (pseudo-)rapidity. However, as the measurement of E_T^{jet} can be a source of systematic errors at low E_T values, the following variable, which does not depend on E_T^{jet}, might be more convenient:

$$x^{p,\gamma}_{LL} = \frac{p_T\,(e^{\pm\eta} + e^{\pm\eta^{jet}})}{2 E^{p,\gamma}} \quad (4)$$

The variable x^γ_{LL} also has the advantage that it has a smoother behaviour for $x^\gamma \to 1$.

3. Numerical studies

Unless stated otherwise, the following input for the numerical results is used: The center of mass energy is $\sqrt{s} = 318\,\text{GeV}$ with $E_e = 27.5\,\text{GeV}$ and $E_p = 920\,\text{GeV}$. The maximal photon virtuality is $Q^2_{max} = 1\,\text{GeV}^2$, and $0.2 < y < 0.7$. For the parton distributions in the proton we take the MRST01[17] parametrisation, for the photon we use AFG04[18] distribution functions and BFG[19] fragmentation functions. For the charged hadron fragmentation functions we use BFGW[23] as default. We take $n_f = 4$ flavours, and for $\alpha_s(\mu)$ we use an exact solution of the two-loop renormalisation group equation, and not an expansion in $\log(\mu/\Lambda)$. The default scale choice is $M = M_F = \mu = p_T$. Jets are defined using the k_T-algorithm.

3.1. The NLO program EPHOX

All numerical results shown here are obtained by the NLO partonic Monte Carlo program EPHOX[20], which allows to obtain integrated cross sections as well as fully differential distributions for the reactions considered here. The program contains the full NLO corrections to all four categories of subprocesses direct-direct, direct-fragmentation, resolved-direct and resolved-fragmentation. It also contains the quark loop box diagram $\gamma g \to \gamma g$ with a flag to turn it on or off. EPHOX can be obtained at http://wwwlapp.in2p3.fr/lapth/PHOX_FAMILY/main.html, together with a comfortable user interface and detailed documentation.

3.2. The gluon distribution in the photon

In order to constrain the gluon distribution $g^\gamma(x^\gamma)$ in the photon, one has to focus on a kinematic region where the gluon content of the resolved photon is large, i.e. on small values of x^γ. According to eqs. (3) and (4), small x^γ corresponds to large values of η and η^{jet}, which means that the forward rapidity region is particularly interesting. Fig. 2 illustrates for the prompt photon plus jet cross section that the gluon distribution $g^\gamma(x^\gamma)$ only becomes important in the very forward region $\eta^\gamma \gtrsim 1.5$ if the rapidity of the jet is integrated over the range $-1 < \eta^{jet} < 2.3$. However, the

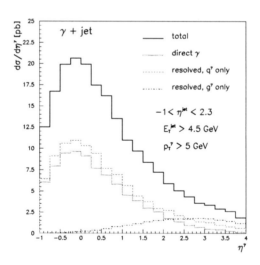

Figure 2. Rapidity range where the gluon in the photon becomes important

relative importance of the gluon distribution $g^\gamma(x)$ can be enhanced by restricting both rapidities, the one of the photon *and* the one of the jet, to the forward region. Fig. 3a shows that with the cuts $\eta^\gamma, \eta^{jet} > 0$, the resolved photon component makes up for a major part of the cross section, but mainly consists of quarks, while for $\eta^\gamma > 0.5$, $\eta^{jet} > 1.5$, the gluon in the photon contributes about 40% to the total cross section, as shown in Fig. 3b.

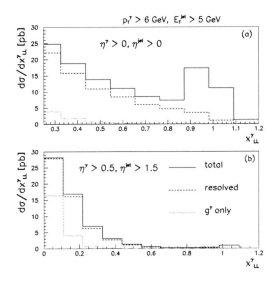

Figure 3. The rapidity cuts $\eta^\gamma > 0.5$, $\eta^{jet} > 1.5$ substantially enhance the relative contribution of the gluon in the resolved photon

3.3. The gluon distribution in the proton

The gluon distribution in the proton is large for small x^p, corresponding to small rapidities. From Fig. 4 one can indeed see that the relative contribution of subprocesses initiated by a gluon from the proton to the $h^\pm$ + jet cross section is about 73% if both rapidities are restricted to the backward region, $-2 < \eta, \eta^{jet} < 0.5$. The analogous is *not* true for the prompt photon + jet cross section, where the gluon contri-

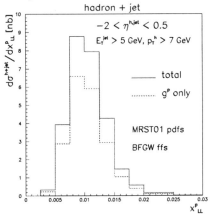

Figure 4. The gluon (from the proton) contribution to the $h^\pm$+jet cross section at small rapidities.

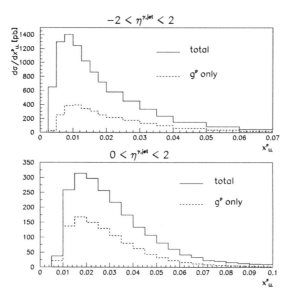

Figure 5. The relative contribution of gluon initiated processes to the γ + jet cross section can be enhanced by the rapidity cuts $0 < \eta^\gamma, \eta^{jet} < 2$. The p_T cuts are $E_T^{jet} > 5\,\text{GeV}$, $p_T^\gamma > 6\,\text{GeV}$.

bution makes up only 13% of the total cross section if the rapidities are restricted to the range $-2 < \eta^\gamma, \eta^{jet} < 0$. This is due to the fact that at small x^p values, x^γ is large, such that direct initial photons should dominate. However, the subprocess $g^p + \gamma \to \gamma(\text{direct}) + \text{jet}$ does not exist at leading order, and the subprocess $g^p + \gamma \to q + \text{jet}$, where the quark subsequently fragments into a photon, is suppressed by isolation. Therefore, in the case of the γ + jet cross section, the subprocess $g^p + q^\gamma \to \gamma + q$ is the dominant one involving g^p whereas in the case of the hadron+jet cross section, $g^p + \gamma \to q + \bar{q}$ is dominant at small x^p. The virtue of this behaviour of the prompt photon cross section is that the region where g^p initiated subprocesses are important is *not* restricted to negative rapidities. As can be seen from Fig. 5, the rapidity cuts $0 < \eta^{\gamma, jet} < 2$ enhance the relative contribution of the gluon g^p to the γ + jet cross section, because they select a region where $g^p + q^\gamma$ initiated subprocesses are important. Note that this rapidity domain is more accessible experimentally than the very backward region.

4. Comparison to HERA data

4.1. Charged hadrons

For the case of single charged hadron production, only data for $h^{\pm}+X$, but not for $h^{\pm}+$jet $+X$ are available so far[4]. The analysis uses a minimum p_T of 3 GeV for the hadron, which is rather close to the non-perturbative regime. This induces a large theoretical uncertainty due to differences in the parametrisation of the fragmentation functions[21,22,23] at low p_T (see Fig. 6), and also a large scale dependence. If a minimum p_T of 7 GeV is chosen for the hadron, the cross section is reduced, but the theoretical uncertainty is much smaller[9].

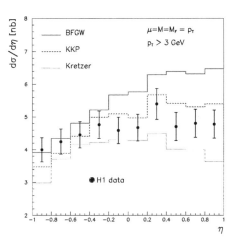

Figure 6. Predictions obtained with different fragmentation functions compared to H1 data.

4.2. Prompt photons (inclusive)

For inclusive prompt photon production, ZEUS data[5,6] as well as preliminary H1 data[7] are available. In Fig. 7[a], one observes that the ZEUS data are above the NLO prediction in the backward region. At large rapidities, there is a trend that theory is above both H1 and ZEUS data. As already mentioned in the introduction, this effect is very likely due to photon isolation:

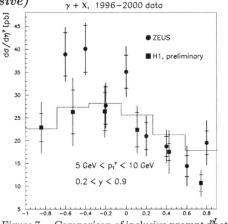

Figure 7. Comparison of inclusive prompt photon data with the NLO QCD prediction.

---

[a]In order to be able compare to the ZEUS data taken at $E_p = 820$ GeV and $0.2 < y < 0.9$, the H1 data taken at $E_p = 920$ GeV and $0.2 < y < 0.7$ have been corrected using PYTHIA[7].

A sizeable amount of hadronic transverse energy in the isolation cone may stem from the underlying event. Therefore, even direct photon events may be rejected, leading to a decrease of the cross section. This fact cannot be simulated by a partonic Monte Carlo, such that the NLO predictions tend to be above the data in kinematic regions where the underlying event activity is expected to be large.

4.3. Prompt photon + jet

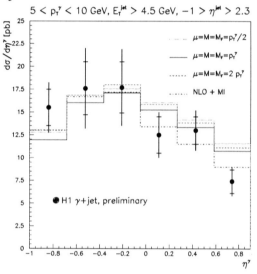

In the rapidity distribution for the γ + jet cross section, the effect mentioned above is again visible at large rapidities. To estimate its impact, H1 recently made an analysis[7,24] where the NLO result is corrected for multiple interaction effects. As can be seen from Fig. 8, the corrections are indeed most pronounced in the forward region where the underlying event activity from the resolved photon remnants is expected to be larger, and they improve the agreement between data and theory. Fig. 8 also shows that the γ+jet cross section is very stable with respect to scale variations.

Figure 8. Comparison of preliminary H1 data to the NLO prediction with different scale choices. NLO+MI denotes the result where the partonic prediction has been corrected for multiple interaction effects.

4.4. Study of intrinsic $\langle k_T \rangle$

The ZEUS collaboration made an analysis on prompt photon + jet data to study the "intrinsic" parton transverse momentum $\langle k_T \rangle$ in the proton[6]. To this aim, the $\langle k_T \rangle$ - sensitive observables $p_\perp$, the photon momentum component perpendicular to the jet direction, and $\Delta\phi$, the azimuthal acollinearity between photon and jet, have been studied. To suppress contributions to $\langle k_T \rangle$ from the resolved photon, the cut $x_\gamma^{obs} > 0.9$ has been imposed. To minimise calibration uncertainties, only normalized cross sections have been considered, as an additional $\langle k_T \rangle$ mainly changes the *shape* rather than the absolute value of the cross section. The best fit with PYTHIA 6.129 lead to the result $\langle k_T \rangle = 1.69 \pm 0.18\text{(stat)}^{+0.18}_{-0.20}\text{(sys)}$ GeV, which includes the shower contribution to $\langle k_T \rangle$. On the other hand, the NLO program EPHOX is able to describe the data without including an additional $\langle k_T \rangle$. This is shown in Fig. 9, where also the minimum E_T^{jet} has been varied in order to estimate the impact of an uncertainty in the determination of the jet energy and of choosing symmetric cuts ($p_T^\gamma > 5$ GeV).

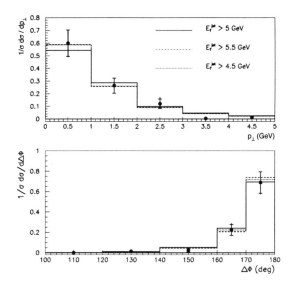

Figure 9. ZEUS data for the normalised cross sections differential in $p_\perp$ and $\Delta\phi$ compared to the EPHOX prediction for different cuts on the minimum jet transverse energies.

5. Conclusions

The reactions $\gamma p \to \gamma + \text{jet} + X$ and $\gamma p \to h^\pm + \text{jet} + X$ offer the possibility to constrain the gluon density in the photon and in the proton. It has been shown that appropriate rapidity cuts can enhance the relative contribution of gluon initiated processes.

Comparing the predictions of the partonic Monte Carlo NLO program EPHOX to HERA data, the following observations can be made:

For inclusive single charged hadron production, the H1 data are well described, but the theoretical uncertainties from fragmentation functions and scale variations are large. These large uncertainties can mainly be attributed to the fact that a minimal p_T of 3 GeV for the hadron is too close to the non-perturbative regime. A $p_T^{\min}$ of 7 GeV improves the stability of the theoretical predictions.

The rapidity distributions for inclusive prompt photon production as well as for γ+jet show an interesting feature: At large rapidities, the NLO prediction always tends to overshoot the data. It has been argued that this behaviour might be attributed to photon isolation: Due to hadronic activity stemming from the underlying event in the isolation cone, even direct photons in the final state may be rejected by the isolation cut, thus leading to a lower cross section than the (partonic) Monte Carlo prediction, especially in the forward region where the probability of resolved photon remnants is higher. A recent H1 analysis which corrects the NLO prediction bin per bin for multiple interaction effects corroborates this explanation.

Finally, the predictions of EPHOX have been compared to a ZEUS analysis on prompt photon+jet data where the intrinsic $\langle k_T \rangle$ of the proton is studied, and it has been found that the NLO prediction describes the data very well without introducing an extra $\langle k_T \rangle$.

Acknowledgements

I would like to thank the organisers of the Ringberg workshop for an interesting and pleasant meeting, and J. Gayler, R. Lemrani and P. Bussey for discussions on the data. I also wish to thank my collaborators M. Fontannaz and J. Ph. Guillet.

References

1. C. Adloff *et al.* [H1 Collaboration], DESY-02-225, hep-ex/0302034.
2. S. Chekanov *et al.* [ZEUS Collaboration], Phys. Lett. B **560** (2003) 7 [hep-ex/0212064].
3. J. Cvach, *in these proceedings*.
4. C. Adloff *et al.* [H1 Collaboration], Eur. Phys. J. C **10** (1999) 363 [hep-ex/9810020].
5. J. Breitweg *et al.* [ZEUS Collaboration], Phys. Lett. B **472** (2000) 175 [hep-ex/9910045];
   J. Breitweg *et al.* [ZEUS Collaboration], Phys. Lett. B **413** (1997) 201 [hep-ex/9708038].
6. S. Chekanov *et al.* [ZEUS Collaboration], Phys. Lett. B **511** (2001) 19 [hep-ex/0104001].

7. H1 Collaboration, submitted to the Int. Europhysics Conference on High Energy Physics, EPS03, July 2003, Aachen (Abstract 093), and to the XXI Int. Symposium on Lepton and Photon Interactions, LP03, August 2003, Fermilab.
8. B. A. Kniehl, G. Kramer and B. Pötter, Nucl. Phys. B **597** (2001) 337 [hep-ph/0011155].
9. M. Fontannaz, J. P. Guillet and G. Heinrich, Eur. Phys. J. C **26** (2002) 209 [hep-ph/0206202].
10. L. E. Gordon and W. Vogelsang, Phys. Rev. D **52**, 58 (1995).
11. L. E. Gordon, Phys. Rev. D **57** (1998) 235 [hep-ph/9707464].
12. M. Fontannaz, J. P. Guillet and G. Heinrich, Eur. Phys. J. C **21** (2001) 303 [hep-ph/0105121].
13. M. Fontannaz, J. P. Guillet and G. Heinrich, Eur. Phys. J. C **22** (2001) 303 [hep-ph/0107262].
14. M. Krawczyk and A. Zembrzuski, Phys. Rev. D **64** (2001) 114017 [hep-ph/0105166].
15. A. Zembrzuski and M. Krawczyk, hep-ph/0309308.
16. M. Fontannaz, G. Heinrich, hep-ph/0312009.
17. A. D. Martin, R. G. Roberts, W. J. Stirling and R. S. Thorne, Eur. Phys. J. C **23** (2002) 73.
18. P. Aurenche, J. P. Guillet and M. Fontannaz, Z. Phys. C **64** (1994) 621; P. Aurenche, J. P. Guillet and M. Fontannaz, new version of AFG, publication in preparation.
19. L. Bourhis, M. Fontannaz and J. Ph. Guillet, Eur. Phys. J. C **2**, 529 (1998).
20. For a detailed description of EPHOX, see ref.[12] and
   http://wwwlapp.in2p3.fr/lapth/PHOX_FAMILY/main.html.
   EPHOX is actually part of a larger family of NLO Monte Carlo programs, containing also programs for hadron-hadron collisions and e^+e^- photoproduction.
21. S. Kretzer, Phys. Rev. D **62** (2000) 054001 [hep-ph/0003177].
22. B. A. Kniehl, G. Kramer and B. Pötter, Nucl. Phys. B **582** (2000) 514 [hep-ph/0010289].
23. L. Bourhis, M. Fontannaz, J. P. Guillet and M. Werlen, Eur. Phys. J. C **19** (2001) 89 [hep-ph/0009101].
24. R. Lemrani [H1 Collaboration], *Talk given at the 11th International Workshop on Deep Inelastic Scattering (DIS 2003), St. Petersburg, Russia, April 2003*, hep-ex/0308066;
   R. Lemrani-Alaoui, *"Prompt photon production at HERA"*, Ph.D. thesis, DESY-THESIS-2003-010, available at
   http://www-h1.desy.de/publications/theses_list.html.

JET PRODUCTION IN DEEP INELASTIC ep SCATTERING AT HERA

C. GLASMAN[*]

Universidad Autónoma de Madrid
Cantoblanco 28049, Madrid, Spain
E-mail: claudia@mail.desy.de

Recent results from jet production in deep inelastic ep scattering at HERA are reviewed. The values of $\alpha_s(M_Z)$ extracted from a QCD analysis of the data are presented.

1. Introduction

Jet production in neutral-current (NC) deep inelastic ep scattering (DIS) provides a test of perturbative QCD (pQCD) calculations and of the parametrisations of the proton parton densities (PDFs). Jet cross sections allow the determination of the fundamental parameter of QCD, the strong coupling constant α_s, and help to constrain the parton densities in the proton.

Up to leading order (LO) in α_s, jet production in NC DIS proceeds via the quark-parton model (QPM) ($Vq \to q$, where $V = \gamma$ or Z^0), boson-gluon fusion (BGF) ($Vg \to q\bar{q}$) and QCD-Compton (QCDC) ($Vq \to qg$) processes. The jet production cross section is given in pQCD by the convolution of the proton PDFs and the subprocess cross section,

$$d\sigma_{\rm jet} = \sum_{a=q,\bar{q},g} \int dx\, f_a(x,\mu_F)\, d\hat{\sigma}_a(x,\alpha_s(\mu_R),\mu_R,\mu_F),$$

where x is the fraction of the proton's momentum taken by the interacting parton, f_a are the proton PDFs, μ_F is the factorisation scale, $\hat{\sigma}_a$ is the subprocess cross section and μ_R is the renormalisation scale.

All the data accumulated from HERA and fixed-target experiments have allowed a good determination of the proton PDFs over a large phase space.

---

[*]Ramón y Cajal Fellow.

Then, measurements of jet production in neutral current DIS provide accurate tests of pQCD and a determination of the fundamental parameter of the theory, α_s.

At high scales, calculations using the DGLAP evolution equations have been found to give a good description of the data up to next-to-leading order (NLO). Therefore, by fitting the data with these calculations, it is possible to extract accurate values of α_s and the gluon density of the proton. However, for scales of $E_T^{\rm jet} \sim Q$, where $E_T^{\rm jet}$ is the jet transverse energy and Q is the exchanged photon virtuality, and large values of the jet pseudorapidity, $\eta^{\rm jet}$, large discrepancies between the data and the NLO calculations have been observed at low x. This could indicate a breakdown of the DGLAP evolution and the onset of BFKL effects. These discrepancies can also be explained by assigning a partonic structure to the exchanged virtual photon or a large contribution of higher order effects at low Q^2.

2. Inclusive jet cross sections

Inclusive jet cross sections have been measured[1] in the Breit frame using the k_T-cluster algorithm in the longitudinally invariant mode. The measurements were made in the kinematic region given by $Q^2 > 125$ GeV^2 and $-0.7 < \cos\gamma < 0.5$, where γ is the angle of the struck quark in the quark-parton model in the HERA laboratory frame. The cross sections refer to jets of $E_{T,\rm B}^{\rm jet} > 8$ GeV and $-2 < \eta_{\rm B}^{\rm jet} < 1.8$, where $E_{T,\rm B}^{\rm jet}$ and $\eta_{\rm B}^{\rm jet}$ are the jet transverse energy and pseudorapidity, respectively, in the Breit frame.

The use of inclusive jet cross sections in a QCD analysis presents several advantages: inclusive jet cross sections are infrared insensitive and better suited to test resummed calculations and the theoretical uncertainties are smaller than for dijet cross sections.

Figure 1 shows the inclusive jet cross section as a function of Q^2 and $E_{T,\rm B}^{\rm jet}$. The dots are the data and the error bars represent the statistical and systematic uncertainties; the shaded band displays the uncertainty on the absolute energy scale of the jets. The measured cross sections have a steep fall-off, by five (four) orders of magnitude within the measured Q^2 ($E_{T,\rm B}^{\rm jet}$) range. The lines are the NLO calculations using DISENT with different choices of the renormalisation scale ($\mu_R = Q$ or $E_{T,\rm B}^{\rm jet}$). The calculations describe reasonably well the Q^2 and $E_{T,\rm B}^{\rm jet}$ dependence of the cross section for $Q^2 > 500$ GeV^2 and $E_{T,\rm B}^{\rm jet} > 15$ GeV. At low Q^2 and low $E_{T,\rm B}^{\rm jet}$, the measurements are above the calculations by about 10%, which is of the same size as the theoretical uncertainties (see below). Therefore, for the extraction of α_s, the phase space was restricted to high Q^2 and high $E_{T,\rm B}^{\rm jet}$.

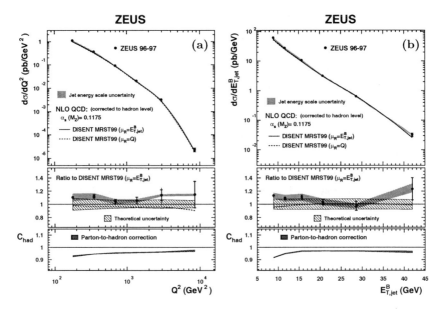

Figure 1. Inclusive jet cross sections[1] as a function of (a) Q^2 and (b) $E^{\text{jet}}_{T,\text{B}}$.

The experimental uncorrelated uncertainties for these cross sections are small, $\sim 5\%$. The uncertainty coming from the absolute energy scale of the jets is also small, $\sim 5\%$. The theoretical uncertainties comprise 5% from the absent higher orders, 3% from the uncertainties of the proton PDFs and 5% from the uncertainty in the value of $\alpha_s(M_Z)$ assumed. The parton-to-hadron corrections are 10% with an uncertainty of 1%.

2.1. Determination of α_s

The method[1] used by ZEUS to extract α_s exploits the dependence of the NLO calculations on $\alpha_s(M_Z)$ through the matrix elements ($\hat\sigma \sim A \cdot \alpha_s + B \cdot \alpha_s^2$) and the proton PDFs ($\alpha_s(M_Z)$ value assumed in the evolution). To take into account properly this correlation, NLO calculations were performed using various sets of PDFs which assumed different values of $\alpha_s(M_Z)$. The calculations were then parametrised as a function of $\alpha_s(M_Z)$ in each measured Q^2 or $E^{\text{jet}}_{T,\text{B}}$ region. From the measured value of the cross section as a function of Q^2 in each region of Q^2, a value of $\alpha_s(M_Z)$ and its uncertainty were extracted using the parametrisations of the NLO calculations.

From the inclusive jet cross section for $Q^2 > 500$ GeV^2, the value

$$\alpha_s(M_Z) = 0.1212 \pm 0.0017 \text{ (stat.) } ^{+0.0023}_{-0.0031} \text{ (exp.) } ^{+0.0028}_{-0.0027} \text{ (th.)}$$

was extracted using the method explained above. The experimental uncertainties are dominated by the uncertainty on the absolute energy scale of the jets (1%). The theoretical uncertainties are: 3% from the absent higher orders, 1% from the PDFs and 0.2% from the hadronisation corrections. This determination is compatible with other independent extractions performed at HERA and with the current world average (see figure 2a). Further precision in the extraction of $\alpha_s(M_Z)$ from inclusive jet cross sections depends upon further experimental and theoretical improvements.

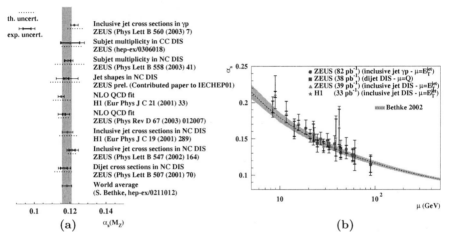

Figure 2. (a) Summary of extracted $\alpha_s(M_Z)$ values at HERA. (b) Summary of extracted $\alpha_s(\mu)$ values as a function of μ at HERA.

The QCD prediction for the energy-scale dependence of α_s has been tested by determining α_s from the measured differential cross sections at different scales[1]. From the measured cross section as a function of $E_{T,B}^{\text{jet}}$, in each region of $E_{T,B}^{\text{jet}}$, a value of $\alpha_s(E_{T,B}^{\text{jet}})$ was extracted. The result, shown in figure 2b (triangles), is compatible with the running of α_s as predicted by QCD (shaded band) over a large range in the scale. Figure 2b also shows other studies of the energy-scale dependence of α_s from HERA: all the results are compatible with each other and with the QCD prediction. This constitutes a test of the scale dependence of α_s between $\mu = 8.4$ and 90 GeV.

3. Parton evolution at low x

Dijet data in DIS may be used to gain insight into the parton dynamics at low x. The evolution of the PDFs with the factorisation scale can be described by the DGLAP evolution equations which sum the leading powers of terms like $\alpha_s \log Q^2$ in the region of strongly ordered transverse momenta k_T. This presciption describes successfully jet production at high Q^2. However, the DGLAP approximation is expected to break down at low x since when $\log Q^2 \ll \log 1/x$, the terms proportional to $\alpha_s \log 1/x$ become important and need to be summed. This is done in the BFKL evolution equations; the integration is taken over the full k_T phase space of the gluons with no k_T-ordering.

Another approach, the CCFM evolution equations with angular-ordered parton emission, is equivalent to the BFKL approach for $x \to 0$ and reproduces the DGLAP evolution equations at large x. Thus, the properties of the dijet system, which depend on the dynamics of the ladder, can be studied to determine whether the cascade has a k_T-ordered or unordered evolution.

Deviations from the DGLAP approach at small x can be tested experimentally. At small x it is expected that parton emission along the exchanged gluon ladder should increase with decreasing x. A clear experimental signature of this effect would be that the two outgoing hard partons are no longer back-to-back and so an excess of events at small azimuthal separations should be observed.

Values of the azimuthal separation of the two hard jets in the γ^*p centre-of-mass frame, $\Delta\phi^*$, different than π can occur in the DGLAP approach only when higher order contributions are included. On the other hand, in the BFKL and CCFM approaches, the number of events with $\Delta\phi^* < \pi$ should increase due to the partons entering the hard process with large k_T.

3.1. *Azimuthal jet separation*

To test the predictions of the different approaches, dijet cross sections have been measured[2] using the k_T-cluster algorithm in the longitudinally inclusive mode in the γ^*p centre-of-mass frame. The measurements were made in the kinematic region given by $5 < Q^2 < 100$ GeV^2 and $10^{-4} < x < 10^{-2}$. The cross sections refer to jets of $E_T^* > 5$ GeV, $-1 < \eta_{\mathrm{LAB}}^{\mathrm{jet}} < 2.5$ and $E_{T,\mathrm{max}}^* > 7$ GeV, where E_T^* is the jet transverse energy in the γ^*p centre-of-mass frame.

Figure 3a shows the measured dijet distribution as a function of $\Delta\phi^*$. A significant fraction of events is observed at a small azimuthal separation. Since a measurement of a multi-differential cross section as a function of x, Q^2 and $\Delta\phi^*$ would be very difficult due to large migrations, the fraction of the number of dijet events with an azimuthal separation between 0 and

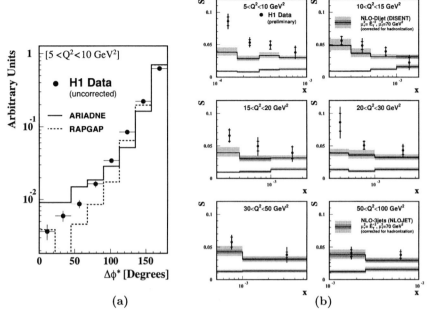

Figure 3. (a) Azimuthal separation between jets[2]. (b) Ratio S as a function of Bjorken x and Q^2 compared with predictions from NLO QCD calculations[2].

α, where α was taken as $\alpha = \frac{2}{3}\pi$, was measured instead. The fraction S, defined as

$$S = \frac{\int_0^\alpha N_{2\text{jet}}(\Delta\phi^*, x, Q^2) d\Delta\phi^*}{\int_0^\pi N_{2\text{jet}}(\Delta\phi^*, x, Q^2) d\Delta\phi^*},$$

is better suited to test small-x effects than a triple differential cross section.

The measured fraction S as a function of Bjorken x in different regions of Q^2 is presented in figure 3b. The data rise towards low x values, especially at low Q^2. The NLO predictions from DISENT, which contain k_T effects only in the first order corrections, are several standard deviations below the data and show no dependence with x. On the other hand, the predictions of NLOJET, which contain k_T effects at next-to-lowest order, provide an accurate description of the data at large Q^2 and large x. However, they fail to describe the increase of the data towards low x values, especially at low Q^2.

Figure 4a shows the data compared with the predictions of RAPGAP with direct only and resolved plus direct processes. A good description of the data is obtained at large Q^2 and large x. However, there is a failure

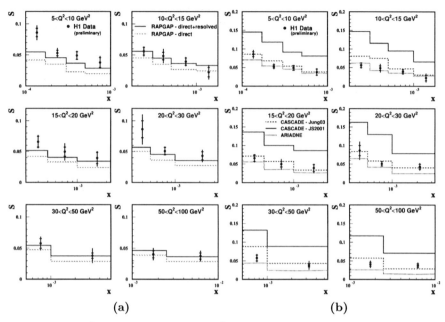

Figure 4. Ratio² S as a function of x and Q^2 compared with predictions from RAPGAP (a) and ARIADNE and CASCADE (b).

to describe the strong rise of the data towards low x, especially at low Q^2, even when including a possible contribution from resolved virtual photon processes, though the description in other regions is improved.

If the observed discrepancies are due to the influence of non-ordered parton emissions, models based on the color dipole or the CCFM evolution could provide a better description of the data. Figure 4b shows the data compared with ARIADNE (dotted lines) and two predictions of CASCADE which use different sets of unintegrated parton distributions. These sets differ in the way the small-k_T region is treated: in Jung2003 the full splitting function, i.e. including the non-singular term, is used in contrast to JS2001, for which only the singular term was considered. The predictions of ARIADNE give a good description of the data at low x and Q^2, but fail to describe the data at high Q^2. The predictions of CASCADE using JS2001 lie significantly above the data in all x and Q^2 regions, whereas those using Jung2003 are closer to the data. Therefore, the measurement of the fraction S is sensitive to the details of the unintegrated parton distributions.

4. Internal structure of jets

The investigation of the internal structure of jets gives insight into the transition between a parton produced in a hard process and the experimentally observable jet of hadrons. The internal structure of a jet depends mainly on the type of primary parton from which it originated and to a lesser extent on the particular hard scattering process. QCD predicts that at sufficiently high E_T^{jet}, where fragmentation effects become negligible, the jet structure is driven by gluon emission off the primary parton and is then calculable in pQCD. The lowest non-trivial order contribution to the jet substructure is given by order $\alpha\alpha_s$ calculations.

The internal structure of the jets can be studied by means of the mean subjet multiplicity. Subjets are resolved within a jet by reapplying the k_T algorithm on all particles belonging to the jet until for every pair of particles the quantity $d_{ij} = \min(E_{T,i}, E_{T,j})^2 \cdot ((\eta_i - \eta_j)^2 + (\varphi_i - \varphi_j)^2)$ is above $d_{\text{cut}} = y_{\text{cut}} \cdot (E_T^{\text{jet}})^2$. All remaining clusters are called subjets. The subjet structure depends upon the value chosen for the resolution parameter y_{cut}.

The mean subjet multiplicity has been measured[3] for jets using the k_T algorithm in the HERA laboratory frame with E_T^{jet} above 15 GeV and $-1 < \eta^{\text{jet}} < 2$, in the kinematic range given by $Q^2 > 125$ GeV^2. Figure 5a shows the mean subjet multiplicity for a fixed value of y_{cut} of 10^{-2} as a function of E_T^{jet}. It decreases as E_T^{jet} increases, i.e. the jets become more collimated. The experimental uncertainties are small (fragmentation model uncertainty $< 1\%$, the uncertainty on the absolute energy scale of the jets is negligible). The detector and hadronisation corrections are $< 10\%$ and $< 17\%$, respectively, for $E_T^{\text{jet}} > 25$ GeV. The data are compared to the LO and NLO predictions of DISENT. The LO calculation fails to describe the data, whereas the NLO calculations provide a good description. These measurements are sensitive to α_s and have been used to extract a value of $\alpha_s(M_Z)$. The result is

$$\alpha_s(M_Z) = 0.1187 \pm 0.0017 \text{ (stat.)} \,^{+0.0024}_{-0.0009} \text{ (exp.)} \,^{+0.0093}_{-0.0076} \text{ (th.)}.$$

This value is compatible with the world average and with previous measurements (see figure 2a).

5. Event shapes

A complementary extraction of α_s, also using the details of the hadronic final state in DIS, comes from the study of the event shape variables, like thrust or jet broadening. Event shape variables are particularly sensitive

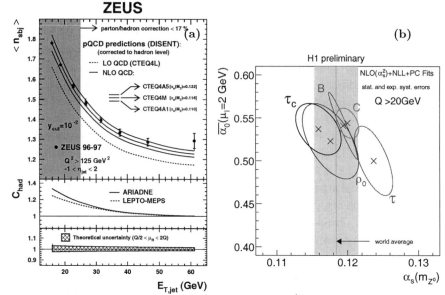

Figure 5. (a) Mean subjet multiplicity[1] as a function of $E_T^{\rm jet}$. (b) $1-\sigma$ contours in the $(\alpha_s, \bar{\alpha}_0)$ plane[4].

to the details of the non-perturbative effects of hadronisation and can be used to test the models for these effects. Recently, new developments with regard to power-law corrections have prompted a revived interest in the understanding of hadronisation from first principles. In this type of analysis, the data are compared to model predictions which combine NLO calculations and the theoretical expectations of the power corrections, which are characterised by an effective coupling $\bar{\alpha}_0$. Previous results supported the concept of power corrections in the approach of Dokshitzer et al. but a large spread of the results suggested that higher order corrections were needed. Now, resummed NLL calculations matched to NLO are available and so it is possible to study event shape distributions instead of only their mean values.

Event shape distributions (thrust, jet broadening, the jet mass ρ and the C parameter) have been measured[4] for particles in the current hemisphere in the kinematic region given by $14 < Q < 200$ GeV and $0.1 < y < 0.7$, where y is the inelasticity variable. Predictions consisting of NLO calculations using DISASTER++, resummed calculations matched to NLO and power corrections have been fitted to the data, leaving α_s and $\bar{\alpha}_0$ as free parameters. A good description of the data by the predictions was obtained at high Q^2, though the description at low Q^2 was poorer.

Figure 5b shows the 1σ-contour results from the fit in the α_s-$\bar{\alpha}_0$ plane. The spread observed in previous studies is much reduced when the resummed calculations are included. A clear anti-correlation between α_s and $\bar{\alpha}_0$ is found for all variables. A universal value for $\bar{\alpha}_0$ of 0.5 at the 10% level was obtained, in agreement with the previous results, but with a smaller spread. There is still a sizeable theoretical uncertainty for both α_s and $\bar{\alpha}_0$, of the order of 5%, which is as large as the experimental uncertainties. This uncertainty comes from the absent higher order corrections.

6. Conclusions

HERA has become a unique QCD-testing machine due to the fact that at large scales considerable progress in understanding and reducing the experimental and theoretical uncertainties has led to very precise measurements of the fundamental parameter of the theory, the strong coupling constant α_s. The use of observables resulting from jet algorithms leads now to determinations that are as precise as those coming from more inclusive measurements, such as from τ decays. To obtain even better accuracy in the determination of QCD, further improvements in the QCD calculations are needed, e.g. next-to-next-to-leading-order corrections.

At low values of x and Q^2, considerable progress has also been obtained in understanding the mechanisms of parton emission, though the interplay between the DGLAP, BFKL and CCFM evolution schemes has still to be fully worked out. Further progress in this respect needs both more experimental and more theoretical work.

Acknowledgments

I would like to thank the organisers for providing a warm atmosphere conducive to many physics discussions and a well organised conference. Special thanks to my colleagues from H1 and ZEUS for their help in preparing this report.

References

1. ZEUS Collaboration, S. Chekanov et al., *Phys. Lett.* **B547**, 164 (2002).
2. H1 Collaboration, Contributed paper N81 to the International Europhysics Conference on High Energy Physics, July 17-23,2003, Aachen.
3. ZEUS Collaboration, S. Chekanov et al., *Phys. Lett.* **B558**, 41 (2003).
4. H1 Collaboration, Contributed paper N111 to the International Europhysics Conference on High Energy Physics, July 17-23,2003, Aachen.

FORWARD JET AND PARTICLE PRODUCTION AT LOW X

L. JÖNSSON[*]

Physics Department, Lund University
Box 118,
S-221 00 Lund, Sweden
E-mail: leif.jonsson@hep.lu.se

In the kinematic region where the partons carry a very small fraction of the proton momentum their dynamics is expected to be characterized by initial state cascades which are non-ordered in virtuality of the propagators. In electron-proton collisions the investigation of jet- and single particle production along the direction of the incoming proton provides high sensitivity to such effects. Some recent experimental results from HERA are presented and compared to next-to-leading order calculations and to different QCD models in order to find evidence for the new effects.

1. Introduction

Quantum Chromodynamics (QCD) is generally regarded to be the correct theory to describe interactions between the quarks, which to our present knowledge are the smallest building blocks of matter. So far QCD has been mainly applicable to explain experimental data from hard scattering processes, where the relevant degrees of freedom are the fundamental particles, quarks and gluons. Although QCD has been successful in describing much data from deep inelastic scattering (DIS) we still face a number of problems like the description of parton dynamics at very small x-values, saturation effects, shadowing phenomena etc. A better understanding of the transition region between short- and long-range physics as well as of the long-range physics itself, and a theoretical description of such processes within the framework of QCD, is one of the most interesting challenges for the future.

Collisions between electrons and protons offer unique possibilities to study the partonic structure of matter. A particular advantage is that in

---

[*]Representing the H1 and ZEUS collaborations

such collisions the hardness of the interaction can be varied by selecting the energy and virtuality of the exchanged photon. With the higher resolution achieved by larger photon virtuality or with larger photon energy, a quark or a gluon can be resolved into a cascade of gradually softer partons. Various approximation schemes have been developed to calculate the evolution of parton cascades. For large photon virtualities the DGLAP evolution scheme [1] is applicable, whereas for very large photon energies the BFKL scheme [2] is relevant. The HERA facility offers the presently highest energies for performing such studies.

One of the most striking observations from the HERA experiments is the sharply rising density of quarks and gluons as the resolving power of the exchanged photon is increased, giving sensitivity to quarks carrying a very small fraction of the parton momentum. This strong rise was rather unexpected, although it was actually predicted by the BFKL formalism. It is clear that this rise of the parton density can not continue beyond limits, since it would ultimately lead to a violation of the unitarity principle, which guarantees that the total reaction probability must be smaller than 1. Thus, for sufficiently small x-values the density of partons becomes so large that they start to recombine. This saturation effect may be within reach at HERA, but it is also possible that the higher energies expected at future colliders are needed.

The dependence of the parton distribution on the energy fraction, x, carried by the parton and of the photon virtuality, Q^2, is one of the most interesting and challenging problems in QCD. This problem is intimately related to such diverse phenomena as shadowing in deep inelastic lepton-nucleon scattering, the amount of energy produced in the central region of heavy ion collisions and perhaps even to the growth of the total hadronic cross section at ultra high energies.

2. QCD models

The basic, or lowest order, parton level process must be corrected for higher order perturbative QCD effects causing the emission of extra partons on a short space-time scale. These can either be calculated by exact matrix elements, which has so far only been possible to low orders in the strong coupling constant, α_s, or approximately by parton cascades based on the iteration (to arbitrary orders in α_s) of the basic quark (q) and gluon (g) radiation processes $q \to qg$, $g \to gg$ and $g \to q\bar{q}$. Fixed order QCD calculations are available up to next-to-leading order (NLO) in the case of inclusive jet production at HERA. It is, however, important to keep in mind that the question to which order in α_s a calculation has been performed depends on the observable. For example leading order (LO) is different in a measurement of inclusive jet production compared to di-jet production. LO is thus the lowest order in which the observable obtains a non-zero value, and NLO is the next higher order. For DIS processes there are four programs existing, which provide matrix element calculations up to NLO. These are DISENT [3], DISASTER++ [4], MEPJET [5] and JetViP [6].

At high energies the interactions leave a large phase space for parton emission. Higher order QCD effects will therefore become important and to account for these it is necessary to use phenomenological parton shower models. There are various models on the market with different formulations of the basic evolution equations for parton branchings, and therefore the models are valid in certain regions of phase space only.

2.1. *The DGLAP model*

The DGLAP evolution equation is of the form

$$\frac{df_j(x,\mu^2)}{d\ln\mu^2} = \frac{\alpha_s(\mu^2)}{2\pi} \sum_i \int_x^1 \frac{dx'}{x'} f_i(x',\mu^2) P_{i\to j,k}(z) \qquad (1)$$

where $f(x,\mu^2)$ is the density of partons carrying a longitudinal momentum fraction x probed at a scale μ^2, and $P_{i\to j,k}(z)$ describes the probability that a parton i is split into two partons j and k with a fraction $z = \frac{x}{x'}$ and $1-z$ of the original parton momentum, respectively. The probability that a gluon splits into two gluons is given by

$$P_{g\to g,g}(z) = \frac{1}{1-z} - 2 + z(1-z) + \frac{1}{z} \qquad (2)$$

where the terms $\frac{1}{1-z}$ and $\frac{1}{z}$ are called singular terms, since they give infinite contributions when $z \to 1$ and $z \to 0$, respectively. Equation (1) thus describes the probability change of finding a parton of type j with momentum fraction x as we increase the scale μ^2. In the DGLAP formalism, the propagator gluons are assumed to be strongly ordered in virtuality

$$\mu^2 \gg |k_n^2| \ldots \gg |k_i^2| \ldots \gg |k_1^2| \gg |k_0^2|, \qquad (3)$$

where k_i is the four momentum of parton i. It can be shown that the ordering in virtuality implies that also the transverse momentum of the propagator partons (at small x) has to be strongly ordered according to

$$\mu^2 \gg |k_{tn}^2| \ldots \gg |k_{ti}^2| \ldots \gg |k_{t1}^2| \gg |k_{t0}^2| \qquad (4)$$

Since the virtualities and transverse momenta of all the gluon propagators are small compared to the hard scale, μ^2, they can be treated as massless and approximated to move in the same direction as the incoming proton (collinear approximation). It can be shown, that the DGLAP evolution gives a resummation of terms of the form $(\alpha_s \ln(\mu^2))^n$ in the expansion of the cross section. Hence, the DGLAP approximation is only valid at large μ^2 where these terms will dominate.

In electron-proton scattering the internal structure of the proton as well as of the exchanged photon can be resolved provided the scale of the hard process is larger than the inverse radius of the proton, $1/R_p^2 \sim \Lambda_{QCD}^2$, and the photon, $1/R_\gamma^2 \sim Q^2$, respectively. Resolved photon processes play an important role in photo-production of high p_t-jets, since $Q^2 = 0$, but they can also give considerable contributions to DIS processes if the scale μ^2 of the hard subprocess is larger than Q^2. This situation can be described by introducing two DGLAP ladders, one from the proton side and another from the photon side [7].

2.2. The BFKL model

The BFKL evolution equation resums the terms $(\alpha_s \ln(\frac{1}{x}))^n$ in the expansion, and is thus only valid at small x. Thus, the evolution is made in increasing $\ln(\frac{1}{x})$, since

$$x_0^2 \gg x_1^2 \ldots \gg x_n^2 \gg x_{Bj}^2 \tag{5}$$

has been assumed. This implies that the emitted gluons will take a large fraction of the propagator momentum. However, there is no restriction in k^2 or k_t^2. This means that the virtualities and transverse momenta of the propagators can take any kinematically allowed value and must not be smaller than the photon virtuality as was the case in the DGLAP description. Consequently the collinear approach is no longer applicable but the matrix elements must be taken off mass shell (the particles can have virtual mass) and unintegrated parton densities, taking the transverse momenta of the propagators into account, have to be used. This is called k_t-factorisation [8]. The BFKL evolution equation is of the form

$$\frac{d\mathcal{G}(x, k_t^2)}{d\ln(\frac{1}{x})} = \int dk_t'^2 \mathcal{G}(x, k_t'^2) \cdot K(k_t^2, k_t'^2) \tag{6}$$

where

$$\mathcal{G}(x, k_t^2) \cong \left.\frac{dg(x, \mu^2)}{d\mu^2}\right|_{\mu^2 = k_t^2} \tag{7}$$

with

$$g(x,\mu^2) \simeq \int_0^{\mu^2} dk_t^2 \mathcal{A}(x, k_t^2, \mu^2). \tag{8}$$

$\mathcal{A}(x, k_t^2, \mu^2)$ is the unintegrated gluon density describing the probability to find a gluon with a longitudinal momentum fraction x and a transverse momentum k_t at a scale μ^2. The function K is the splitting kernel equivalent to P in (1). To distinguish the different parton distributions, the following notation has been used: $g(x, \mu^2)$ is the parton distribution in the collinear approach, while $\mathcal{G}(x, k_t^2)$ and $\mathcal{A}(x, k_t^2, \mu^2)$ are the one- and two-scale distributions in the k_t-factorization approach.

2.3. *The CCFM model*

The CCFM [9] evolution equation is valid both at large and small x, since it resums terms of both the form $(\alpha_s \ln(\frac{1}{x}))^n$ and $(\alpha_s \ln(\frac{1}{1-x}))^n$. This means that at large x the CCFM evolution will be DGLAP-like, and at small x it will be BFKL-like. The CCFM evolution includes angular ordering in the initial state cascade, to account for colour coherence effects, which means that the emission angles of the partons with respect to the incoming proton increases as one moves towards the quark box,

$$\Xi \gg \xi_n \gg \ldots \xi_1 \gg \xi_0, \tag{9}$$

where the maximum allowed angle Ξ is set by the hard quark box.

The original CCFM splitting function $\tilde{P}$ was defined as

$$\tilde{P}_g(z, k_t^2, \bar{q}) = \frac{\bar{\alpha}_s(\bar{q}_i^2)}{1 - z_i} + \frac{\bar{\alpha}_s(k_{ti}^2)}{z_i} \Delta_{ns}(z_i, k_{ti}^2, \bar{q}_i^2), \tag{10}$$

where

$$\bar{q}_i = \frac{p_{ti}}{1 - z_i} = x_{i-1}\sqrt{s\xi_i}$$

is the rescaled transverse momenta of the emitted gluons and $z_i = \frac{x_i}{x_{i-1}}$. The CCFM splitting function differs from the DGLAP splitting function in the sense that it only includes the singular terms. Another difference is the additional function Δ_{ns}, called the non-Sudakov form factor. The non-Sudakov form factor originates from the fact that, in CCFM and BFKL, all virtual corrections in the gluon vertex are automatically taken into account. This is called the Reggeization of the gluon vertex.

At very high energies the $1/z$-term in the splitting function will dominate and it is sufficient to include the singular terms. However, for the energies at present colliders it is still necessary to take the non-singular terms into account in the splitting function, which then takes the form:

$$\tilde{P}_g(z, k_t^2, \bar{q}) = \bar{\alpha}_s(\bar{q}_i^2) \left(\frac{z_i}{1-z_i} + \frac{z_i(1-z_i)}{2} \right)$$
$$+ \bar{\alpha}_s(k_{ti}^2) \left(\frac{(1-z_i)}{z_i} + \frac{z_i(1-z_i)}{2} \right) \Delta_{ns} \quad (11)$$

2.4. The Colour Dipole Model

In the Colour Dipole Model (CDM) [10] the emissions are assumed to originate from colour dipoles streched between the partons. The primordial dipole is spanned between the scattered quark and the proton remnant. The emission of a gluon by this dipole produces a 'kink' in the dipole leading to secondary dipoles, which may in turn produce further dipoles by emitting more gluons. Since the dipoles are radiating independently, the parton cascade will not be ordered in transverse momentum and therefore exhibit a BFKL-like behaviour. The fact that the proton remnant is an extended object will reduce the phase space for gluon emission because the gluon can only access a fraction of the momentum carried by the proton remnant. In a resolved photon case the photon can also be regarded as an extended object, described by its parton density functions just like the proton, and thus a similar suppression is obtained on the photon side.

3. Inclusive jet production

In the region of low x-values the interacting parton frequently produces a cascade of emissions before it interacts with the virtual photon. Due to the strong ordering in virtuality, the emissions of the DGLAP evolution are very soft close to the proton direction, whereas BFKL emissions can produce large transverse momenta in this region. Thus, deviations from the DGLAP parton evolution scheme are expected to be most visible in a region close to the direction of the incoming proton. The ZEUS collaboration has studied inclusive jet production in DIS at photon virtualities, $Q^2 > 25\ GeV^2$, and x-values down to $7 \cdot 10^{-4}$ in order to investigate whether deviations from the DGLAP picture could be observed, which could then be better described by the BFKL model.

The jet search was performed by applying the longitudinally invariant k_t-algorithm to the information from the calorimeter cells. Jets were accepted provided they had a transverse energy $E_{t,jet} > 6\ GeV$ and were reconstructed inside the rapidity range $-1 < \eta_{jet} < 3$ in the laboratory system.

The data sample obtained provided the basis for two independent analyses. The first one used the full phase space as defined by the cuts given above. In the second analysis the so called hadronic angle, γ_h, was calculated from a calorimetric measurement of the hadronic final state. Only events where the momentum vector of the hadronic system was pointing in the backward direction ($\cos \gamma_h < 0$) and the jet (jets) was reconstructed in the forward part of the detector ($\eta_{jet} > 0$) were accepted in this analysis. The data sample of the first analysis was dominated by single jet events, whereas the hadronic angle restriction removed such events and enhanced the contribution from di-jet and multi-jet events in the second sample. To distiguish the two samples the first one is referred to as the 'inclusive phase space' and the second as the 'di-jet phase space'.

The differential cross section for the two samples was measured as a function of the transverse jet energy, the jet pseudorapidity, the photon virtuality and Bjorken-x. The results were compared to the predictions from NLO calculations in the DGLAP approach as given by the DISENT program and from the phenomenological descriptions given by the DGLAP evolution as implemented in the LEPTO generator [11] and by CDM in the ARIADNE program [12]. Radiative corrections were accounted for by using the DJANGO program [13], which provides an interface between the QED generator HERACLES [14] and LEPTO or ARIADNE. Hadronisation was performed using the Lund string model [15].

The results from the 'inclusive phase space' (Fig. 1) show that the description of data as η_{jet} increases becomes worse. This can be understood from the fact that the α_s^o contribution goes to zero at fixed Q^2 and x (alternatively y), since the value of η_{jet} is fixed in an α_s calculation by $\frac{1}{2} \ln \frac{Q^2(1-y)}{4E_e^2 y^2}$. Experimentally a certain range in Q^2 is covered which obviously influences the maximum value of η_{jet}. In the kinematic region chosen by ZEUS the maximum η_{jet} is around 1 and this is also where NLO calculations start deviating from data. Both CDM (ARIADNE) and the DGLAP model (LEPTO) are in good agreement with data. The QCD models as well as NLO calculations are able to reproduce the Bjorken-x distribution down to values of about $2 \cdot 10^{-3}$ but at lower values the DGLAP model and even more the NLO calculations strongly disagree with data. All this indicates that calculations in NLO are not enough and that higher order corrections are necessary.

The jet cross section in the 'di-jet phase space' seems to be better de-

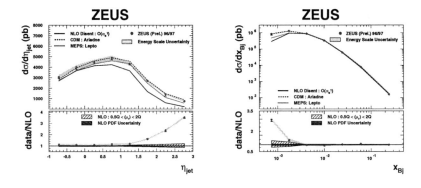

Figure 1. $d\sigma/d\eta_{jet}$ and $d\sigma/dx$ for the 'inclusive phase space' data.

scribed by NLO calculations and by the QCD models both when it is plotted as a function of η_{jet} and as a function of Bjorken-x (Fig. 2). On the other hand we can observe that the scale uncertainty has blown up which excludes any clear conclusions.

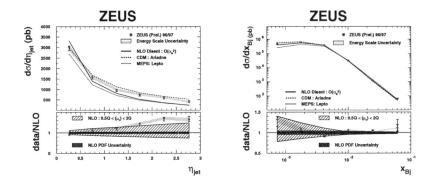

Figure 2. $d\sigma/d\eta_{jet}$ and $d\sigma/dx$ for the 'di-jet phase space' data.

It should be recalled that NLO for inclusive jet production is order α_s^1, whereas NLO for di-jet production is order α_s^2.

4. Forward jet production

A measurement of the forward jet production cross section at small Bjorken-x, as proposed by Mueller and Navelet [16], has since long been regarded as the most promising test of perturbative parton dynamics. The idea is to select events with a jet close to the proton direction having the virtuality of the propagator closest to the proton approximately equal to the virtuality of the exchanged photon. This will prevent an evolution with strong ordering in virtuality as is the case in the DGLAP evolution. The additional requirement that the forward jet takes a large fraction of the proton momentum, $x_{jet} = E_{jet}/E_p$, such that $x_{jet} \gg x_{Bj}$ opens up for an evolution where the propagators are strongly ordered in the longitudinal momentum fraction like in the BFKL scheme. Experimentally this is realized by demanding the transverse momentum of the forward jet to be of the same order as Q^2 and x_{jet} to be larger than a preselected value which still gives reasonable statistics.

The H1 experiment has recently measured the forward jet cross section from a sample of events obtained by applying the cuts $E_{e'} > 10\ GeV$, $156^o < \Theta_e < 175^o$, $0.1 < y < 0.7$ and $5 < Q^2 < 75\ GeV$. Jets were reconstructed using the inclusive k_t-algorithm in the Breit frame. In order to fulfill the conditions discussed above to suppress DGLAP and enhance BFKL evolution, it was required that $0.5 < p_t^2/Q^2 < 2$ and $x_{jet} > 0.035$. The jet angle had to be larger than 7° with respect to the proton direction, which ensures that the jet is well inside the acceptance of the detector. Further the transverse energy of the jet had to be $p_{t,jet} > 3.5\ GeV$ to stay in the perturbative region.

The cross section of forward jet production is shown as a function of x in Fig. 3. Data are compared to the predictions of the DGLAP model with and without resolved virtual photons (DGLAP 'resolved' and DGLAP 'direct', respectively) as implemented in the RAPGAP Monte Carlo program [17]. Also shown are the predictions from CDM as given by ARIADNE and of the CCFM evolution, which has been parametrized in the CASCADE generator [18]. Three different unintegrated gluon densities were used in CASCADE. That of JS2001 was determined through a fit to the early $F_2(x, Q^2)$-data from HERA taken in 1994. The soft region was avoided by applying a cut $k_t^{cut} > 0.25\ GeV$ and collinear emissions were excluded by setting $Q_o > 1.4\ GeV$. In the CCFM evolution k_t can take any kinematically allowed value i.e. also values smaller than k_t^{cut}. In such a case no real emission is allowed in CASCADE and the evolution in angle continues until the k_t-condition is fulfilled. For the versions J2003-1 and -2 the unintegrated gluon densities have been determined from fits to the $F_2(x, Q^2)$-data obtained at HERA in 1994 and 1996/97. The input parameters of the model was defined such as to fit the data in the region $x > 5 \cdot 10^{-3}$ and $Q^2 > 4.5~GeV^2$. J2003-1 has $k_t^{cut} = Q_o = 1.33~GeV$, whereas J2003-2 has $k_t^{cut} = Q_o = 1.18~GeV$. The difference between these two sets is that set 1 only includes singular terms in the splitting function whereas set 2 also takes the non-singular terms into account.

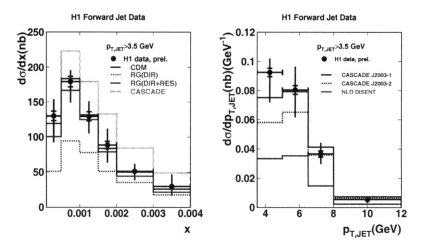

Figure 3. $d\sigma/dx$ for forward jet production compared to various QCD models (CDM, DGLAP 'direct' (RG(DIR)), DGLAP 'resolved' (RG(DIR+RES)), CCFM (CASCADE JS2001, J2003-1, J2003-2) and NLO calculations (NLO DISENT)

From Fig. 3 it is seen that DGLAP 'direct' is undershooting the data and that the contribution from resolved photons gives agreement with data. CDM describes the data well while CASCADE (JS2001) gives too high a cross section. Further the NLO calculations by DISENT give cross sections which are a factor two smaller than data. CASCADE version J2003-1 reproduces data well and J2003-2, with non-singular terms in the splitting fuction, gives a reasonable agreement although it is somewhat on the low side.

5. Forward pion production

Following the idea behind the forward jet measurement it is also possible to select high energy single particles in the forward direction and require the criteria of Mueller and Navelet to be fulfilled. The advantage of using single particles is that angles closer to the proton direction can be reached compared to jets, which have a certain lateral extension. On the other hand the cross section for events with a specific high energy forward particle is lower than the jet cross section and the sensitivity to fragmentation effects becomes higher.

The H1 collaboration has performed a measurement on single forward π^o-meson production, where the π^o-meson is identified through its dominant decay into two photons. The photons were measured in the calorimeter but due to the high energy of the pion, the photons could not be separated but were detected as one single cluster. The analysis was restricted to the kinematic range defined by the cuts $0.1 < y < 0.6$ and $2 < Q^2 < 70\ GeV^2$. The polar angle of the pion was required to be in the range $5^o < \Theta_\pi < 25^o$ with $x_\pi = E_\pi/E_p > 0.01$.

Inclusive forward π^o cross sections for $p_{t,\pi^o} > 2.5\ GeV$, measured in the proton-photon centre-of-mass system, are shown in Fig. 4 as a function of x for three different regions in Q^2. It should be noticed that the measurements extend down to $x = 4 \cdot 10^{-5}$. The results show that the prediction of DGLAP 'direct' (DIR) again is well below the data, whereas a reasonable description is obtained if the resolved photon contribution (DIR+RES) is added. Somewhat unexpectedly all sets of CASCADE (only JS2001 is shown) fall below the data at small x-values. One explanation for this could be that CASCADE produces mostly gluon jets in this region, while RAPGAP also produces quark jets. Since the fragmentation of quarks into pions has a harder spectrum the selection criteria will enhance the quark-jet contribution. It should, however, be pointed out that there is no direct contradiction since the discrepancies in the π^o cross section occur in a region of x which is mostly not covered by the forward jet measurement. A comparison was also made to a modified LO BFKL calculation [19] which was supplemented by a consistency constraint to mimic higher orders in the perturbative expansion. There is good agreement between this calculation and the data.

The parton cascades from the different evolution schemes should result in transverse energy flow distributions, which are different. The π^o data have been used to study the transverse energy flow with respect to the pion in the hadronic centre-of-mass system. The results are presented for three different rapidity ranges and compared to the predictions of the QCD models in Fig. 5. The experimental distribution in the rapidity range closest to the proton direction indicates that the energy compensation occurs over

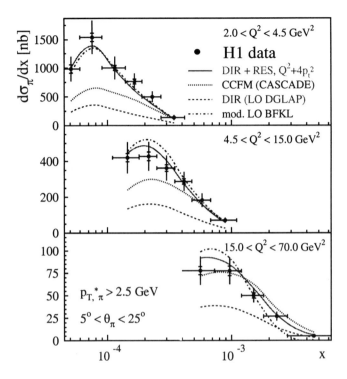

Figure 4. $d\sigma/dx$ for forward pion production compared to various QCD models (DGLAP 'direct' (DIR), DGLAP 'resolved' (DIR+RES), CCFM (CASCADE JS2001) and a modified BFKL calculation (LO BFKL).

the full length of the parton ladder, which is in disagreement with the predictions of long range compensation from DGLAP. The best description of data is given by DGLAP 'resolved'.

6. Conclusions

The production of jets and single particles in the forward region has been studied in order to search for deviations from the standard DGLAP description of the parton dynamics. Significant discrepancies between data and the simple DGLAP predictions were observed but if the exchanged photon is allowed to interact via its partonic content (resolved virtual photons) good agreement is obtained with most data. The CCFM evolution scheme, which combines the properties of DGLAP and BFKL, give similar agreement and

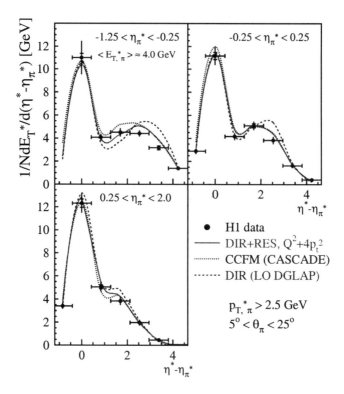

Figure 5. $1/NdE_t/d(\eta - \eta_\pi)$ for forward pion production compared to various QCD models (DGLAP 'direct' (DIR), DGLAP 'resolved' (DIR+RES), CCFM (CASCADE JS2001).

has the advantage to contain less free parameters. The final conclusion is, however, that so far no clear evidence for new parton dynamics has been observed.

7. Acknowledgments

I would like to express my gratitude to the organising committee for the invitation to give this talk and for creating such a nice atmosphere during the conference. M. Klein, T. Greenshaw, P. Newman, H. Jung and J. Butterworth are acknowledged for good advises and comments in the preparation of the talk.

References

1. V. Gribov and L. Lipatov, *Sov. J. Nucl. Phys.* **15**, 438 and 675 (1972)
   L. Lipatov, *Sov. J. Nucl. Phys.* **20**, 94 (1975)
   G. Altarelli and G. Parisi, *Nucl. Phys.* **B 126**, 298 (1977)
   Y. Dokshitzer, *Sov. Phys. JETP* **46** 641 (1977)
2. E. Kuraev, L. Lipatov and V. Fadin, *Sov. Phys. JETP* **44**, 443 (1976)
   E. Kuraev, L. Lipatov and V. Fadin, *Sov. Phys. JETP* **45**, 199 (1977)
   Y. Balinsky and L. Lipatov, *Sov. J. Nucl. Phys.* **28**, 822 (1978)
3. S. Catani, M.H. Seymour, *Phys. Lett.* **B 378**, 287 (1996), *Nucl. Phys.* **B 485**, 291 (1997)
4. D. Graudenz, PSI.PR-97-20, hep-ph/979240
5. E. Mirkes, D. Zeppenfeld, *Phys. Lett.* **B 380**, 23 (1996), *Acta Phys. Polon.* **B 27**, 1392 (1996)
6. B. Pötter, *Comp. Phys. Comm.* **119**, 45 (1999)
   B. Pötter, *Comp. Phys. Comm.* **133**, 105 (2000)
7. H. Jung, L. Jönsson, H. Küster, *Eur. Phys. J.* **C 9**, 383 (1999)
8. S. Catani, M. Ciafaloni, F. Hautmann, *Nucl. Phys.* **B 366**, 135 (1991)
   J. Collins, R. Ellis, *Nucl. Phys.* **B 360**, 3 (1991)
   L. Gribov, E. Levin, M. Ryskin, *Phys. Rep.* **100**, 1 (1983)
   E.M. Levin, M.G. Ryskin, Y.M. Shabelski, A.G. Shuvaev, *Sov. J. Nucl. Phys.* **53**, 657 (1991)
9. M. Ciafaloni, *Nucl. Phys.* **B 296**, 49 (1988)
   S. Catani, F. Fiorani and G. Marchesini, *Phys. Lett.* **B 234**, 339 (1990)
   S. Catani, F. Fiorani and G. Marchesini, *Nucl. Phys.* **B 336**, 18 (1990)
   G. Marchesini, *Nucl. Phys.* **B 445**, 49 (1995)
10. G. Gustafson, *Phys. Lett.* **B 175**, 453 (1986)
    G. Gustafson, U. Pettersson *Nucl. Phys.* **C 306**, 746 (1988)
11. G. Ingelman, A. Edin and J. Rathsman, *Comp. Phys. Comm.* **101**, 108 (1997)
12. L. Lönnblad, *Comp. Phys. Comm.* **71**, 15 (1992)
    L. Lönnblad, *Z. Pys.* **C 65**, 285 (1995)
13. K. Charchula, G.A. Schuler, H. Spiesberger, *Comp. Phys. Comm.* **81**, 381 (1994)
14. A. Kwiatkowski, H. Spiesberger, H.-J. Möhring *Comp. Phys. Comm.* **69**, 155 (1992)
15. T. Sjöstrand *Comp. Phys. Comm.* **82**, 74 (1994)
16. A.Mueller, *Nucl. Phys. B (Proc. Suppl.)* **18C**, 125 (1990)
    A. Mueller, *J. Phys.* **G 17**, 1443 (1991)
    A. Mueller, H. Navelet, *Nucl. Phys.* **B 282**, 727 (1987)
17. H. Jung, http://www.quark.lu.se/ hannes/rapgap/
18. H. Jung, *Eur. Phys. J.* **C 19**, 351 (2001)
    H. Jung, *Comp. Phys. Comm.* **143**, 100 (2002)
19. J. Kwiecinski, A. Martin, J. Outhwait, *Eur. Phys. J.* **C 9**, 611 (1999)

SINGLE HADRON PRODUCTION IN DEEP INELASTIC SCATTERING

M. MANIATIS

*II. Institut für Theoretische Physik, Universität Hamburg,
Luruper Chaussee 149, 22761 Hamburg, Germany
E-mail: markos@mail.desy.de*

The NLO-QCD correction to single hadron production in deep inelastic scattering is calculated. We require the final state meson to carry a non-vanishing transversal momentum, thus being sensitive to perturbative QCD effects. Factorization allows us to convolute the hard scattering process with parton densities and fragmentation functions. The predictions are directly comparable to experimental results at the HERA collider at DESY. The results are sensitive to the gluon density in the proton and allow us to test universality of fragmentation functions.

1. Introduction

The predictive power of QCD lies in the factorization theorem. In deep inelastic scattering (DIS) factorization in short and long distance parts allows us to describe the observed hadrons as a convolution of the partonic processes with non-perturbative parton densities and fragmentation functions [1]. Single meson production in electron proton scattering

$$e^-(k) + P(p_a) \to e^-(k') + h(p_b) + X \quad (1)$$

occurs partonically already in the absence of strong interactions ($\mathcal{O}(\alpha_S^0)$), where one parton of the proton (a quark) interacts with the leptonic current and fragments into a meson (h) (naive parton model).

Since we are interested in perturbative QCD effects we require the meson to carry a non-vanishing transversal momentum with respect to the centre-of-mass frame of virtual vector boson coming from the electron and initial proton ($p_{b\perp} > 0$). Thus at partonic level at least two final state partons are required to balance the transversal momentum. The leading order processes with non-vanishing transversal momentum of a fragmenting parton into a meson ($\mathcal{O}(\alpha_S)$) are

(a) $\gamma^* + q \to q + g$

(b) $\gamma^* + g \to q + \bar{q}$,

where the virtual photon originates from the electron current. If the virtuality of the photon $Q^2 := -q^2$ ($q := k - k'$) is not too large compared to the squared Z-mass, the contribution from Z boson exchange is suppressed and may be neglected. Dealing then with C-invariant processes we do not have to calculate separately the processes with interchanged quarks and antiquarks. Factorization, proven generally for DIS processes [1], describes the events as a convolution of the hard scattering processes with parton densities and fragmentation functions.

The investigation of single hadron production is interesting for several reasons: First of all it is a test of perturbative QCD and factorization. The predictions depend beside the perturbative partonic calculation essentially on universal parton densities and fragmentation functions. Especially fragmentation functions, fitted to electron positron annihilation data, may be tested, in particular their universality.

Further, the predictions allow for a direct comparison with experimental data, in particular there is no need to use any kind of Monte Carlo procedure. Thus we may expect very meaningful results.

The predictions are due to process (b) directly sensitive to the gluon density in the proton and may allow us to draw conclusions concerning the gluon density in the proton. From the experimental side precise data are available from the HERA collider at DESY. For instance π mesons were measured in the forward region (with small angles with respect to the proton remnant) based on events detected in the H1 detector [2] and charged hadrons were measured at the ZEUS experiment [3].

In 1978 the process (1) with non-vanishing transversal momentum of the hadron ($p_{b\perp}$) was calculated by Méndez [4] at tree level accuracy ($\mathcal{O}(\alpha_S)$). Since QCD corrections are typically large and we are confronted with precise experimental data it is desirable to compare these data with predictions of at least next-to-leading order (NLO) ($\mathcal{O}(\alpha_S^2)$) accuracy. Also NLO-QCD predictions were computed with the assumption of purely transversal photons, neglecting the longitudinal degrees of freedom of the exchanged virtual photon [5].

Here we present a NLO-QCD calculation based on the dipole subtraction formalism [6]. In contrast to the more conventional phase space slicing method there is no need to introduce any unphysical parameter to cut the phase space in soft, respectively collinear regions. Also all cancellations of infrared singularities occur before any numerical phase space integration is performed. Thus we may present numerically very stable predictions.

2. Calculation

The differential cross section for process (1) reads as a convolution with the parton densities and fragmentation functions as

$$\frac{d\sigma^h}{d\bar{x}dyd\bar{z}d\phi} = \sum_{ab} \int_{\bar{x}}^1 \frac{dx}{x} \int_{\bar{z}}^1 \frac{dz}{z} f_a(\frac{\bar{x}}{x},Q^2) \frac{d\sigma^{ab}}{dxdydzd\phi} D_b^h(\frac{\bar{z}}{z},Q^2), \qquad (2)$$

where, as usual, the variables x, y, and z are defined as $x = \frac{Q^2}{2p_a q}$, $y = \frac{p_a q}{p_a k}$, $z = \frac{p_a p_b}{p_a q}$ with respect to the partonic momenta and the bar quantities $\bar{x}$, $\bar{y}$, $\bar{z}$ with respect to the momenta of the hadrons (with $\bar{y} = y$). The angle ϕ denotes the azimuthal angle between the planes defined on one hand by the directions of the leptons and on the other hand by the momenta of final state hadron and virtual vector boson in the centre-of-mass frame of vector boson and initial parton. The sum is over the different initial (a) and final (b) state partons and f_a and D_b^h denote the corresponding parton density respectively fragmentation function. The hard scattering process may be written as a contraction of a lepton tensor ($l^{\mu\nu}$) with a hadron tensor ($H_{\mu\nu}^{ab}$):

$$\frac{d\sigma^{ab}}{d\bar{x}dyd\bar{z}d\phi} = \frac{\alpha^2}{16\pi^2} y \frac{1}{Q^4} l^{\mu\nu} H_{\mu\nu}^{ab}. \qquad (3)$$

If we consider the centre-of-mass system of virtual photon and initial parton, both unpolarized, there cannot be any dependence on the azimuthal angle ϕ. Integrating out this angle dependence we find a decomposition into a transversal and a longitudinal part of the virtual photon:

$$\frac{d\sigma^{ab}}{d\bar{x}dyd\bar{z}d\phi} = \frac{\alpha^2}{8\pi} y \frac{1}{Q^4} \left\{ Q^2 \frac{2y - y^2 - 2}{2y^2} g^{\mu\nu} + 2Q^4 \frac{y^2 - 6y + 6}{\hat{s}^2 y^4} p_a^\mu p_a^\nu \right\} H_{\mu\nu}^{ab}. \qquad (4)$$

The computation of the correction to process (1) were carried out in the subtraction formalism [6]. The general idea of the subtraction formalism is to subtract from the real correction an artificial counterterm which has the same pointwise singular behaviour in $D = 4 - 2\epsilon$ dimensions as the real correction itself. Thus the limit $\epsilon \to 0$ can be performed and the real phase space integral can be evaluated numerically. The artificial counterterm is constructed in a way that it also can be integrated over the one-parton subspace analytically leading to ϵ poles. Adding these terms to the virtual part of the correction these poles cancel all singularities in the virtual part analytically and the remaining integration over the phase space can be carried out numerically. The advantage compared to phase space slicing methods is that all singularities cancel *before* any numerical integration is performed. Also there is no need to introduce an unphysical cut parameter which in phase space slicing methods separates soft, respectively collinear phase space regions from the remaining hard region.

In our case of a convolution of a partonic cross section with distribution functions additional kinematic constraints have to be taken into account. The momenta of the partons which enter the convolution have to be kept fixed (called *identified* partons). This leads to modified artificial counterterms. Its analytical integration gives collinear singularities which cancel the singularities of the non-perturbative distribution functions yielding scheme and scale dependent convergent parts.

The whole calculation was done with the help of the algebra package Form [8]. At $\mathcal{O}(\alpha_S^2)$ the real correction is given by the squared matrix elements of the diagrams

(a) $\gamma^* + q \to q + g + g$
(b) $\gamma^* + g \to q + \bar{q} + g$
(c) $\gamma^* + q \to q + q + \bar{q}$,

where in turn each final state parton serves as an observed hadron. In process (c) we have to consider two flavours in the fermion traces. This yields 4 diagrams for every pair of different flavours in contrast to 8 diagrams in the case of one uniform flavour. Care must be taken to adjust statistical factors properly. The artificial counterterm was constructed as described by the subtraction formalism. The phase space integral over the 3-particle final state can be performed yielding a finite real contribution. The virtual contribution of the correction, i.e. the interference term of the Born matrix elements and the one-loop matrix elements is computed. Here we encounter 2-point, 3-point, and 4-point tensor integral contributions which were reduced to scalar integrals via tensor reduction [9]. The scalar integrals, containing ultraviolet and infrared singularities, were computed analytically in dimensional regularization. The analytic expressions were compared with the literature [7]. The virtual contribution was renormalized in a mixed scheme, where the wave functions were renormalized on-shell and the strong coupling constant in the $\overline{MS}$ scheme, yielding an ultraviolet finite virtual contribution. The infrared singularities cancel exactly the contributions given by the integrated artificial counterterm in the subtraction formalism.

From the subtraction of remaining singularities into the parton densities and the fragmentation functions we obtain also finite remainders which depend on the factorization scheme and on the factorization scale. In this context we choose the $\overline{MS}$ scheme.

Thus we end up with three contributions, the real part, the virtual part, and the part related to identified partons. All these finite parts have to be integrated over the 2- respectively 3- particle final state phase space. To this purpose a C-routine was written to perform these integrations numerically.

3. Results

In the convolution (2) we use the parton densities published by the CTEQ collaboration [10]. Herein the parton distribution set, called CTEQ5M, where M denotes the $\overline{MS}$ scheme, matches our conditions with the assumption of 5 light quarks.

We adopt the KKP fragmentation functions to our calculation [11]. Since these fragmentation functions are fitted to e^+e^- data with high accuracy and applied here to a DIS process the comparison with experimental data will serve as a good check of universality.

We use the value of the strong coupling constant at the Z-scale as given by the Particle Data Group [12] $\alpha_S(M_Z) = 0.118$ and evolve this value with the NLO evolution equation. In the LO approximation we evolve this value with the LO evolution equation for consistency reasons.

In Fig. 1 we show the differential cross section $\frac{d\sigma^h}{d\bar{x}dyd\bar{z}}$ as a function of $p_{b\perp}$ for fixed values of $\bar{x} = 0.1$, $y = 0.1$, $\bar{z} = 0.5$, and $Q^2 = 360$ GeV^2, where $p_{b\perp}$ is defined in the centre-of-mass frame of vector boson and initial parton. We set the renormalization scale (μ_R) equal to both the initial (μ_{F_i}) and the final (μ_{F_f}) fragmentation scales $\mu := \mu_R = \mu_{F_i} = \mu_{F_f}$, where the initial scale is related to the parton densities and the final scale to the fragmentation functions.

The steep fall of the cross section over several orders of magnitude makes it hard to resolve the NLO contribution (full line) in contrast to the LO prediction (dashed line) graphically. The lower part of the figure shows the ratio of the NLO correction over the LO result and displays more clearly the higher order effects. The correction changes from $+10$ % for small $p_{b\perp}$ to -10 % for large $p_{b\perp}$. For $p_{b\perp} \approx 1$ GeV we get an even larger correction due to collinear configurations.

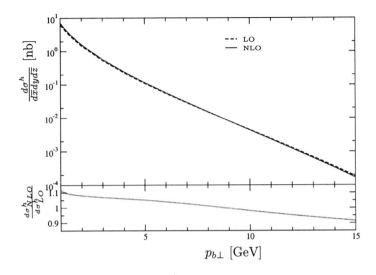

Figure 1. Differential cross section versus $p_{b\perp}$ for $\bar{x} = 0.1$, $y = 0.1$, and $\bar{z} = 0.5$ with $Q^2 = 360$ GeV^2. The lower part shows the corresponding K-factor $\frac{d\sigma^h_{NLO}}{d\sigma^h_{LO}}$ of the differential cross sections at NLO and LO.

The theoretical uncertainty of the predictions of a fixed order calculation is mainly given by the renormalization and fragmentation scale dependence. The theoretical uncertainties due to the parton densities and fragmentation functions are not considered here. Another source of uncertainty lies in the factorization theorem itself since it predicts correct results with an error of $\mathcal{O}(\Lambda^2_{QCD}/p^2_{b\perp})$ which may be become large for very low $p_{b\perp}$.

We check the scale dependence of the cross section for the kinematics already depicted in context with Fig. 1 but with a fixed transversal hadron momentum of $p_{b\perp} = 5$ GeV. In Fig. 2 the differential cross section is shown where we vary the common scale μ^2 over two orders of magnitude with respect to the reference scale $\mu_0^2 = Q^2$. The scale dependence varies in this range about +33 % (+27 %) for very low scales to -20 % (-18 %) for rather high scales at LO (NLO). Thus there is only an unexpectedly slight scale dependence reduction at NLO compared to LO.

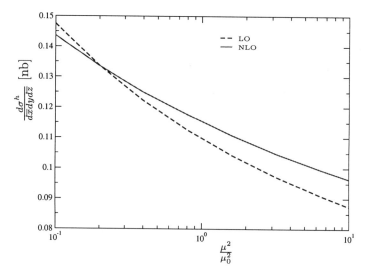

Figure 2. Scale dependence of the differential cross section for $\bar{x} = 0.1$, $y = 0.1$, and $\bar{z} = 0.5$ with $Q^2 = 360$ GeV^2 and a fixed $p_{b\perp} = 5$ GeV. The scale $\mu^2 := \mu_R^2 = \mu_{F_i}^2 = \mu_{F_f}^2$ is varied with respect to the reference scale $\mu_0^2 = 360$ GeV^2.

4. Conclusion

The calculation of single hadron production in deep inelastic scattering to $\mathcal{O}(\alpha_S^2)$ was presented. Owing to the factorization theorem this calculation was performed as a convolution of the hard scattering process with universal and process independent parton densities and fragmentation functions. This allows for meaningful predictions, directly comparable to experimental data.

Results were shown for specific kinematic values. A correction of the order of several percent is predicted depending on the transversal momentum of the observed hadron. The scale dependence of the correction gives slightly reduced theoretical uncertainties compared to LO predictions. Varying the renormalization and factorization scales over two orders of magnitude yields theoretical uncertainties of about ±25 % at NLO.

An extensive study of NLO effects will be published elsewhere which in particular will include a detailed comparison of predictions with experimental data from the HERA experiments H1 and ZEUS.

Note added

Just after finishing the calculation, a preprint appeared on NLO calculations of hadron production with non-vanishing transversal momentum [13]. In this paper matrix elements were adopted from the DISENT program package and the phase space slicing method was applied to handle singularities.

Acknowledgments

Thanks to Gustav Kramer and Bernd A. Kniehl for proposing this project and collaboration. We are very grateful to Michael Klasen and Dominik Stöckinger for many helpful discussions on the calculation. We want to thank Michael Spira for his advises to apply the subtraction formalism in the rather involved case of *identified* partons. For comments on the manuscript we thank Ingo Schienbein.

References

1. J. C. Collins, D. E. Soper and G. Sterman, Adv. Ser. Direct. High Energy Phys. **5** (1988) 1.
2. C. Adloff *et al.* [H1 Collaboration], Phys. Lett. B **462** (1999) 440.
3. M. Derrick *et al.* [ZEUS Collaboration], Z. Phys. C **70** (1996) 1.
4. A. Mendez, Nucl. Phys. B **145** (1978) 199.
5. P. Büttner, "Inklusive Meson-Produktion in tief-inelastischer Elektron-Proton-Streuung in nächstführender Ordnung der QCD," DESY-thesis 1999-004.
6. S. Catani and M. H. Seymour, Nucl. Phys. B **485** (1997) 291 (Erratum-ibid. B **510** (1997) 503).
7. D. Graudenz, Phys. Rev. D **49** (1994) 3291.
8. J. A. M. Vermaseren, arXiv:math-ph/0010025.
9. G. Passarino and M. J. G. Veltman, Nucl. Phys. B **160** (1979) 151.
10. H. L. Lai *et al.* [CTEQ Collaboration], Eur. Phys. J. C **12** (2000) 375.
11. B. A. Kniehl, G. Kramer and B. Potter, Nucl. Phys. B **582** (2000) 514.
12. K. Hagiwara *et al.* [Particle Data Group Collaboration], Phys. Rev. D **66** (2002) 010001.
13. P. Aurenche, R. Basu, M. Fontannaz and R. M. Godbole, arXiv:hep-ph/0312359.

PHOTO- AND ELECTROPRODUCTION OF SINGLE HADRONS AND RESONANCES

F. CORRIVEAU

McGill University,
Department of Physics,
3600 University Street,
Montréal, Qc,
Canada, H3A 2T8
E-mail: corriveau@physics.mcgill.ca

Production of single hadrons and resonances emerges as a promising field for the study of Quantum Chromodynamics at HERA. Four recent studies are presented here to illustrate the scope and reach of the technique: light mesons and resonances test the universality of hadroproduction, strange particles challenge the hadronisation models or directly probe the quark content of the proton sea while the first observation of $K - K$ resonant states offers a glueball candidate.

1. Introduction

The process by which quarks and gluons convert into colorless hadrons is not well described by Quantum Chromodynamics (QCD) processes. This is particularly true for light quarks, where perturbative QCD does not apply. As a consequence, phenomenological hadronisation models such as the Lund String Model or the Cluster Fragmentation Model are widely used to represent the processes.

Production studies of particles such as some of the light mesons or baryons containing a strange quark, or light resonances, should therefore contribute to the understanding of the hadronisation mechanisms. In many ways, HERA lags behind LEP in several of such measurements, but its different kinematical reach and the presence of quarks in the initial state strongly enhances its potential.

This paper will cover four examples of photo- and electroproduction processes: neutral hadronic resonances, strange particles, the use of strange mesons to probe the proton sea content and the study of $K_s^0 K_s^0$ resonances in the search for special states of matter.

2. Neutral Hadronic Resonances at HERA

This first extensive set of measurements done by the H1 experiment at HERA on light hadronic mesons and resonances used the 39 pb^{-1} of the 2000 data, covering the laboratory rapidity region of $|y_{lab}|<1$ with an average energy of 210 GeV in the photon-proton system. The production rates of η, ρ^0, $f_0(980)$ and $f_2(1270)$ were investigated[1].

A large fraction of the η's decay in a pair of photons which can be identified by clusters in the liquid argon detector, as illustrated in figure 1. The other states are predominantly decaying in $\pi^+\pi^-$ pairs, easily detected as charged tracks in the jet chambers. However, as shown in figure 2, the low-mass background is considerable and special care had to be taken in its combinatorial representation and in the assumptions regarding reflections from ω and K^* states. The ρ^0 peak is shifted to low values from the expectations, but this is also partly due to the acceptance limitations in that range and introduces but a negligible bias on the f_0 and f_2 resonant peaks.

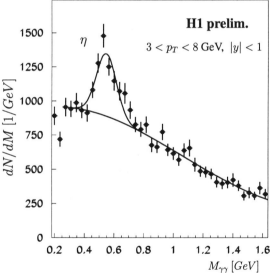

Figure 1. The two-photon mass spectrum. The curves were obtained from a fit to a Gaussian distribution and a polynomial background function.

When the single isospin data is corrected for spin and plotted as function of the sum of the meson mass and its transverse momentum (figure 3), one observes a striking scaling as the same behaviour of light, long-lived hadrons is reproduced, thus testing the universality of hadroproduction.

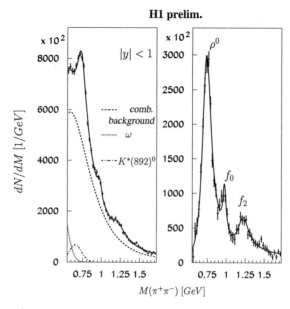

Figure 2. The $\pi^+\pi^-$ invariant mass spectrum before and after subtraction of the background and of the estimated low-energy contributions from reflections. The curves contain relativistic Breit-Wigner shapes from the fit.

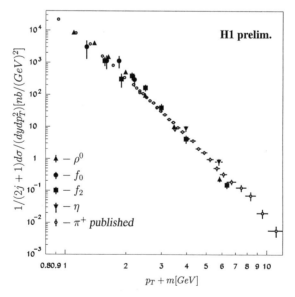

Figure 3. The differential photoproduction cross sections of light mesons as function of the sum of the meson mass and its transverse momentum p_T.

3. Strangeness Production

Strange quarks may be produced by flavour excitation, QCD Compton, gluon-splitting, boson-gluon fusion, various hadronisation processes, decay from higher mass states or diffractive processes, most of which are not always readily differentiated. Strange particle production becomes very useful because of the associated low masses, copious rates and unambiguous signals. K_s^0's and Λ's have been studied previously at HERA in deep inelastic scattering (DIS) and photoproduction, but only from very early data sets[2,3,4,5], and no cross-section had as yet been determined.

Using the 60 pb^{-1} of the ZEUS 1999-2000 data[6,7] in the DIS regime ($50 < Q^2 < 500$ GeV^2), and restricting the particle p_T range to the 0.5–5 GeV interval, figure 4 shows the quality of the particle identification for the main decays of the particles. Table 1 compares the results to HERWIG and the Lund Color Dipole Model (under ARIADNE) for two configurations of the strangeness suppression factor $\lambda_s = P(s)/P(d)$, whereby $P(d) = P(u)$. The measurements fall between the values of 0.2 and 0.3 in Lund model while HERWIG fails to predict the total cross sections. The ratio between Λ and $\overline{\Lambda}$ remains however unity.

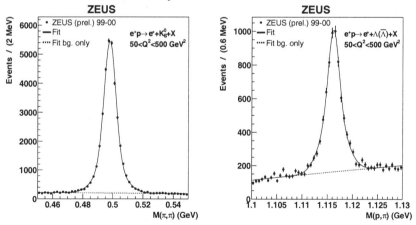

Figure 4. The K_s^0 (left) and Λ (right) invariant mass spectra.

In figure 5, drawn in the laboratory frame, even a renormalized HERWIG simulation does not reproduce the shapes of the distributions. Moreover, the effect of changing the value of λ_s is shown to be non-uniform according to the type of particle or its transverse momentum. From the pseudorapidity (η) plot, there is an indication of increased baryon to meson production in the forward region.

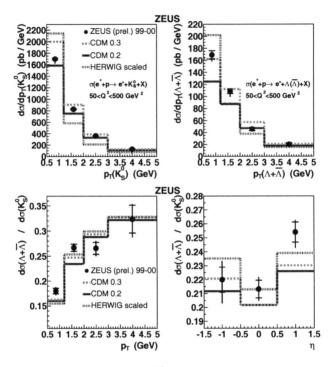

Figure 5. Differential cross sections for K_s^0 and Λ as function of the transverse momentum p_T and the pseudorapidity η in the laboratory frame.

Table 1. Preliminary cross-sections and ratios thereof for K_s^0 and Λ.

	$\sigma(K_s^0)$ [pb]	$\sigma(\Lambda + \bar{\Lambda})$ [pb]	$\sigma(\Lambda + \bar{\Lambda})/\sigma(K_s^0)$
ZEUS (prelim.)	$2454 \pm 18^{+32}_{-102}$	$567 \pm 12^{+13}_{-34}$	$0.231 \pm 0.005^{+0.005}_{-0.006}$
CDM ($\lambda_s = 0.3$)	2762	603	0.218
CDM ($\lambda_s = 0.2$)	2257	483	0.214
HERWIG	1854	1329	0.717

	$\sigma(\Lambda)$ [pb]	$\sigma(\bar{\Lambda})$ [pb]	$\sigma(\Lambda)/\sigma(\bar{\Lambda})$
ZEUS (prelim.)	$292 \pm 9^{+7}_{-18}$	$279 \pm 9^{+12}_{-18}$	$1.05 \pm 0.05^{+0.05}_{-0.05}$
CDM ($\lambda_s = 0.3$)	302	301	1.00
CDM ($\lambda_s = 0.2$)	240	243	0.99
HERWIG	661	668	0.99

This last observation motivated further investigation in the Breit Frame, where the radiation from the struck quark is naturally separated from the proton remnant (figure 6). The variable x_p represents the particle scaled momentum. The current region is equivalent to one hemisphere of a e^+e^- collider system and the results confirm the low sensitivity to the λ_s parameter. In the target region, associated to the proton remnant, the data tend towards the higher value $\lambda_s = 0.3$, which may suggest that this parameter could be related to the gluon density. A single parameter might not be adequate to estimate the production rates of strange particle over such kinematical ranges. Studies of the heavier strange particles Σ and Ξ are currently being done in ZEUS to further extend the scope of this study.

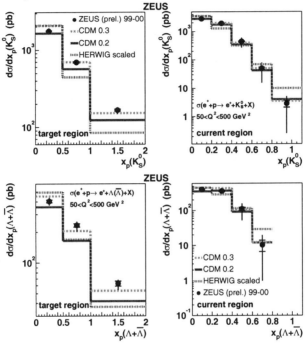

Figure 6. Differential cross sections for K_s^0 and Λ as function of the scaled momentum x_p in the Breit Frame.

4. Strange Content of the Sea

The ϕ-meson is another strange particle with a clean decay mode of a pair of kaons at the vertex which can be used to study not only hadron production in yet another channel, but also to probe directly the content of the proton sea. Detection of ϕ's in the 1995-1997 ZEUS data sample[8] reveals a high background but exhibits a clear signal in the laboratory frame, which becomes even more enhanced in spite of low statistics in the current region of the Breit Frame (figure 7).

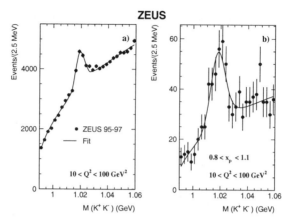

Figure 7. Invariant mass plot of K^+K^- pairs in the laboratory and in the Breit Frame.

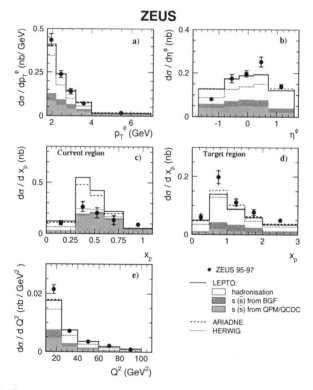

Figure 8. Differential ϕ-meson cross-sections as function of the kinematic variables p_T, η, x_p (current and target regions) and Q^2. Contributions of subprocesses of Boson-Gluon Fusion, Quark Parton Model or QCD Compton are also indicated.

The cross-section measured in the selected kinematical range, $\sigma(e^+p \to e^+\phi X) = 0.57 \pm 0.022^{+0.010}_{-0.008}$ nb agrees very well with the LEPTO (0.0501 nb) and ARIADNE (0.509 nb) simulations while establishing a preliminary value of 0.22 ± 0.02 for λ_s in that range. The plots of figure 8, whereby the Monte-Carlo inputs can be controlled, illustrate some of the observations: the hard QCD processes give very significant relative contributions in the current region of the Breit Frame while vanishing in the target region. This, coupled to the fact that their fraction increases as function of p_T or x_p, points to important involvement of the sea s-quarks.

By concentrating on the leading ϕ-mesons ($x_p > 0.8$), one realizes that uncertainties in the QCD processes and hadronisation are very small at high p_T, that gluon emissions become negligible and that the photon scattering on s-quarks is a QED process, i.e. well described. The main uncertainties left arise from the modellisations themselves and the allowed λ_s range, which is taken to be 0.2–0.3. Under such conditions, the leading ϕ-mesons in figure 9 unambiguously require s-quark scattering, thus showing direct evidence of the proton strange sea contribution to hadron production at low values of the scaled momentum x.

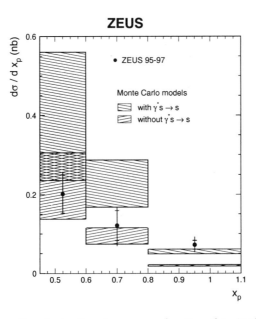

Figure 9. Cross sections for leading ϕ-mesons as function of x_p in the current region of the Breit Frame. The hatched bands represent the full range of uncertainties in the simulations.

5. $K_s^0 K_s^0$ Resonances

With the maturity of the experiments and the recent developments in the field, there is now considerable interest in "meson spectroscopy" at HERA. The striking example of the $K_s^0 K_s^0$ resonant states in DIS with the ZEUS experiment[9,10] illustrates the potential for discoveries.

Kaons are again chosen in this search because of their high detection efficiency, good acceptance and low background in the central rapidity region. The K_s^0 momentum threshold is chosen to be 200 MeV. A special cut on the opening angle θ_{KK} between the 2 kaons is applied to reject the $f_0/a_0(980)$ state. 2553 candidates are found. Their invariant mass plot of the system is displayed in figure 10 and fitted with three relativistic modified Breit-Wigner distributions and one background function.

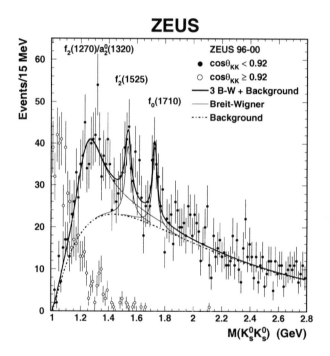

Figure 10. Invariant mass plot of the $K_s - K_s$ pairs. The lines identify the residual combinatorial background and the peaks used in the overall fit to the data (full dots). The open dots describe the already rejected f_0/a_0 background.

Three regions are identified. Below 1500 MeV, the contributions from the $f_2(1270)$ and $a_2^0(1320)$ merge over the acceptance threshold. The θ_{KK}

cut is shown to leave negligible possible contributions under the remaining two peaks. The first of these, at 1537±9 MeV and with a width of ~50 MeV, is consistent with the known $f_2'(1525)$ state. The second peak, at 1726±7 MeV, contains 79^{+29}_{-23} events and has a measured width of 38^{+20}_{-14} MeV. This width is in open contradiction from the observations from BES, Belle and L3, which published 167, 323 and 200 MeV respectively. A large number of permutations and assumptions has been tested in the fits to the data, the other state widths, the background shapes, etc., but results in this high mass range were proven to be rather insensitive to them. A further check in the Breit Frame revealed that 93% of the candidates are found in the target region, i.e. associated with the proton remnant. The expected sizeable initial state gluon radiation thus strengthens the hypothesis that the state could be a glueball candidate.

6. Summary

First measurements of inclusive photoproduction cross-sections of the light resonances η, ρ^0, $f_0(980)$ and $f_2(1270)$ have been reported. Strangeness production studies in deep inelastic scattering reveal details of hadronisation processes and inconsistencies in some of the currently available simulation packages. ϕ-mesons were further used to provide direct evidence of the strangeness content of the proton sea. Finally, in a first case of complex resonant system of K_s^0 pairs, two high mass states have been clearly identified, offering one glueball candidate.

Acknowledgments

This work has been partly supported by the National Sciences and Engineering Research Council of Canada and by DESY.

I would like to thank the organisers of this workshop for creating such a stimulating environment in a truly congenial setting.

References

1. The H1 Collaboration, presented by A. Kropivnitskaia, *Measurements of the Inclusive Photoproduction of η, ρ^0, $f_0(980)$ and $f_2(1270)$ Mesons at HERA*, International Europhysics Conference on High Energy Physics, EPS03, Aachen, July 17–23, 2003.
2. The ZEUS Collaboration, M. Derrick et al., *Neutral Strange Particle Production in Deep Inelastic Scattering at HERA*, DESY-95-084 (April 1995), Zeitschrift f. Physik **C68**, 29–42 (1995).

3. The H1 Collaboration, S. Aid et al., *Strangeness Production in Deep Inelastic Positron-Proton Scattering at HERA*, DESY-96-122 (June 1996), *Nucl. Phys.* **B480**, 3 (1996).
4. The H1 Collaboration, C. Adloff et al., *Photoproduction of K^0 and Λ at HERA and a Comparison with Deep Inelastic Scattering*, DESY-97-095 (May 1997), *Zeitschrift f. Physik* **C76**, 213 (1997).
5. The ZEUS Collaboration, J. Breitweg et al., *Charged Particles and Neutral Kaons in Photoproduced Jets at HERA*, DESY-97-229 (November 1997), *Eur. Phys. Journal.* **C2**, 77–93 (1997).
6. The ZEUS Collaboration, presented by A. Ziegler, *Strange Particle Production in DIS at HERA*, International Workshop in Deep Inelastic Scattering, DIS2003, St. Petersburg, April 23-27, 2003.
7. The ZEUS Collaboration, presented by B. Levtchenko, *Measurements of Strange Particle Production in Deep Inelastic $e^+p \to e^+X$ Scattering at HERA*, International Europhysics Conference on High Energy Physics, EPS03, Aachen, July 17–23, 2003.
8. The ZEUS Collaboration, S. Chekanov et al., *Observation of the Strange Sea in the Proton via Inclusive ϕ-Meson Production in Neutral Current Deep Inleastic Scattering at HERA*, DESY-02-184 (October 2002), *Phys. Lett.* **B553**, 141–158 (2002).
9. The ZEUS Collaboration, presented by B. Levtchenko, *Observation of $K_s^0 K_s^0$ Resonances in Deep Inelastic Scattering at HERA*, International Europhysics Conference on High Energy Physics, EPS03, Aachen, July 17–23, 2003.
10. The ZEUS Collaboration, S. Chekanov et al., *Observation of $K_s^0 K_s^0$ resonances in deep inelastic scattering at HERA*, DESY-03-098 (July 2003), accepted for publication by *Phys. Lett. B*.

GLUONIC MESON PRODUCTION

PETER MINKOWSKI*

Institute for Theoretical Physics, Univ. of Bern,
CH-3012 Bern, Switzerland
E-mail: mink@itp.unibe.ch

WOLFGANG OCHS

Max-Planck-Institut für Physik (Werner Heisenberg-Institut)
Föhringer Ring 6
D-80805 München, Germany
E-mail: wwo@mppmu.mpg.de

The existence of glueballs is predicted in QCD, the lightest one with quantum numbers $J^{PC} = 0^{++}$, but different calculations do not well agree on its mass in the range below 1800 MeV. Several theoretical schemes have been proposed to cope with the experimental data which often have considerable uncertainties. Further experimental studies of the scalar meson sector are therefore important and we discuss recent proposals to study leading clusters in gluon jets and charmless B-decays to serve this purpose.

1. Introduction

The existence of glueballs, i.e. bound states of two or more gluons, are among the early predictions of QCD. There is general agreement in that the lightest glueball should be in the scalar channel with $J^{PC} = 0^{++}$. The experimental situation is still controversial as the properties of scalar resonances are not well known. This can be illustrated by looking at the Particle Data Group[1] listing on the lightest scalar mesons (status end 2003). Glueballs can be searched for among the 5 isoscalar states with mass below 1800 MeV, listed in Table 1. The nature of two broad states $f_0(600)$ (also called σ) and $f_0(1370)$ is still controversial. Whereas quite a number of decay modes are "seen", no branching ratios are quoted for any one of these

---

*Work partially supported by grant 2-4570.5 of the Swiss National Science Foundation.

Table 1. Information on isoscalar mesons with $J^{PC} = 0^{++}$ from the Particle Data Group.

states width [MeV]	$f_0(600)$ 600-1000	$f_0(980)$ 40-100	$f_0(1370)$ 300-500	$f_0(1500)$ 109 ± 7	$f_0(1710)$ 125 ± 10
hadronic modes "seen"	1	2	11	13	3
ratios of rates published	-	1	13	18	5
"used" by PDG	-	0	0	5	2
from more than 1 exp.	-	0	0	2	0

particles. Only few ratios of rates are considered as acceptable of which altogether only two have been measured by two experiments independently.

Besides the decay branching ratios the production properties are important for the identification of gluonic mesons. Glueballs should be produced with enhanced rate in a "gluonic environment", such as central hadronic production (double Pomeron exchange), radiative J/ψ decay, proton antiproton annihilation, decay of excited Quarkonia into their respective ground states, whereas their production should be suppressed in $\gamma\gamma$ collisions.

There is not yet an agreement on the mass of the lightest 0^{++} glueball and its mixing with other states, it is even debated whether it has been seen at all. It is rather clear that the evidence for the scalar glueball will emerge only from a thorough experimental study of the low mass scalar channels and the identification of the scalar $q\bar{q}$ nonet in a parallel effort. Various channels should be probed with the aim to obtain more precise information on the production and decay of the scalar resonances.

A recent analysis of $K\bar{K}$ mass spectra at HERA[2] with a prominent peak near the mass of 1700 MeV (presumably $f_0(1710)$) has shown the potential for hadron spectroscopy at HERA. In this report we will discuss the recent proposal to search for glueballs in the leading part of the gluon jet[3] and first results[4] (see also discussions in[7,8,9]), which may be applicable at HERA. Furthermore we report on the possibility for glueball searches and results in B decays.[10]

2. QCD Expectations

The basic triplet of binary glueball states which can be formed by two "constituent gluons" corresponds to the three invariants which can be built from the bilinear expressions of gluon fields and carry quantum numbers $J^{PC} = 0^{++}$, 0^{-+} and 2^{++}.[11] Quantitative results are derived today from the QCD lattice calculations or QCD sum rules, both agree that the lightest glueball has quantum numbers $J^{PC} = 0^{++}$.

Lattice calculations in quenched approximation[12,13,14,15] (without light sea quark-antiquark pairs) suggest the lightest glueball to have a mass in the range 1400-1800 MeV.[16] Results from unquenched calculations still suffer from systematic effects, the large quark masses of the order of the strange quark mass and large lattice spacings. Typically, present results on the glueball mass are about 20% lower than the quenched results.[17,18] Another interesting result would be the mass of the light scalar $\bar{q}q$ mesons. A recent result[18] suggests the mass (1.0 ± 0.2) GeV for the scalar a_0; this is well consistent with the mass of $a_0(980)$ but in view of the systematic uncertainties it is not yet possible to exclude $a_0(1450)$ as the lightest isovector scalar particle.

Results on glueballs have also been obtained from QCD sum rules. Recent calculations[19] for the 0^{++} glueball yield a mass consistent with the quenched lattice result but in addition require a gluonic state near 1 GeV. A strong mixing with $q\bar{q}$ is suggested resulting finally in the broad σ and narrow $f_0(980)$. Similar results with a low glueball mass around 1 GeV are obtained also in other calculations.[20] On the other hand, it is argued[21] that the sum rules can also be saturated by a single glueball state with mass 1.25 ± 0.2 GeV.

In conclusion, there is agreement in the QCD based calculations on the existence of a 0^{++} glueball but the mass and width of the lightest state is not yet certain and phenomenological searches should allow a mass range of about 1000-1800 MeV.

There is also another class of gluonic mesons, the hybrid $q\bar{q}g$ states, both in the theoretical and experimental analysis, but we will not further consider these here (see, for example, review[23]).

3. Spectroscopy of Scalar Mesons - Phenomenology

In the mass range below 1800 MeV the PDG lists two isovectors $a_0(980)$ and $a_0(1450)$ as well as the strange $K_0^*(1430)$, furthermore, there is a possibility of a light strange particle $\kappa(800)$. From these particles and the isoscalars in Tab. 1 one should build the relevant multiplets, one (or more) $q\bar{q}$ nonets and a glueball. There are various schemes for the spectroscopy of the light scalars. We emphasize two different routes in the interpretation of the data, which are essentially different in the classification of the states, both with some further possibilities in details.

Route I: "Heavy" $q\bar{q}$ multiplet (above $f_0(980)$) and "heavy" glueball
One may start from the glueball assuming a mass around 1600 MeV

as found in the quenched lattice calculations emphasized above. In the isoscalar channel there are the states $f_0(1370)$, $f_0(1500)$ and $f_0(1710)$ nearby in mass which are assumed to be mixtures of the two members of the nonet and the glueball. The multiplet of higher mass then includes furthermore the uncontroversial $q\bar{q}$ state $K^*(1430)$ and also the nearby $a_0(1450)$. After the original proposal[22] several such mixing schemes have been considered (review[23]) using different phenomenological constraints. There are schemes[22] with the largest gluon component residing in $f_0(1500)$ and others[13] where this role is taken by $f_0(1710)$. In these schemes $f_0(980)$ and $a_0(980)$ are sometimes superfluous, they are taken as multiquark bound states[24] or $K\bar{K}$ molecules[25] and then are removed from $q\bar{q}$ spectroscopy. An attractive possibility is the existence of an additional light nonet, which includes $\sigma/f_0(600)$, κ, $a_0(980)$ and $f_0(980)$ either as $q\bar{q}$ or $qq\bar{q}\bar{q}$ bound states. Such schemes appear in theories of meson meson scattering in a realization of chiral symmetry, for an outline, see review[26].

Route II: $q\bar{q}$ multiplet including $f_0(980)$ and "light" glueball

Alternatively, one may start from an identification of the lightest nonet and then look for the glueball among the remaining states.[27] Several approaches agree on a similar nonet with $f_0(980)$ as the lightest member,[28,27,29,19] although with different intrinsic structure. Also $K^*(1430)$ belongs to this nonet whereas the identification of the other members differs. There are arguments[27,28] for a strong flavour mixing similar to the pseudoscalar sector with the correspondence $f_0(980) \leftrightarrow \eta'$ near flavour singlet and $f_0(1500) \leftrightarrow \eta$ near flavour octet. The isovector could be $a_0(980)$[27] or $a_0(1450)$.[28] There is no room for κ in these schemes.

The remaining light scalars, the broad σ and $f_0(1370)$, are then candidates for the lightest glueball. The interpretation of σ which is related to the strong $\pi\pi$ scattering up to 1 GeV is subject to intense discussions and controversies.[30] In our phenomenological analysis[27] we consider both states as a single object with a width of about 500-1000 MeV.

Our arguments in favour of the glueball hypothesis include: the strong central production in pp collisions, the appearence in the Quarkonium decays $\psi', \psi'' \to J/\psi \pi\pi$, also in corresponding Y', Y'' decays, in $p\bar{p}$ annihilation and the suppression in $\gamma\gamma$ collisions;[31] on the other hand, contrary to expectation, there is no strong signal in radiative J/ψ decays. A glueball in this mass region is also located in the sum rule analyses[19,20,21] and the K matrix fits to a variety of production processes.[29] Alternatively, this broad $\pi\pi$ "background" has been viewed as due to non-gluonic exchange processes.[23]

As a common feature of the above schemes the lightest glueball is not expected to appear as a single narrow resonance but rather as a phenomenon spread over a mass range of around 500 MeV. Either it is mixed with several moderately narrow isoscalar resonances in a range from 1300-1800 MeV (route I) or it appears mainly as a broad state with a large width by itself (route II). In view of these different possibilities it is important to improve our knowledge on gluonic interactions and to look for further possibilities of glueball production.

4. Gluonic Meson Production in Gluon Jets

New information on gluonic mesons can be obtained from the comparative study of the leading particle systems in quark and gluon jets. The possible appearance of isoscalar particles in the leading system of a gluon jet has already been considered long ago.[7] More specifically, the production of glueballs in the fragmentation region at large Feynman x has been considered.[8] The search for glueballs applying a rapidity gap selection of events and charge distributions in quark and gluon jets has been suggested recently.[3]

There is a well established fragmentation phenomenology for quark jets: a particle which carries the primary quark as valence quark is produced with larger probability at high momentum fraction x than other particles. For example, the u-quark will produce more π^+ than π^- at large x. A natural extension of this phenomenology applies for gluon jets: particles with large Feynman x are predominantly those which carry the initial gluon as valence gluon, these are glueballs or hybrids if they exist.

Whether the idea can be transferred from quark to gluon jets in this way depends on the hadronization mechanism at the distances of about 1 fm, where the colour confinement forces become important, but there is no firm approach to deal with these non-perturbative processes. At the end of a parton cascade the valence quark will form a hadron by recombining with an anti-quark corresponding to an interaction between the colour triplets. In a simple case a primary pair of energetic q and $\bar{q}$ in a colour singlet state is neutralised by a soft $\bar{q}q$ pair, see Fig. 1a. For gluons there are in general two different possibilities. A pair of energetic valence gluons can be colour neutralized again by colour triplet forces (two soft $q\bar{q}$ pairs) or by colour octet forces (a pair of soft gluons), see Fig. 1b. Whereas the standard hadronization models support only the colour triplet neutralization, the second mechanism would allow glueball production.

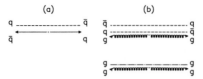

Figure 1. The colour neutralization (a) of an initial $q\bar{q}$ pair by $\bar{q}q$ and (b) of an initial gg pair by either double colour triplet $q\bar{q}$ or by colour octet gg.

It is not obvious to what extent the two types of neutralization mechanisms are realized in a given process. The octet neutralization could in general be a rare process enhanced only for particular kinematic configurations. We argued[3] that in a gluon jet with a large rapidity gap the octet mechanism should become visible if it exists. For large rapidity gaps the more complex multi-quark pair exchanges through the gap will be suppressed.

The relative importance of the colour octet mechanism can be tested by studying the distribution of electric charge of the leading hadronic cluster beyond the gap. These charges should approach for large gaps a limiting distribution corresponding to the minimal number of partons traversing the gap for the considered neutralization mechanism.

There is a clear distinction between the triplet and octet mechanisms: the leading charge in the minimal triplet ($q\bar{q}$) configuration is $Q = 0, \pm 1$ whereas for the octet configuration (gg) it is $Q = 0$. Note that this result is independent of the existence of glueballs. On the other hand, if no extra $Q = 0$ component is observed, then there is no evidence for the octet mechanism whose existence is a precondition for the production of glueballs: glueballs could exist in nature but not be produced in gluon jets.

An exploration of these possibilities has been performed in e^+e^- annihilation at LEP, first with the DELPHI data[4] and recently also with OPAL[5] and ALEPH data.[6] For illustration we show in Fig. 2 the results by DELPHI. The charge distribution has been obtained for the leading cluster, as defined by a rapidity gap $\Delta y = 2$, in quark and gluon jets. The data with this selection do not yet show the limiting charge distribution as can be concluded from the presence of charges $Q = \pm 2, \pm 3$ at the level of 10%. The data for quark jets are in a good agreement with the JETSET Monte Carlo[32] which is based on the triplet neutralization mechanism whereas in gluon jets there is a significant excess of events with charge $Q = 0$ above the MC expectation with a significance of about 4σ. A similar excess is found by OPAL, again for JETSET, but an even larger effect for ARIADNE[33]

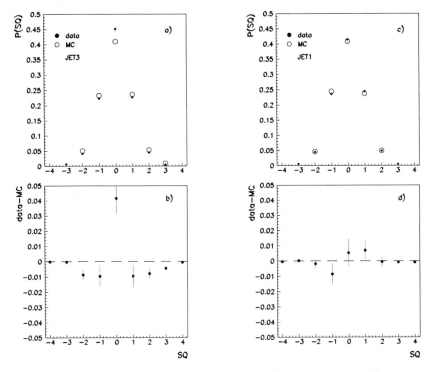

Figure 2. Sum of charges of leading system for a,b) gluon jets and c,d) quark jets (preliminary results by DELPHI[4]) showing the significant excess of data for gluon jets over the expectations from the JETSET MC.

and a smaller effect for HERWIG.[34] The same excesses in comparison to JETSET and ARIADNE have also been observed by ALEPH where the effect is also shown to increase with the size of the rapidity gap and appears mainly at the small multiplicities ≤ 4. These results show the inadequacy of MC's to describe the charge distribution of leading clusters in gluon jets. This may be taken as a hint towards the presence of the octet neutralization mechanism.

The next question to ask is whether the events with a leading cluster of charge $Q = 0$ show indeed a sign of a glueball which would be a natural explanation of the observed excesses. The mass spectra shown by DELPHI for quark jets are again in good agreement with the MC expectations. The mass distribution for events with charge $Q = 0$ in gluon jets in the mass region below 2 GeV show an excess near $f_0(980)$ and possibly in the mass region around 1400 MeV.[a] OPAL observes a moderate excess in the region 1.0-2.5 GeV of 2σ significance with a coarse binning whereas ALEPH did not study mass spectra (however, their finding of the most sgnificant excess at low multiplicities indicates the main effect at low mass as well).

At present one cannot claim a significant signal in the mass spectra in view of insufficient statistics or the preliminary nature of the data. We want to stress therefore the importance of further dedicated studies.

If the finding of a significant excess of $f_0(980)$ in gluon jets is confirmed it would be a strong argument in favour of a large flavour singlet component of this meson[28,27,19] which could couple to two gluons and disfavour a flavour octet assignment.[36] It would also support the picture with the correspondence of $f_0(980)$ and η' both with gg coupling but not being a glueball; similar effects are discussed in charmless B-decays.[10] Also we note that such an excess is not what would be naturally expected from a 4 quark model for this state. Furthermore, there is the possibility of the broad 0^{++} glueball state above 1 GeV which may be indicated by the above findings.

For further improvements of the analysis it seems important to obtain a better separation of the leading cluster from the other particles in the event. This is not immediately garanteed if the rapidity gap is only applied to the particles inside one jet and not in the full event. This can be achieved if the gap is required for the gluon jet in a frame with a symmetric configuration and the same angle between the gluon and the two quark jets.[37] Then larger rapidity gaps can possibly be achieved with reasonable statistics which would reduce the background from non-minimal configurations.

5. Evidence for Scalar Glueball in Charmless B-Decay

The interest in charmless B-decays with strangeness has been stimulated through the observation of a large decay rate $B \to \eta' K$ and $B \to \eta' X$.[38] It has been suggested that these decays, at least partially, proceed through the b-quark decay $b \to sg$. This decay could be a source of mesons with large gluon affinity.[39,40,41,42] In consequence, besides η' also other gluonic states, in particular also glueballs could be produced in a similar way.

The total rate $b \to sg$ has been calculated perturbatively in leading[43] and next-to-leading order[44]

$$\text{Br}(b \to sg) = \begin{cases} (2-5) \times 10^{-3} & \text{in LO (for } \mu = m_b \ldots m_b/2) \\ (5 \pm 1) \times 10^{-3} & \text{in NLO} \end{cases} \quad (1)$$

---

[a]The significance of these excesses is still under experimental investigation.[35]

The energetic massless gluon in this process could turn entirely into gluonic mesons by a nonperturbative transition after neutralization by a second gluon. Alternatively, colour neutralization through $q\bar{q}$ pairs is possible as well (see also Sec. 4 and Fig. 1). This is to be distinguished from the short distance process $b \to s\bar{q}q$ with virtual intermediate gluon which has to be added to the CKM-suppressed decays $b \to q_1 \bar{q}_2 q_2$. These quark processes with s have been calculated and amount to branching fractions of $\sim 2 \times 10^{-3}$ each.[45,46,44]

Recently, the BELLE collaboration[47,48] has studied charmless decays $B \to Khh$ with $h = \pi, K$ showing a strong signal of the decay $B \to K f_0(980)$ with decay rate comparable to $B \to K\pi$. Similar results have been obtained by BaBar.[49,50] These results show that scalar particles are easily produced and open the possibility to identify the scalar nonet related to $f_0(980)$ and to determine its properties.[10]

There is another interesting feature in the decays $B^+ \to K^+\pi^+\pi^-$ and $B^+ \to K^+K^-K^+$ observed by the BELLE collaboration.[47] The latter channel shows a broad enhancement in the K^+K^- mass spectrum in the region $1.0 - 1.7$ GeV. The flat distribution in the Dalitz plot of these events suggests this object to be produced with spin $J = 0$. Its contribution can be parametrized as scalar state $f_X(1500)$ with mass $M = 1500$ MeV and $\Gamma = 700$ MeV. More recent preliminary data[48] of higher statistics indicate a narrower substructure above the broad "background". In the $\pi\pi$ channel there is also some background under $f_0(980)$ which drops above 1400 MeV.

As discussed elsewhere[10] we consider the enhancements in $\pi\pi$ and $K\overline{K}$ observed by BELLE as a new manifestation of the broad scalar glueball discussed above under route II. Whereas the center of the peak in $\pi\pi$ is closer to 1 GeV, it is shifted to higher mass in the $K\overline{K}$ channel. The glueball decays with equal rates into $u\bar{u}$, $d\bar{d}$ and $s\bar{s}$ and also into $gb\ gb$. The strange quarks produce dominantly $K\overline{K}$, the nonstrange quarks produce $\pi\pi$ but also resonances in the 4π channel (like $\rho\rho$ or $\sigma\sigma$). We therefore interpret the drop in the $\pi\pi$ spectrum as consequence of the opening of the 4π channel for glueball decay. An important test of our glueball hypothesis is the verification of the expected branching ratios; with the published data there is no contradiction with the hypothesis.

Using the published B branching ratios[47] we have estimated the total production rate of the scalar glueball to be

$$\text{Br}(B^+ \to gb(0^{++}) + X_s) \sim 1.2 \times 10^{-3}, \qquad (2)$$

adding the gluonic part of $f_0(980)$ and η' production we estimate

$$\mathrm{Br}(B^+ \to gb(0^{++}) + f_0 + \eta' + X_s) \sim (1.5 \pm 0.5) \times 10^{-3} \qquad (3)$$

which is of the same size as the leading order result for the process $b \to sg$ in (1) and about 1/3 of the full rate obtained in NLO. Further gluonic contributions are expected from other glueballs, in particular, from the parity partner 0^{-+} so that the total production rate of the decay $b \to sg$ could be saturated by gluonic mesons and gluonic production of flavour singlet mesons.

6. Summary

1. Recent results from QCD calculations confirm the existence of glueballs, the lightest one with quantum numbers 0^{++}. The results for its mass vary in a range from 1000 to 1800 MeV.

2. There are different scenarios in the phenomenological interpretation of the data: the glueball has a mass around 1500 MeV and mixes with the isoscalar members of the nonet into observed isoscalars in the range 1300 - 1800 MeV. Alternatively, the glueball is around 1 GeV or a bit heavier with a large width of 500-1000 MeV. In any case, there is no single narrow resonance representing the glueball.

3. Many relevant data (decay branching ratios of f_0's, production rates and phases of f_0's in various production and decay channels have often large errors or are controversial. An improvement of this experimental situation by better measurements is essential for further progress. We think here in particular of improved measurements on the various f_0's (and gb) in central pp (double Pomeron) production, J/ψ decays ($\to f_0\omega$, $f_0\phi$; $\gamma 2\pi$, $\gamma 4\pi$) and D-decays ($f_0\pi$).

A crucial role is played by $f_0(980)$ which could be either a member of the higher mass or lower mass nonet. Its properties and flavour composition are still controversial.

4. An attractive new possibility to search for glueballs lies in the comparison of leading clusters in gluon and quark jets. Also new information on flavour singlet mesons can be obtained (η', $f_0(980)$?). First results from LEP show an additional neutral component in gluon jets not expected from standard MC's. The origin of this excess has to be understood in terms of hadronic states. Such studies are possible at HERA.

5. Charmless B-decays are an attractive source of glueballs and other gluonic mesons from the decay $b \to sg$ as well. Recent results have been

interpreted in terms of a broad scalar glueball decaying into $K\bar{K}$ in the mass range 1000-1600 MeV.

The recent results from B decays and gluon jets are quite promising. They allow to contrast quark and gluon structures in the same experiment. As we hope, there is a large potential to establish the light scalar $q\bar{q}$ nonet with its mixing and the lightest scalar glueball.

References

1. K.Hagiwara et al. (Particle Data Group), *Phys. Rev.* **D66**, 010001 (2002); URL: http://pdg.lbl.gov.
2. S. Chekanov et al. (ZEUS Collaboration), *Phys.Lett.* **B578**, 33 (2004).
3. P. Minkowski and W. Ochs, *Phys. Lett.* **B485**, 139 (2000).
4. B. Buschbeck and F.Mandl (DELPHI Collaboration), in Proc. of the Int. Symp. on Multiparticle Dynamics (ISMD 2001), Sept. 2001, Eds. Bai Yuting et al. (World Scientific, Singapore 2002), p.50.
5. G.Abbiendi et al. (OPAL Collaboration), arXiv:hep-ex/0306021, subm. to *Eur. Phys. J.C.* (2003).
6. ALEPH Collaboration (contact G. Rudolph), Contr. to Int. Europhysics Conf. on HEP, July 2003, Aachen, Germany, CONF 2003-005.
7. C. Peterson and T.F. Walsh, *Phys. Lett.* **B91**, 455 (1980).
8. P. Roy, K. Sridhar, *JHEP* **9907**, 013 (1999).
9. H. Spiesberger and P. Zerwas, *Phys. Lett.* **B481**, 236 (2000).
10. P. Minkowski and W. Ochs, arXiv:hep-ph/0304144.
11. H. Fritzsch and P. Minkowski, *Nuovo Cim.* **30A**, 393 (1975).
12. G. Bali et al., *Phys. Lett.* **B309**, 378 (1993).
13. J. Sexton, A. Vaccarino and D. Weingarten, *Phys. Rev. Lett.* **75**, 4563 (1995).
14. C.J. Morningstar and M. Peardon, *Phys. Rev.* **D60**, 034509 (1999).
15. B. Lucini and M. Teper, *JHEP* **0106**, 050 (2001).
16. G. Bali, arXiv:hep-lat/0308015.
17. A. Hart and M. Teper, *Phys. Rev.* **D65**, 034502 (2002).
18. A. Hart, C. McNeile and C. Michael, *Nucl. Phys. B (Proc. Suppl.)* **119**, 266 (2003).
19. S. Narison, *Nucl. Phys.* **B509**, 312 (1998); *Nucl. Phys. B (Proc. Suppl.)* **121**, 13 (2003).
20. E. Bagan and T.G. Steele, *Phys. Lett.* **B243**, 413 (1990); T.G. Steele, D. Harnett and G. Orlandini, arXiv:hep-ph/0308074.
21. H. Forkel, arXiv:hep-ph/0312049.
22. C. Amsler and F.E. Close, *Phys. Rev.* **D53**, 295 (1996); *Phys. Lett.* **B353**, 385 (1995).
23. E. Klempt, *"Meson Spectroscopy"*, PSI Zuoz Summer School, Aug. 2000, arXiv:hep-ex/0101031.
24. R.L. Jaffe, *Phys. Rev.* **D 15** 267, 281 (1977).
25. J. Weinstein and N. Isgur, *Phys. Rev.* **D27**, 588 (1983).
26. F.E. Close and N.A. Törnqvist, *Phys. G* **28**, R249 (2002).

27. P. Minkowski and W. Ochs, *Eur. Phys. J.* **C9**, 283 (1999).
28. E. Klempt, B.C. Metsch, C.R. Münz and H.R. Petry, *Phys. Lett.* **B361**, 160 (1995).
29. V.V. Anisovich and A.V. Sarantsev, *Eur.Phys.J.* **A16**, 229 (2003).
30. W. Ochs, arXiv:hep-ph/0311144.
31. P. Minkowski and W. Ochs, in *Proc. Workshop on Hadron Spectroscopy*, Frascati, March 1999, Italy, Eds. T. Bressani et al. (Frascati Physics Series XV, 1999) p.245.
32. T. Sjöstrand, *Comp. Phys. Comm.* **82**, 74 (1994).
33. L. Lönnblad, *Comp. Phys. Comm.* **71**, 15 (1992).
34. G. Marchesini et al., *Comp. Phys. Comm.* **67**, 465 (1992); G. Corcella et al., *JHEP* **0101**, 010 (2001).
35. B. Buschbeck, private communication.
36. V.V. Anisovich, V.A. Nikonov and L. Montanet, *Phys. Lett.* **B480**, 19 (2000).
37. G. Abbiendi et al., (OPAL Collaboration) arXiv:hep-ex/0310048, subm. to *Phys.Rev.D*.
38. B.H. Behrens et al. (CLEO Collaboration), *Phys. Rev. Lett.* **80**, 3710 (1998); T.E. Browder et al., *Phys. Rev. Lett.* **81**, 1786 (1998).
39. D. Atwood and A. Soni, *Phys. Rev. Lett.* **79**, 5206 (1997).
40. H. Fritzsch, *Phys. Lett.* **B415**, 83 (1997).
41. X.-G. He, W.-S. Hou and C.-S. Huang, *Phys. Lett.* **B429**, 99 (1998).
42. A.S. Dighe, M. Gronau and J.L. Rosner, *Phys. Lett.* **B367**, 357 (1996); *Phys. Rev. Lett.* **79**, 4333 (1997).
43. M. Ciuchini, E. Franco, G. Martinelli, L. Reina and L. Silvestrini, *Phys. Lett.* **B334**, 137 (1994).
44. C. Greub and P. Liniger, *Phys. Rev.* **D63**, 054025 (2001).
45. G. Altarelli and S. Petrarca, *Phys. Lett.* **B261**, 303 (1991).
46. A. Lenz, U. Nierste, and G. Ostermaier, *Phys. Rev.* **D56**, 7228 (1997).
47. A. Garmash et al. (BELLE Collaboration), *Phys. Rev.* **D65**, 092005 (2002).
48. K. Abe et al., BELLE-CONF-0225, arXiv:hep-ex/0208030.
49. B. Aubert et al. (BaBar Collaboration), arXiv:hep-ex/0308065.
50. B. Aubert et al., BaBar-Conf-02/009, arXiv:hep-ex/0206004.

4
Heavy-Flavour Production

OPEN CHARM AND BEAUTY PRODUCTION

SANJAY PADHI

(ON BEHALF OF THE H1 AND ZEUS COLLABORATIONS)

Deutsches Elektronen-Synchrotron,
Notkestrasse 85, 22607 Hamburg, Germany
E-mail: spadhi@mail.desy.de

New results in open heavy flavour production in ep collisions are presented. Aspects sensitive to the various stages of the production mechanism are compared to the experimental data. Based on the measurements and the comparison with leading-order parton shower Monte Carlo models and next-to-leading order QCD predictions, conclusions related to data alone as well as model initiated issues are drawn.

1. Introduction

Open heavy quark production at HERA offers a novel way of testing both perturbative and non-perturbative aspects of quantum chromodynamics (QCD). For the quark to be "heavy", its mass[a] has to be larger than the QCD scale $\Lambda_{QCD} \sim 200 - 300$ MeV. Hence throughout this article, "heavy quark" will refer to a charm or a bottom quark. The following steps in heavy quark production can be distinguished (See Fig. 1 a):

- Hard subprocess: basic $2 \to 2$ process, e.g $\gamma g \to Q\bar{Q}$, boson gluon fusion (BGF) to form the heavy quark $Q\bar{Q}$ pair.
- Initial/Final state parton radiation.
- Fragmentation: transition of final state coloured partons to colourless hadrons.
- Unstable hadrons decay according to their branching ratios.

The first two stages are based on perturbative QCD (pQCD) calculations, while the last two rely on non-perturbative models. The cross section of a generic collision of two hadrons producing heavy quark pairs at HERA can

---

[a]The top quark, however, is too heavy and decays weakly before hadronizing.

be written as:

$$\sigma(S) = \sum_{i,j} \int dx_1 \int dx_2 f_i^\gamma(x_1,\mu_F) f_j^p(x_2,\mu_F) \hat{\sigma}_{ij}(x_1 x_2 S, m_Q^2, \mu_R^2) \quad (1)$$

where the cross section is a convolution of parton densities, $f_i^\gamma(x_1,\mu_F)$ and $f_j^p(x_2,\mu_F)$ of the photon and proton with the short distance cross section $\hat{\sigma}_{ij}$. The fractional hadronic momenta carried by the interacting partons are represented by x_1 and x_2 with μ_F and μ_R as the factorisation and renormalisation scales. The partonic cross section presented above further needs to be convoluted with a fragmentation function to obtain the hadronic cross section.

Two different collinear schemes and their combination are used for heavy quark production. In the massive-quark or fixed-order next-to-leading order scheme (FO NLO), the heavy quark Q is on-mass-shell and it only appears in the final state, but not as an active parton inside the incoming photon or proton. The calculation is expected to be valid for $p_T \sim m_Q$. However in the massless-quark scheme or zero-mass variable-flavour-number scheme (VFNS), the quark Q, (with $p_T > m_Q$) is treated as massless, such that it appears as an active parton of the incoming photon or proton, with a non perturbative parton density function. The attempt made to combine the results from FO NLO with suitably subtracted VFNS and with perturbative fragmentation functions (PFFs), resulted in the fixed order next-to-leading logarithm (FONLL) scheme[1]. The FONLL thus is expected to give more reliable predictions for heavy quark production.

1.1. *Experimental techniques*

The reconstruction of heavy quark mesons in the two HERA experiments H1 and ZEUS is based on either mass or lifetime tags. Due to the large b-quark mass H1 and ZEUS consider muons from semi-leptonic b-decays, which usually have high values of transverse momentum p_T^{rel}, with respect to the closest b quark initiated jet axis. The muons coming from charm and light quark decays have low values of p_T^{rel} as shown in Fig. 1(b). Also, the decay length of beauty is longer than that for charm and other light quarks. Hence the silicon vertex detector in H1 was used to tag the lifetime (see Fig. 1(c)) to distinguish beauty from other quarks.

The fraction of events from b-decays in a specific data sample[2] extracted on a statistical basis by adjusting the relative fractions of the simulated bottom, charm and light quarks decays to the measured p_T^{rel} and impact

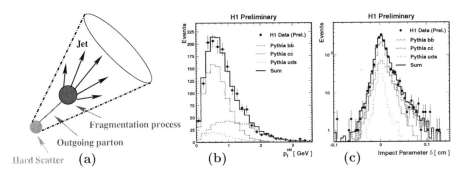

Figure 1. (a) Representation of the various stages of production and fragmentation of heavy quarks. The distribution of (b) the muon p_T^{rel} relative to the jet axis and (c) the impact parameter δ, of the muon track, are compared with the Monte Carlo simulations.

parameter δ distribution[b] is respectively found to be $f_b = 28.8 \pm 2.8\%$ and $f_b = 28.0 \pm 4.2\%$ in the H1 analysis. The two independent methods are found to be consistent in predicting the beauty fraction.

The charmed hadrons on the other hand are generally observed by forming the invariant mass of the tracks identified with a specific decay channel. In the presence of a vertex detector backgrounds can be reduced by measuring the displaced vertices from the primary interaction point. The recent measurement[3] by H1 for the charmed meson D^+, with a tag on the decay significance length, shows an improvement in signal to background ratio by a factor of $\sim \mathcal{O}(50)$.

In the following sections, results of open heavy quark production from HERA are discussed. Beauty, due to its large mass, is expected to have a faster convergence in the perturbative expansion in powers of the strong coupling constant α_s. Thus QCD calculations are expected to provide more reliable predictions for beauty than for charm, and will be discussed in the next section followed by the experimental results for open charm production.

2. Measurements of open beauty cross sections

One of the primary limitations of experimental measurements is the restriction of kinematic region, which is usually defined by the detector ac-

---

[b]The impact parameter δ is calculated in the plane transverse to the beam axis. Its magnitude is the distance of the closest approach of the track to the primary event vertex.

ceptance. Measurements performed in the restricted region are often extrapolated to the full phase space using a particular theoretical framework (either Monte Carlo models or next-to-leading order calculations), in order to compare between different experiments. As long as the model assumptions are not absorbed in the experimental results, but in the corresponding QCD predictions, the published measurements can be considered to be "safe".

On the other hand in the case of extrapolation to an unmeasured kinematic region, the extrapolation done should not only be quoted, but should also be compared within the same theoretical framework. For example the extrapolated cross section to an unmeasured phase space using a given Monte Carlo simulation cannot be reliably compared to the cross section predicted by a NLO calculation. Thus only those beauty measurements are considered here, where the measurements are made in the visible kinematic region without extrapolation.

The measurement of the beauty cross section by ZEUS[4] in the deep inelastic scattering (DIS) regime, using the p_T^{rel} method, as a function of Q^2 is shown in Fig. 2(a). The total visible cross section σ was determined in the kinematic range $Q^2 > 2$ GeV^2, inelasticity between $0.05 < y < 0.7$, with at least one jet in the Breit frame with $E_{T,jet}^{Breit} > 6$ GeV and pseudorapidity $\eta_{jet}^{Lab} < 2.5$, and with a muon between $30° < \theta^\mu < 160°$ and $p_T^\mu > 2$ GeV. The measured cross section is

$$\sigma = 38.7 \pm 7.7 \text{ (stat.)}^{+6.1}_{-5.0}\text{(syst.) pb}.$$

about a factor of 1.4 above the NLO QCD predictions of $28.1^{+5.3}_{-3.5}$ pb. It is, however, consistent with the measurement within the large theoretical and experimental uncertainties. The errors on the prediction were estimated by varying the factorisation and renormalisation scales by a factor of 2 and the b-quark mass between 4.5 and 5.0 GeV. CASCADE, based on the CCFM evolution equations, is in agreement with the measured cross section. The RAPGAP Monte carlo simulation based on DGLAP on the other hand lies below the data.

In order to obtain a unified picture on the beauty production H1[2] and ZEUS[5] recently measured the beauty cross section in the photoproduction regime within almost identical phase spaces. Fig. 2(b) shows the ratio of the measured cross section over theoretical expectations as a function of p_T^μ obtained either using the p_T^{rel} and δ method (H1) or the p_T^{rel} method (ZEUS). The low virtuality ($Q^2 < 1$ GeV^2), dijet photoproduction samples, are considered with $E_T^{jet1(2)} > 7(6)$ GeV, $|\eta^{jet}| < 2.5$ and

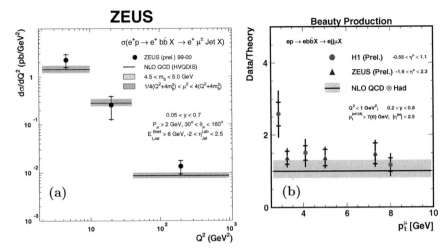

Figure 2. Beauty differential cross section (a) in DIS as a function of Q^2 with at least one reconstructed jet in the Breit frame in association with a muon and (b) the ratio of measured cross section over theoretical expectation, in the photoproduction regime as a function of p_T^μ, for events with two jets and a muon with $p_T^\mu > 2.5$.

$0.2 < y < 0.8$. The differential cross section as a function of the transverse momentum of the associated muons with $p_T^\mu > 2.5$ GeV and the pseudorapidity $-0.55 < \eta^\mu < 1.1$ (H1) or $-1.6 < \eta^\mu < 2.3$ (ZEUS) are found to be in agreement within the two experiments. The FO NLO QCD calculation is in reasonable agreement with the data within the large experimental and theoretical uncertainties, which are of similar size, except for the lowest p_T^μ H1 measured bin.

3. Measurements of open charm cross sections

Inclusive production of the charmed mesons D^+, D^0, D_s and D^+ in DIS was measured[3] with the H1 detector, using an integrated luminosity of 48 pb^{-1}. In these measurements the charm lifetime properties are exploited using a two layer silicon vertex detector. Differential cross sections are measured for the D^+ and D^0 mesons and compared with the AROMA MC predictions in the kinematic range $2 < Q^2 < 100$ GeV^2, $0.05 < y < 0.7$, $p_t(D) > 2.5$ GeV and $|\eta(D)| < 1.5$. Results for the D^+ channel are shown in Fig. 3(A). The distributions are well described both in shape and normalisation at LO with a parton shower QCD simulation. The lower shaded band indicates the AROMA beauty contribution, scaled by a factor of 4.3, according to the H1 published values[6].

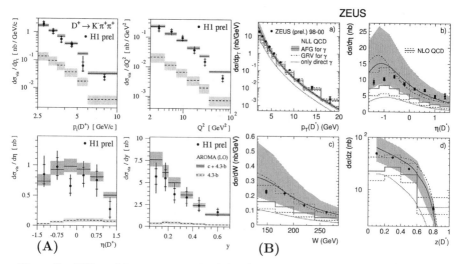

Figure 3. Differential cross section for (A) D^+ as a function of $p_t(D^+)$, Q^2, $\eta(D^+)$ and electron inelasticity y in the DIS regime, compared with the AROMA predictions and (B) inclusive D^* photoproduction events, as a function of a) $p_T(D^*)$, b) $\eta(D^*)$, c) $W(D^*)$ and d) $z(D^*)$. FO NLO predictions with nominal parameters are given by the solid histogram. The VFNS predictions (labeled as NLL) are shown by the solid curves. The GRV photon parameterisation and direct photon predictions are given by dash-dotted and dotted lines.

Inclusive photoproduction of D^* mesons has been measured[7] with the ZEUS detector in the kinematic region $Q^2 < 1$ GeV^2, γp centre-of-mass energies $130 < W < 280$ GeV, $1.9 < p_T(D^*) < 20$ GeV and $|\eta(D^*)| < 1.6$ using an integrated luminosity of 79 pb^{-1}. The measured differential cross sections were compared with FO NLO, VFNS and FONLL QCD predictions. In Fig. 3(B) the differential cross section as a function of a) $p_T(D*)$, b) $\eta(D^*)$, c) $W(D^*)$ and d) $z(D^*)$ are compared with the FO NLO and VFNS QCD calculations, where $z(D^*)$ is the fraction of the photon energy carried by the charmed meson in the proton rest frame. The precision of the data is enormously better than the theoretical uncertainties. The uncertainties for VFNS are larger than for the FO NLO calculations, which were obtained by varying the charm mass and the renormalisation scales simultaneously. The central FO NLO predictions are below the data, mainly in the proton direction ($\eta > 0$) and the low z region. The VFNS prediction is closer to the data, in particular it is better than FO NLO for $d\sigma/dz$ and

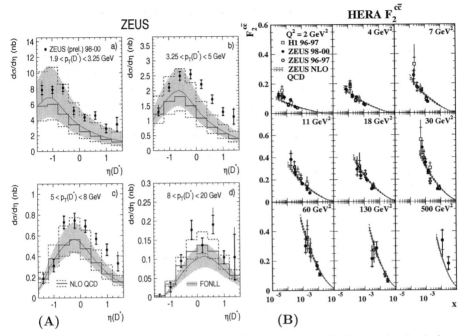

Figure 4. (A) Differential cross section $d\sigma/d\eta$ for inclusive D^* photoproduction in four p_T regions, compared to FO NLO (histograms) and FONLL (curves) predictions. (B) Charm contribution to the proton structure function for different Q^2 at a function of x.

for positive pseudorapidity. The direct photon processes alone in VFNS cannot describe the data distribution and hence a significant resolved contribution is required, which as expected shows sensitivity to the variation in the photon structure function parameterisations.

Given the precision of the ZEUS data, the differential cross sections for different $p_T(D^*)$ regions as a function of $\eta(D^*)$ are compared to the FO NLO and FONLL predictions in Fig. 4(A). This comparison allows to identify the region of phase space where the discrepancy between the data and the calculations can be localized. The data is close to the upper limit of the uncertainties of the predictions and is significantly above both the FO NLO and FONLL predictions at medium p_T and positive η. As expected due to the inherited properties from FO NLO, the FONLL is close to the former at low p_T, but is surprisingly below FO NLO at large p_T.

3.1. *Measurements sensitive to structure functions and hard scattering*

The photon acts as a pointlike probe of the proton in DIS, where the BGF (Fig. 5(e)) process dominates at HERA. This provides an unique opportunity to study the charm contribution $F_2^{c\bar{c}}$, to the proton structure F_2, which in turn is sensitive to the gluon density of the proton. The production of D^* mesons has been studied in DIS at HERA[8] in the kinematic region $1.5 < Q^2 < 1000$ GeV^2, $0.02 < y < 0.5$, $1.5 < p_T(D^*) < 15$ GeV and $|\eta(D^*)| < 1.5$. Predictions from NLO QCD are found to be in reasonable agreement with the measured differential cross section as a function of Q^2 and x. Hence this calculation was then used to extract $F_2^{c\bar{c}}$ using:

$$F_{2,\text{meas}}^{c\bar{c}} = \frac{\sigma(x,Q^2)_{\text{meas}}}{\sigma(x,Q^2)_{\text{theo}}} F_{2,\text{theo}}^{c\bar{c}}. \qquad (2)$$

The extrapolation factors to the full phase space vary between 4.7 and 1.5 at low and high Q^2 respectively. The extracted $F_2^{c\bar{c}}$ are shown compared with the NLO QCD prediction in Fig. 4(B). A remarkable confirmation of the gluon density obtained from scaling violations of the inclusive F_2 can be observed within 10% accuracy. The data shows a steep rise at low x, indicating a large contribution of the gluon density in the proton. Since at low Q^2, the uncertainties of the data are comparable to those from the PDF fit, the measured differential cross section $\sigma(x,Q^2)$ should be used in future fits to constrain the gluon density of the proton.

In an analogous way, dijet photoproduction at HERA allows to study the structure of the photon. In leading order (LO) QCD the underlying sub-processes (Fig. 5) can be divided into either direct photon or resolved photon processes. In direct photon processes the photon participates in the hard scatter predominantly via the boson-gluon fusion process. Hence the fraction of the photon energy participating in the formation of two highest transverse energy jets, experimentally defined as x_γ^{obs}, is close to 1. On the other hand in resolved photon processes the photon acts like a source of incoming partons (quarks and gluons) and only a fraction of its momentum participates in the hard scatter, giving rise to $x_\gamma^{\text{obs}} < 1$.

ZEUS recently measured such a photon energy fraction associated with at least one charm quark for dijet photoproduction events[9]. Fig. 6(a) shows the differential cross section as a function of x_γ^{obs} in comparison with PYTHIA, HERWIG and CASCADE. Not only a rise at high x_γ^{obs} is observed, there is a substantial tail at low x_γ^{obs}, which requires a significant ($\approx 40\%$) LO-resolved contribution to describe the data. The low x_γ^{obs}

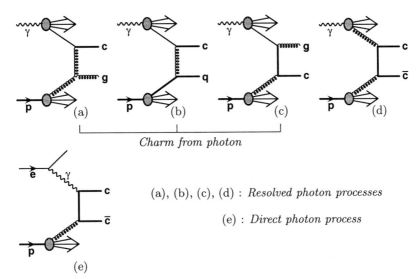

Figure 5. Various LO sub-processes with charm, dominant in the HERA region of phase space.

data are consistent with the presence of resolved photon processes, and can also be simulated in CASCADE via initial-state radiation based on CCFM evolution. However, CASCADE overestimates the high x_γ^{obs} region.

A similar study in both the γp and DIS regimes[10] with two high E_T^{jet} jets was done in order to probe the structure of the virtual photon. The ratio of the low and high x_γ^{obs} regions is shown in Fig. 6(b) and compared to CASCADE and AROMA expectations. The AROMA model using the DGLAP evolution scheme lies below the data, whereas CASCADE with CCFM evolution is much closer to the data. The fact that the ratio does not change significantly with Q^2 is in marked contrast to the case in which the presence of charm is not required[11], suggesting that the suppression of the low x_γ^{obs} cross section due to the charm mass and the suppression due to photon virtuality as the scales are not independent.

In order to probe the dynamics of the hard scattering in the sub-process (see Fig. 5) and in particular to study the charm content of the photon, the differential cross section $d\sigma/d|\cos\theta^*|$ as shown in Fig. 7(a), as a function of $|\cos\theta^*|$ was measured[9], where θ^* is the angle between the jet-jet axis and the beam direction in the dijet rest frame. The distribution was measured for resolved-enriched ($x_\gamma^{obs} < 0.75$) and direct-enriched ($x_\gamma^{obs} > 0.75$) samples. The cross section for the sample enriched with resolved photon events exhibits a more rapid rise towards high values of $|\cos\theta^*|$ than does

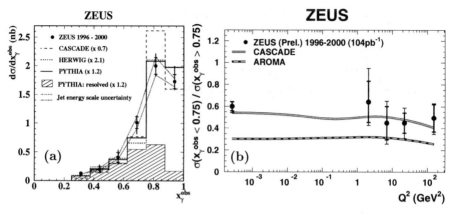

Figure 6. (a) Differential cross section $d\sigma/dx_\gamma^{\rm obs}$ for photoproduction events compared with MC simulations. Each MC distribution is normalised to the data, as indicated in the brackets. (b) Ratio of low to high x_γ^{obs} for events with a D^* compared to the predictions of AROMA and CASCADE, which do not include photon structure.

the cross section for the direct-enriched sample. The measured differential cross section in comparison to the LO partonic matrix elements reflects the different spins of the quark and gluon propagators in the dominant diagrams. Consequently the subprocess $gg \to c\bar{c}$ cannot be the dominant resolved photon process for the charm dijet events.

Independent of any model assumptions, the Rutherford scattering cross section for a q-exchange, $(\approx (1 - |\cos\theta^*|)^{-1})$ should show a mild rise, in contrast to the g-exchange diagram $\approx (1-|\cos\theta^*|)^{-2}$. To quantify this rise, the measured cross section for the direct and resolved enriched samples were fitted to a function $1/(1 - |\cos\theta^*|)^\kappa$. The resultant values of κ are:

$$\begin{aligned}\kappa(x_\gamma^{\rm obs} < 0.75) &= 1.735 \pm 0.176;\\ \kappa(x_\gamma^{\rm obs} > 0.75) &= 0.743 \pm 0.107\end{aligned} \quad (3)$$

Given the fact that the shapes are expected to be distorted due to the additional parton radiation and hadronisation corrections, the results are consistent with the Rutherford scattering expectations.

The two jets are then distinguished into a charm-initiated jet (D^* jet), and the other jet in $\eta - \phi$ space (the variable $\Delta R_i \equiv \sqrt{(\phi_{jet_i} - \phi_{D^*})^2 + (\eta_{jet_i} - \eta_{D^*})^2}$) is used with the D^* jet having the smallest $\Delta R_{i(1=1,2)} < 1.0$. Thus the sign of the unfolded $\cos\theta^*$ distribution is given by the direction of the D^* meson. It is positive for the proton direction and negative for the photon direction.

Fig. 7(b) shows the differential cross section as a function of $\cos\theta^*$ for the

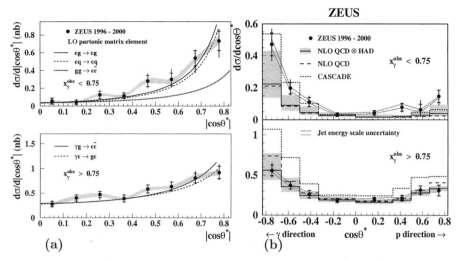

Figure 7. (a) Differential cross section $d\sigma/d|\cos\theta^*|$ for samples enriched in resolved and direct photon events, compared to leading order $2 \to 2$ matrix element with at least a charm in the final state. All matrix element distributions are normalised to the first data bin. (b) Differential cross section as a function of $\cos\theta^*$ for samples enriched in resolved and direct photon events compared with FO NLO predictions. The NLO uncertainty after hadronisation correction is given by the shaded band.

resolved- and direct-enriched samples. The $\cos\theta^*$ distribution for direct-enriched sample is almost symmetric, as expected from the BGF process. The slight asymmetry can be explained by the PYTHIA estimation[9] of the resolved process contribution to the direct enriched sample. The strong rise in the cross section towards the photon direction for the resolved-enriched sample is clear evidence for charm originating from the photon with a gluon propagator. Although the shape for low x_γ^{obs} events is reproduced by both CASCADE and the FO NLO predictions, the large uncertainties in the FO NLO prediction localized in the photon direction indicate non-convergent terms in the perturbative expansion due to collinear emissions of partons causing a large scale dependence. These terms then need to be absorbed either in the fragmentation or in the charm distribution functions of the photon and proton. Of course, to perform this absorption a charm contribution in the structure function is needed.

3.2. Charm fragmentation

The non-perturbative aspect of QCD was studied in terms of charm fragmentation. The data comparison to the theoretical predictions in all hadronic colliders is done based on the assumption of universality in charm fragmentation. This assumption then allows to use the parameterisations of the fragmentation characteristics obtained in e^+e^- annihilations in the calculations for charm production in ep scattering. The measurements of these fragmentation characteristics at HERA will not only permit the verification of the universality of charm fragmentation, but also can significantly contribute to the understanding of the non-perturbative process.

The H1 and ZEUS measurements in the DIS[3] and photoproduction[12] regime respectively, were used to obtain preliminary results on the transfer of the quark's energy to a given meson. The fragmentation fractions of the charm quark to the various mesons are shown in Table 1. In addition H1 and ZEUS measured cross sections, which were used to test the isospin invariance of the fragmentation process by calculating the ratio of neutral ($c\bar{u}$) to charged ($c\bar{d}$) D meson production. The measured ratio $R_{u/d}$ is consistent with 1, confirming the isospin invariance. The s quark production is measured to be suppressed by a factor ≈ 3.5 as shown by γ_s. Both H1 and ZEUS rule out the naive spin counting prediction of 0.75 for the fraction of D mesons produced in a vector state P_V. The results in the table are compared with the e^+e^- world average values[13].

Table 1. The fractions of c quarks hadronising to a specific charmed meson. The results for $R_{u/d}$, γ_s and P_V are also provided, compared to the world average.

	H1 (DIS)	ZEUS (γp)	Combined e^+e^-
$f(c \to D^+)$	$0.20 \pm 0.02^{+0.04+0.03}_{-0.03-0.02}$	$0.249 \pm 0.014^{+0.004}_{-0.008}$	0.232 ± 0.010
$f(c \to D^0)$	$0.66 \pm 0.05^{+0.12+0.09}_{-0.14-0.05}$	$0.557 \pm 0.019^{+0.005}_{-0.013}$	0.549 ± 0.023
$f(c \to D_s^+)$	$0.16 \pm 0.04^{+0.04+0.05}_{-0.04-0.05}$	$0.107 \pm 0.009^{+0.005}_{-0.005}$	0.101 ± 0.009
$f(c \to \Lambda_c^+)$		$0.076 \pm 0.020^{+0.017}_{-0.001}$	0.076 ± 0.007
$f(c \to D^{*+})$	$0.26 \pm 0.02^{+0.06+0.03}_{-0.04-0.02}$	$0.223 \pm 0.009^{+0.003}_{-0.005}$	0.235 ± 0.007
$R_{u/d}$	$1.26 \pm 0.20^{+0.11+0.04}_{-0.11-0.04}$	$1.014 \pm 0.068^{+0.024}_{-0.031}$	1.00 ± 0.09
γ_s	$0.36 \pm 0.10^{+0.01+0.08}_{-0.01-0.08}$	$0.266 \pm 0.023^{+0.014}_{-0.012}$	0.26 ± 0.03
P_V	$0.69 \pm 0.045^{+0.004+0.009}_{-0.004-0.009}$	$0.554 \pm 0.019^{+0.008}_{-0.004}$	0.60 ± 0.03

The final uncertainty in the non-perturbative part of the QCD calculations is the fragmentation function, which affects both the shape and the absolute normalisation of the predicted cross section. It was recently measured by the ZEUS collaboration[14]. Charm associated jet events were used in the photoproduction regime in order to measure the jet energy fraction carried by a D^* meson defined as $z = (E + p_\parallel)_{D^*}/(E + p_\parallel)_{jet}$, where the $p_\parallel$ is the longitudinal momentum of the D^* relative to the jet axis. The relative cross section shown in Fig. 8(a) is compared to the PYTHIA MC simulations with different values of the free parameter ϵ within the Peterson fragmentation function[15]. Strong sensitivity to the parameter ϵ, varied between 0.01 to 0.1 was found. The fit using χ^2 minimisation yields $\epsilon = 0.064 \pm 0.006^{+0.011}_{-0.008}$ in comparison to 0.053 from LO fits[16] to the LEP data. Although there are independent ways of measuring the fragmentation function, in Fig. 8(b) the OPAL and ARGUS results with different z definition are compared. They are similar in shape to the HERA data. The low-z peak in the OPAL data arises from a significant gluon-splitting due to the much higher centre-of-mass energy.

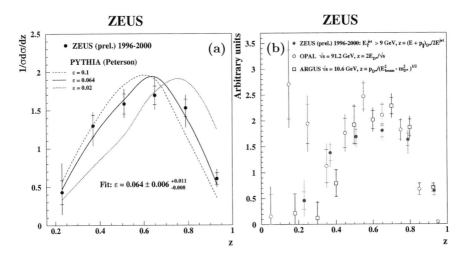

Figure 8. Relative cross section $1/\sigma d\sigma/dz$ for the data compared (a) with PYTHIA predictions for different values of the parameter ϵ in the Peterson fragmentation function and (b) to e^+e^- measurements, where the data sets are normalised to 1/(bin width) for $z > 0.3$.

The ZEUS fragmentation measurements are in good agreement with e^+e^- annihilations, indicating the universality of charm fragmentation.

4. Conclusions

Open heavy flavour measurements provide an important tool for understanding both perturbative and non-perturbative aspects of QCD. The QCD calculations do a fair job in describing the data, but fail within certain kinematic regions, and have large theoretical uncertainties. Beauty measurements made in almost identical kinematic phase spaces by the H1 and ZEUS experiments agree with each other. Very precise open charm measurements are made in both DIS and photoproduction.

Although the first HERA open charm signal consisted of 48±11 events[17], with the enormous amount of available statistics, measurements sensitive to each aspect of the convolution (PDF$_p$ ⊗ PDF$_\gamma$ ⊗ hard scatter ⊗ fragmentation) have been made. The recent trend at HERA not only includes the open beauty measurements in the visible range, but also structure function sensitive measurements like $F_2^{c\bar{c}}$, evidence of charm originating from the photon, multiscale issues with charm in deeply inelastic virtual photon scattering and of course new measurements of the charm fragmentation confirming the evidence of its universality.

References

1. M. Cacciari et al., *JHEP* **0103**, 6 (2001).
2. H1 Coll., Contributed paper to EPS 2003 conference, Abstract 117, Aachen, Germany, July 2003.
3. H1 Coll., Contributed paper to EPS 2003 conference, Abstract 096, Aachen, Germany, July 2003.
4. ZEUS Coll., Contributed paper to ICHEP02 conference, Abstract 783, Amsterdam, The Netherlands, July 2002.
5. ZEUS Coll., Contributed paper to ICHEP02 conference, Abstract 785, Amsterdam, The Netherlands, July 2002.
6. H1 Coll., S. Aid et al.*Phys. Lett.* **B 467**, 156 (1999).
7. ZEUS Coll., Contributed paper to ICHEP02 conference, Abstract 786, Amsterdam, The Netherlands, July 2002.
8. H1 Coll., Contributed paper to EPS2003 conference, Abstract 98, Aachen, Germany, July 2003; ZEUS Coll., S. Chekanov et al. DESY-03-115 (August 2003).
9. ZEUS Coll., *Phys. Lett.* **B 565**, 87 (2003).
10. ZEUS Coll., Contributed paper to EPS01 conference, Abstract 495, Budapest, Hungary, July 2001.
11. ZEUS Coll., Contributed paper to ICHEP00 conference, Abstract 1039, Osaka, Japan, July 2000.
12. ZEUS Coll., Contributed paper to EPS 2003 conference, Abstract 564, Aachen, Germany, July 2003.
13. L. Gladilin, Preprint hep-ex/9912064 (1999).
14. ZEUS Coll., Contributed paper to ICHEP02 conference, Abstract 778, Amsterdam, The Netherlands, July 2002.
15. C. Peterson et al., *Phys. Rev.* **D27**, 105 (1983).
16. P. Nason and C. Oleari, *Nucl. Phys.* **B 565**, 245 (2000).
17. ZEUS Coll., *Phys. Lett.* **B 349**, 225 (1995).

OPEN HEAVY-FLAVOUR PHOTOPRODUCTION AT NLO

I. SCHIENBEIN
DESY,
Notkestrasse 85,
22603 Hamburg, Germany
E-mail: schien@mail.desy.de

I review theoretical approaches to open heavy flavour photoproduction and discuss theoretical aspects of a new calculation in a massive theory with $\overline{MS}$ subtraction based on the factorization theorem of John Collins with heavy quarks.

1. Introduction

Heavy quarks are those with masses $m_h \gg \Lambda_{QCD}$ such that $\alpha_s(m_h^2) \propto \ln^{-1}(\frac{m_h^2}{\Lambda_{QCD}^2}) \ll 1$ and hence, according to this definition, the charm, bottom and top quarks (c, b, t) are heavy whereas the up, down and strange quarks (u, d, s) are light. A lot of reasons can be found in the literature why heavy quark production is interesting. Most importantly, such processes are fundamental elementary particle processes which take place with substantial rates at high energy colliders. Therefore a good phenomenological understanding of heavy quark production is very important. Moreover, there are more theoretical reasons: The heavy quark mass m_h acts as a physical long-distance cut-off such that heavy quark production is an ideal testing ground of perturbative QCD. Very often in addition to m_h another large scale is involved in the hard scattering process such that one has to deal with multi-scale processes which complicate the perturbative analysis.

In this talk I will restrict to open charm photoproduction, $\gamma p \to D^\star X$, and describe theoretical aspects of a new calculation in a massive variable flavour number scheme (VFNS) based on the factorization theorem of John Collins including heavy quark masses [1]. This work is along the lines of previous studies of the process $\gamma\gamma \to D^\star X$ for the direct contribution [2] and the single-resolved part [3]. Note also that the latter constitutes the direct contribution to $\gamma p \to D^\star X$ [4] considered here. A calculation of the resolved contribution is in progress [5] which will allow to perform a complete NLO

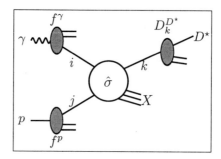

Figure 1. Open charm photoproduction, $\gamma p \to D^\star X$, according to Eq. (1).

analysis of inclusive photoproduction of $D^\star$ mesons in a massive VFNS which extends the successful massless calculation [6] by terms of the order $O(m_c^2/p_T^2)$ where p_T is the transverse momentum of the $D^\star$ meson. A different approach to construct a massive VFNS has been adopted in Refs. 7, 8 ('FONLL').

2. Factorization Formula

Generically, the theoretical description of open charm photoproduction is based on the following factorization formula [see Fig. 1]

$$\mathrm{d}\sigma(\gamma p \to D^\star X) = \sum_{i,j,k} \int dx_1 \, dx_2 \, dz \, f_i^\gamma(x_1,\mu_f) \, f_j^p(x_2,\mu_f) \times$$
$$\mathrm{d}\hat{\sigma}(ij \to kX) \, D_k^{D^\star}(z,\mu_f') + O(\alpha_s^{n+1}, \tfrac{\Lambda^2}{p_T^2}), \quad (1)$$

where Λ is a typical hadronic scale of a few hundred MeV, p_T is the transverse momentum of the $D^\star$ meson and where $f_i^\gamma(x_1,\mu_f)$ and $f_j^p(x_2,\mu_f)$ are (universal) parton distribution functions (PDFs) of the photon and the proton, respectively. The partons of the proton are gluons and quarks, $j = g, q, \ldots$ ($q = u, d, s$) while the partons of the photon are $i = g, q, \ldots$ ('resolved contribution') and $i = \gamma$ ('direct contribution'[a]). The ellipses indicate that the question whether heavy quarks are included as partons depends on the factorization scale μ_f and the heavy flavour scheme to be discussed in the next section. Furthermore, the fragmentation of the final state parton $k = g, q, \ldots$ into the $D^\star$ meson is described by (universal) fragmentation functions (FFs) $D_k^{D^\star}(z,\mu_f')$. The ellipses again refer to

---

[a] $f_\gamma^\gamma(x,\mu_f) = \delta(1-x)$

the heavy quark treatment. [b] Finally, the hard scattering cross sections $d\hat{\sigma}(ij \to kX)(\mu_f, \mu'_f, \alpha_s(\mu), [\frac{m_c}{p_T}])$ are pertubatively calculable and depend on the factorization scales μ_f and μ'_f and the renormalization scale μ. As has been proved to all orders of perturbation theory in Ref. 1 heavy quark mass effects can be consistently included in the hard part. The square bracket around $\frac{m_c}{p_T}$ indicates that, depending on the heavy flavour scheme, these terms are included or neglected. Of course, in order to obtain a physically meaningful (scheme independent) result the hard part and the PDFs and FFs have to be employed in the same factorization scheme.

Before turning to a discussion of heavy flavour schemes it is instructive to consider the theoretical uncertainties of predictions based on Eq. (1). The sources are:

- Uncertainties in the PDFs and FFs.
- The error of the factorization formula which is of the order $O(\frac{\Lambda^2}{p_T^2})$.
- The order of the perturbative calculation of the hard part. An NLO calculation includes terms of the order $O(\alpha_s^2)$ ($n = 2$ in Eq. (1)). Usually the error due to higher order terms is estimated by varying the factorization and renormalization scales.

3. Heavy Flavour Schemes

In this section I will give a basic overview over the main heavy flavour schemes used in the literature. For a more detailed review see, e.g., Ref. 9. References to NLO calculations of $\gamma p \to D^\star X$ based on Eq. (1) and using a given heavy flavour scheme can be found in the corresponding section headlines.

3.1. *Conventional massless Parton Model: ZM-VFNS* [6]

The conventional (massless) parton model is also known as 'zero mass variable flavour number scheme' (ZM-VFNS). In this approach the charm mass m_c is set to zero in the calculation of the hard scattering cross section. The number of partons is variable depending on the factorization scale with the

---

[b]It should be noted that in any case charm quarks (heavy quarks) will be produced in the final state of the partonic subprocess ($k = c$). However, if the final state charm quark is not treated as a parton it is nevertheless phenomenologically necessary to include a scale-independent fragmentation function $D_c^{D^\star}(z)$. Such a function is in principle not related to the factorization formula in Eq. (1) and it is not clear whether it has a universal meaning. See also Sec. 3.

following finite transformations (valid up to NLO) at the so-called transition scale $Q_0 = m_c$: [c]

$$n_f = 3 \to n_f = 4$$
$$\alpha_s^{(3)} \to \alpha_s^{(4)} = \alpha_s^{(3)} + O(\alpha_s^3)$$
$$f_i^{(3)} \to f_i^{(4)} = f_i^{(3)} + O(\alpha_s^2) \qquad (2)$$

and the perturbative boundary condition for the charm quark

$$f_c^{(4)}(x, Q_0^2 = m_c^2) = 0 \ . \qquad (3)$$

The advantage of this approach is that it is simple and large collinear logarithms $\ln \frac{\mu^2}{m_c^2}$ are resummed to all orders in perturbation theory in the evolved distributions $f_c(x, \mu^2)$ and $D_c^{D^*}(x, \mu^2)$. On the other hand, charm mass terms $\frac{m_c^2}{p_T^2}$ are neglected in the hard part, such that theoretical predictions are good only for $p_T^2 \gg m_c^2$.

3.2. ($n_f = 3$)-Fixed Flavour Number Scheme (3-FFNS)
[11,12]

In this approach $m_c \neq 0$ and the number of flavours is fixed to $n_f = 3$. The partonic subprocesses involve gluons and light quarks as partons in the initial state whereas no charm parton distribution is taken into account ($f_c(x, \mu_f) = 0$). On the other hand charm quarks are produced in the partonic subprocesses and hence appear in the final state. Here it is phenomenologically unavoidable to include a non-perturbative function, for example of Peterson type, describing the transition from the charm quark to the observable D^* meson. It should be noted that this fragmentation function is *not* related to the fragmentation functions in the factorization theorem in Eq. (1) which are universal and factorization scale dependent.

In the fixed flavour scheme perturbative calculations are more involved due to $m_c \neq 0$. The advantage is that $\frac{m_c^2}{p_T^2}$ terms are included and a correct threshold behaviour is guaranteed which is especially important in deep inelastic scattering. On the other hand, the fixed order logarithms $\ln \frac{p_T^2}{m_c^2}$ can become large for $p_T^2 \gg m_c^2$ such that a resummation of these large logarithms is necessary. Therefore, calculations within the FFNS are ex-

---

[c]For a discussion of the general case $Q_0 = O(m_c)$ see, e.g., Ref. 10. Therein, also references to the literature of the matching conditions of the PDFs and α_s can be found.

pected to be good mainly in the kinematical region $p_T^2 \simeq m_c^2$.[d] A major drawback is that the charm fragmentation function $D_c^{D^*}(z)$ is not based on the factorization theorem and hence not universal. Therefore there is no deep theoretical reason to fix this function in one experiment and to use it making predictions for other experiments. Finally, it is noteworthy that it is not possible to include a non-perturbative charm component in the proton in this scheme.

3.3. Massive VFNS (GM-VFNS) [7,8,5]

The massive VFNS (GM-VFNS) is a variable flavour number scheme as introduced in Sec. 3.1 which includes heavy quark mass effects in the hard scattering cross sections/coefficients. Besides the usual gluon and light quark parton distributions a charm quark parton distribution is taken into account. In order to avoid double counting collinear logarithms $\ln \frac{\mu^2}{m_c^2}$ are subtracted from the partonic subprocesses and are resummed to all orders with help of the DGLAP evolution equations ($\to f_c \neq 0$). The same is true for the final state, where collinear logarithms are subtracted from the hard part and are resummed to all orders by introducing a charm quark fragmentation function $D_c^{D^*}(z, {\mu'_f}^2)$.

This scheme is most ambitious aiming at combining the virtues of both the massless VFNS and the FFNS and therefore it is technically more involved. A calculation requires to consider all subprocesses occurring already in the FFNS with $m_c \neq 0$ and to subtract the collinear parts in order to obtain 'IR-safe' hard parts. Technically this is equivalent to mass factorization within a massive regularization scheme. In addition the subprocesses with charm quarks in the initial state (occurring also in the massless VFNS) have to be included. However, all desired features are fulfilled in this approach: The large collinear logarithms $\ln \frac{\mu^2}{m_c^2}$ are resummed in terms of evolved charm quark PDFs and FFs $f_c(x, \mu^2)$ and $D_c^{D^*}(x, \mu^2)$, respectively, and charm quark mass terms $O(\frac{m_c^2}{p_T^2})$ are included in the hard parts. Therefore, calculations in this scheme are expected to be valid for both $p_T^2 \simeq m_c^2$ and $p_T^2 \gg m_c^2$ such that for $p_T^2 \lesssim m_c^2$ the transition to the 3-FFNS is achieved and asymptotically, for $p_T^2 \gg m_c^2$, the ZM-VFNS is reproduced. Again it should be noted that as in the massless VFNS a non-perturbative charm component in the proton can be included.

---

[d]Of course it is a matter of debate when conditions like $p_T^2 \gg m_c^2$ are actually satisfied and the answer certainly also depends on the process under consideration.

4. Theoretical Aspects of Open Charm Photoproduction in a massive VFNS

In this section some theoretical issues relevant for open charm photoproduction in a massive VFNS will be addressed. One of the key ingredients in establishing such a scheme is the removal of collinear would-be singularities in order to avoid double counting. As already mentioned, this is technically nothing else but mass factorization within a massive regularization scheme and will be discussed in Sec. 4.1. Then, in Sec. 4.2 the residual implementation freedom concerning finite mass terms will be discussed which is the main reason for the different approaches to massive VFNS existing in the literature. Our procedure adopted for establishing a massive VFNS which fixes these ambiguities and which is applicable to all kinds of processes will be described.

4.1. *Mass Factorization*

Before turning to mass factorization in a massively regularized theory it is helpful to recap some facts about $\overline{MS}$ mass factorization in a dimensionally regularized theory. As is well-known, mass divergences ($\equiv$ 'collinear divergences') occur when an internal propagator gets on its mass-shell due to a *collinear* splitting process as exemplified in Fig. 2. The associated

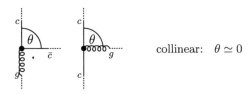

Figure 2. Collinear splitting processes $g \to c\bar{c}$ and $c \to cg$.

regularized collinear 'divergences' are subtracted from the (partonic) cross sections and are absorbed into PDFs and FFs. Along with the genuine (regularized) poles *finite terms* can be subtracted leading to different factorization schemes; among them the $\overline{MS}$ scheme considered here. It should be noted that even for the same factorization scheme the *subtraction terms* do depend on the *regularization procedure*. Usually the $\overline{MS}$ (factorization/renormalization) scheme is introduced or defined within *dimensional regularization* ($m_c = 0$) such that collinear divergences occur as $\frac{1}{\epsilon}$ poles.

As an example consider Fig. 3. The collinear part to be subtracted via

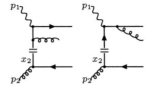

Figure 3. Diagrammatical representation of the subtraction term in Eq. (4).

mass factorization is given by

$$d\sigma_{\rm SUB}(\mu_f{}^2,\mu^2) = \int_0^1 dx_2\, G^1_{cg}(x_2) d\hat\sigma^{(0)}_{\gamma c\to cg}(p_1,x_2 p_2) \equiv G^1_{cg}\otimes d\hat\sigma^{(0)}_{\gamma c\to cg} \quad (4)$$

with the universal function $[n = 4 - 2\epsilon]$

$$G^1_{cg}(x,\mu_f{}^2,\mu^2) = \tfrac{\alpha_s(\mu^2)}{2\pi}\Big[P^{(0)}_{cg}(x)(-\tfrac{1}{\epsilon}+\gamma_E-\ln 4\pi)(\tfrac{\mu^2}{\mu_f{}^2})^\epsilon + \underbrace{g^1_{cg}(x)}_{\overline{\rm MS}=0}\Big] \quad (5)$$

containing the $\tfrac{1}{\epsilon}$ pole and the finite part $g^1_{cg}(x)$ being zero in the $\overline{\rm MS}$ scheme. A few remarks are in order: (i) The variable x_2 in Eq. (4) can be calculated in terms of the external Mandelstam variables, $x_2 = x_2(s,t,u)$, since the Born cross section is proportional to the Delta-function $\delta(\hat s+\hat t+\hat u)$ where $\hat s,\hat t,\hat u$ are obtained in the usual way from the 4-momenta $p_1, x_2 p_2$ entering the Born subprocess. (ii) The Born cross section in Eq. (4) is usually taken to be n dimensional which generates finite terms when multiplied with the $\tfrac{1}{\epsilon}$ pole in the function G^1_{cg}. (iii) Finally it is interesting to keep track of the logarithms involved in the mass factorization procedure. The collinear part of the unsubtracted partonic cross section is proportional to $-\tfrac{1}{\epsilon}(\mu^2/s)^\epsilon = -\tfrac{1}{\epsilon}+\ln\tfrac{s}{\mu^2}$ from which $-\tfrac{1}{\epsilon}(\mu^2/\mu_f{}^2)^\epsilon = -\tfrac{1}{\epsilon}+\ln\tfrac{\mu_f{}^2}{\mu^2}$ is subtracted [see Eq. (5)] generating a logarithm $\ln\tfrac{s}{\mu_f{}^2}$ in the IR-safe hard part which depends on the factorization scale.

The above example can be systematically generalized using universal functions of the form

$$G^1_{ij}(x,\mu_f{}^2,\mu^2) = \tfrac{\alpha_s(\mu^2)}{2\pi}\Big[P^{(0)}_{ij}(x)(-\tfrac{1}{\epsilon}+\gamma_E-\ln 4\pi)(\tfrac{\mu^2}{\mu_f{}^2})^\epsilon + \underbrace{g^1_{ij}(x)}_{\overline{\rm MS}=0}\Big]. \quad (6)$$

In the case of heavy quark production it is natural to use the heavy quark mass m as collinear regulator such that the collinear poles appear as logarithms $\ln\tfrac{s}{m^2}$. The *massive* $\overline{\rm MS}$ subtraction terms $d\sigma_{\rm SUB}(\mu_f,m)$ are identified by the condition

$$\lim_{m\to 0}\left(d\sigma(m)-d\sigma_{\rm SUB}(m)\right)\stackrel{!}{=}d\hat\sigma(\overline{\rm MS}) \quad (7)$$

As before the subtraction terms can be factorized into universal functions $G^1_{ij}(x,\mu_f{}^2,\mu^2)$ and $D^1_{ij}(x,\mu_f{}^2,\mu^2)$ for the initial and final state, respectively, and Born cross sections: $d\sigma^1_{\text{SUB}}(\mu_f{}^2,\mu^2) = G^1(x) \otimes d\hat{\sigma}^{(0)}$, $d\hat{\sigma}^{(0)} \otimes D^1(z)$. The hard (IR safe) cross sections entering the factorization formula in Eq. (1) are then obtained by removing the mass singular parts from the partonic cross section

$$d\hat{\sigma}^{(1)} = d\sigma^1 - d\sigma^1_{\text{SUB}} \,. \tag{8}$$

Setting $\mu_f = \mu$, the massively regularized functions G^1_{ij} (initial state) are given in the $\overline{\text{MS}}$ scheme by

$$\begin{aligned}
G^1_{cg}(x,\mu^2) &= \tfrac{\alpha_s(\mu)}{2\pi} P^{(0)}_{cg}(x) \ln \tfrac{\mu^2}{m^2}\,, \\
G^1_{cc}(x,\mu^2) &= \tfrac{\alpha_s(\mu)}{2\pi} P^{(0)}_{cc}(x) \ln \tfrac{\mu^2}{m^2}\,, \\
G^1_{gg}(x,\mu^2) &= \tfrac{\alpha_s(\mu)}{2\pi} \ln \tfrac{\mu^2}{m^2} \delta(1-x)\,,
\end{aligned} \tag{9}$$

and the functions D^1_{ij} (final state) read

$$\begin{aligned}
D^1_{cg}(z,\mu^2) &= \tfrac{\alpha_s(\mu)}{2\pi} P^{(0)}_{cg}(z) \ln \tfrac{\mu^2}{m^2}\,, \\
D^1_{cc}(z,\mu^2) &= C_F \tfrac{\alpha_s(\mu)}{2\pi} \left[\tfrac{1+z^2}{1-z}(\ln \tfrac{\mu^2}{m^2} - 2\ln(1-z) - 1)\right]_+ \,.
\end{aligned} \tag{10}$$

The function $D^1_{cc}(z,\mu^2)$ has been derived for the first time from the process $\gamma^* \to c\bar{c}g$ with a timelike photon [13] and has also been found by taking the massless limit of the process $\gamma^* c \to cg$ with a spacelike photon [14] and also in Ref. 2 for the process $\gamma\gamma \to c\bar{c}g$ demonstrating the universality of this function. All other functions not listed here are zero to this order in the coupling constant α_s. Furthermore, analogous results for processes involving photon splittings can be found by some obvious replacements ($g \to \gamma$, $\alpha_s \to \alpha$ and appropriate modifications of colour factors) in Eqs. (9) and (10).

4.2. Implementation Freedom and Adopted Procedure

Even fixing the factorization scheme to $\overline{\text{MS}}$ leaves some freedom in the implementation of a massive VFNS as has been discussed for the case of deep inelastic scattering (DIS) in Ref. 9. This can be most easily seen from the condition in Eq. (7) which fixes the subtraction term $d\sigma_{\text{SUB}}(\tfrac{\mu}{m}, \tfrac{m}{p_T})$ only up to terms $\tfrac{m}{p_T}$ vanishing in the limit $m \to 0$. The precise treatment of such terms $\tfrac{m}{p_T}$ is not prescribed by factorization.

In our approach to open charm photoproduction the following procedure has been adopted [2,3]:

- Derive the $m \to 0$ limit of the massive 3-FFNS calculation only keeping m as regulator in logarithms $\ln \frac{m^2}{s}$. Here special care is required in order to recover certain distributions ($\delta(1-w)$, $(\frac{1}{1-w})_+$, $(\frac{\ln(1-w)}{1-w})_+$) occurring in the massless $\overline{\text{MS}}$ calculation.
- Compare the massless limit with the corresponding contribution in the (massless) $\overline{\text{MS}}$ calculation in order to identify the subtraction terms by the unique prescription

$$\mathrm{d}\sigma_{\text{SUB}} = \left(\lim_{m \to 0} \mathrm{d}\sigma(m)\right) - \mathrm{d}\hat{\sigma}(\overline{\text{MS}}) . \quad (11)$$

By this procedure no mass terms are removed from the partonic cross section apart from the collinear logarithms $\ln \frac{m^2}{s}$.
- Include contributions with charm quarks in the *initial state* in the massless approach [15]. This rule is of great practical importance and also has been employed in the recent work [16] of the CTEQ collaboration.

5. Summary

In this talk I have discussed theoretical issues relevant for heavy flavour photoproduction, $\gamma p \to D^\star X$, in a massive VFNS. A first study including mass effects for the direct contribution has been performed in Ref. 4. These results indicate an improved description of HERA data at small transverse momenta of the $D^\star$ meson. A calculation of the resolved contribution including mass effects is in progress allowing for a complete NLO analysis of open heavy flavour production in a massive VFNS rigorously based on the factorization theorem with massive partons by John Collins [1] which is the theoretical basis to incorporate information on the FFs into $D^\star$ mesons from other reactions, as for example $e^+e^- \to D^\star X$.

Acknowledgments

I am grateful to the organizers for the kind invitation to this very well organized workshop at such an inspiring location and to B. A. Kniehl, G. Kramer and H. Spiesberger for their collaboration.

References

1. J. C. Collins, Phys. Rev. **D58**, 094002 (1998).
2. G. Kramer and H. Spiesberger, Eur. Phys. J. **C22**, 289 (2001).
3. G. Kramer and H. Spiesberger, Eur. Phys. J. **C28**, 495 (2003).

4. G. Kramer and H. Spiesberger, *Inclusive photoproduction of D^* mesons with massive charm quarks*, hep-ph/0311062.
5. B. A. Kniehl, G. Kramer, I. Schienbein, and H. Spiesberger, work in progress.
6. J. Binnewies, B. A. Kniehl, and G. Kramer, Phys. Rev. **D58**, 014014 (1998), and references therein.
7. M. Cacciari, S. Frixione, and P. Nason, JHEP **03**, 006 (2001).
8. S. Frixione and P. Nason, JHEP **03**, 053 (2002).
9. W.-K. Tung, S. Kretzer, and C. Schmidt, J. Phys. **G28**, 983 (2002).
10. S. Kretzer, Phys. Lett. **B471**, 227 (1999).
11. S. Frixione, M. L. Mangano, P. Nason, and G. Ridolfi, Phys. Lett. **B348**, 633 (1995).
12. S. Frixione, P. Nason, and G. Ridolfi, Nucl. Phys. **B454**, 3 (1995).
13. B. Mele and P. Nason, Nucl. Phys. **B361**, 626 (1991).
14. S. Kretzer and I. Schienbein, Phys. Rev. **D59**, 054004 (1999).
15. M. Krämer, F. I. Olness, and D. E. Soper, Phys. Rev. **D62**, 096007 (2000).
16. S. Kretzer, H. L. Lai, F. I. Olness, and W. K. Tung, *CTEQ6 parton distributions with heavy quark mass effects*, hep-ph/0307022.

EXPERIMENTAL RESULTS ON EXCLUSIVE VECTOR MESONS AND HEAVY QUARKONIUM

P. THOMPSON

School of Physics and Astronomy,
University of Birmingham,
B15 2TT, UK.
E-mail: pdt@hep.ph.bham.ac.uk

This contribution reviews recent experimental results on the production of light and heavy vector mesons from the H1 and ZEUS collaborations during HERA-I data taking. The data are compared in detail with the predictions of theoretical models based on perturbative QCD.

1. Introduction

The study of vector meson production at HERA provides an experimentally clean process which allows the nature of the strong interaction at high energy to be investigated. HERA is a unique facility which allows simultaneous control of the different kinematic scales: the squared mass of the vector meson, $M_{\rm VM}^2$, the virtuality of the exchanged photon, Q^2, and the four-momentum transferred at the proton vertex, t. The studies at HERA involve identifying transitions from long range, or soft, to short distance, or hard, behaviour as a function of the various scales, in order to test the applicability of perturbative QCD (pQCD). It is hoped the study of hard diffraction processes will lead to a better understanding of the vacuum-exchange and the strong interaction.

The kinematics for exclusive, or diffractive, vector meson production $ep \rightarrow e({\rm VM})Y$ are described in terms of the ep centre-of-mass-energy squared $s = (k+p)^2$, the virtuality of the photon $Q^2 = -q^2 = -(k-k')^2$, the square of the centre-of-mass energy of the initial photon-proton system $W^2 = (q+p)^2$ and the four-momentum transfer squared $t = (p-p_Y)^2$. Here k (k') is the four-momentum of the incident (scattered) lepton and q is the four-momentum of the virtual photon. The four-momentum of the incident proton is denoted by p and p_Y is the four-momentum of the system Y, which represents either an elastically scattered proton or a dissociated proton system.

2. Vector Meson Dominance and Regge Theory

In the Vector Meson Dominance (VMD) model the photon is assumed to fluctuate into a vector meson at a large distance before the target with the vector meson subsequently undergoing a soft scattering from the proton target. In the model, the vector meson retains the helicity of the photon and, therefore, s-channel helicity conservation (SCHC) is satisfied. The VMD model can be combined with Regge phenomenology, which has been successful in parameterising soft hadronic elastic and total cross sections. At high energy, in Regge theory, the cross section is dominated by the "soft pomeron" trajectory with an intercept $\alpha_{I\!P}(0) = 1.08$ and a slope $\alpha'_{I\!P}(t) = 0.25$ GeV^2 [1]. The intercept describes the experimentally observed weak energy dependence $\sim W^{0.2}$ of the total and elastic hadron-hadron cross sections using $\sigma_{tot} \propto W^\delta$ where $\delta = 2(\alpha_{I\!P}(0) - 1)$. The soft pomeron is characterised by scattering at small angles and exhibits an exponential t dependence $d\sigma/dt \propto e^{b(W)t}$, where $b(W)$ is the slope parameter. The foward scattering shows a logarithmic shrinkage with increasing W given by $b = b_0 + 4\alpha'_{I\!P} \log(W/W_0)$.

3. The Dipole Picture and QCD

An increasingly popular approach to describe diffractive interactions is to consider the scattering of $q\bar{q}$ fluctuations of the virtual photon, as colour dipoles scattering off the proton target in the proton rest frame. The probability of the photon to fluctuate into a $q\bar{q}$ pair may be parameterised in QCD using the vector meson wave function. The interaction of the dipole with the proton is described at lowest order by the exchange of two gluons in a colour singlet state. In the leading logarithmic (LL) approximation, this process is described by the effective exchange of a gluonic ladder and the cross section is related to the gluon density within the proton. In QCD, the transverse separation of the dipole is given by $r \sim \frac{1}{(z(1-z)Q^2 + m_q^2)^{1/2}}$, where m_q is the mass of the quark and z is the longitudinal momentum fraction of the photon carried by the quark. Therefore, at large Q^2 or large m_q^2 the dipole may resolve short distances within the proton suggesting that perturbative QCD may be applicable. In contrast, at small Q^2 and small m_q^2 the dipole will be large and the process is likely to contain a significant non-perturbative component.

There are many different approaches combining the dipole approach and pQCD [2,3,4,5,6]. However, the majority of the models have common features, and include the following. A fast rise of the cross section with W ($\sigma \sim W^{0.8}$) due to the rise of the proton gluon density at low x. At asymptotically large Q^2, the cross section for longitudinally polarised photons is expected to dominate. The Q^2 dependence in the longitudinally polarised cross section is expected to be slower than $1/Q^6$ due to the effect of the gluon density. The transverse cross section is expected to have a $1/Q^8$ dependence, although endpoint effects in the wavefunctions of light vector mesons mean that there may be large non-perturbative contributions. At low $|t|$, the $|t|$-dependence is expected to be universal and the two gluon form-factor leads to $\mathrm{d}\sigma/\mathrm{d}t \propto e^{-4|t|}$, with little or no shrinkage expected.

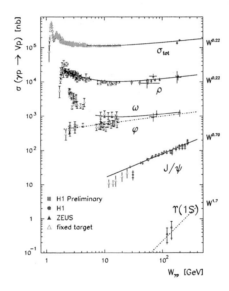

Figure 1. The cross section versus W of the elastic photoproduction of ρ, ω, ϕ, J/ψ and Υ vector mesons.

4. Elastic Vector Meson Photoproduction and Energy Dependence

In figure 1 the cross section for vector meson photoproduction is shown as a function of the centre-of-mass energy W, integrated over low values of $|t|$. The light vector mesons show an energy dependence compatible with the soft pomeron, whereas the J/ψ shows a steeper behaviour $\sim W^{0.7}$. The J/ψ

energy dependence is qualitatively described by pQCD models [4,5] which use the mass of the J/ψ as a perturbative scale. The J/ψ data may be used to study the W dependence at different values of t, in order to measure the pomeron trajectory for the process. The intercept and the slope are found to be $\alpha_{\mathbb{P}}(0) \sim 1.2$ and $\alpha'_{\mathbb{P}} \sim 0.1$ GeV^2 [7,8], respectively, incompatible with the trajectory of the soft pomeron derived from hadron-hadron scattering.

5. Energy Dependence in Elastic Electroproduction

The variation of the energy dependence with Q^2 has been studied with increasing accuracy for the ρ meson [9,10]. The W dependence is found to become steeper with increasing Q^2 and tends toward the same value as J/ψ photoproduction, indicating that Q^2 is a possible scale for the hard process. As yet, there are no quantitative models to which the data can be compared. The variation of the W dependence with Q^2 has also been studied in J/ψ production [11]. The energy dependence is found to be similar at all values of Q^2. The J/ψ data are qualitatively described by pQCD models in which the steep energy dependence arises from the gluon density of the proton [4,6]. The pomeron trajectory for J/ψ production was measured with increasing precision at $\langle Q^2 \rangle = 6.8$ GeV^2, and is found to be compatible with that measured in photoproduction.

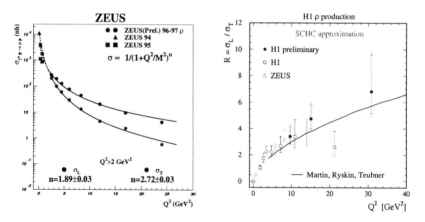

Figure 2. (left) The cross sections σ_L and σ_T and (right) the ratio $R = \sigma_L/\sigma_T$ as a function of Q^2 for elastic ρ electroproduction

6. Q^2 Dependence in Elastic Electroproduction

In pQCD, the cross sections for longitudinally and transversely polarised photons are expected to have different Q^2 dependences. The Q^2 dependence of the two components has been studied in ρ production [9,10]. In figure 2, the data are parameterised according to $1/(1+Q^2/M^2)^n$. For the longitudinal cross section a value of $n = 1.89 \pm 0.03$ is found which is consistent with the prediction of pQCD. The parameterisation is found to be inadequate for $Q^2 < 5$ GeV^2, which is thought to be attributable to wave function effects. The ratio $R = \sigma_L/\sigma_T$ is also shown in the figure. The increase of the ratio with Q^2 is compatible with calculations [4] based on wave particle duality using the details of the wavefunction. The Q^2 dependence of the cross section in J/ψ production has also been studied [11], although the determination of R, which is obtained from helicity frame measurements, is not possible due to limited statistics. The functional form $(Q^2 + M^2)^{-n}$ is found to fit the J/ψ data, although the χ^2 of the fit improves considerably as the minimum Q^2 of the fit is increased. Models based on pQCD are able to give a reasonable description of the J/ψ data.

7. t-dependence at low t

The t-dependence at low $|t|$ is found to be well described by an exponential form $\sim e^{bt}$. The b-slope of the ρ decreases with increasing Q^2, whereas the J/ψ b-slope is constant with Q^2. At large Q^2 or large M_{VM}^2 the b-slope has the value $b \sim 4.5$ GeV^2, suggestive of an underlying t-dependence which is a feature of pQCD models.

8. Helicity Measurements

The measurement of helicity angles gives information on the spin density matrix elements [12], which in turn are related to the helicity transition amplitudes. Precise measurements for ρ and ϕ elecroproduction [13] have established small violations of SCHC in r_{00}^5 indicating that the dominant single helicity flip amplitude is T_{01}. The data are well described by a pQCD model [14] based on two gluon exchange in which the SCHC violation is attainable if the longitudinal momentum of the photon is shared asymmetrically by the $q\bar{q}$ pair. The measurements have been extended [15] as a function of $|t|$ in events where the final state proton dissociates into a low mass state. In figure 3, combinations of spin-density matrix elements show increasing SCHC violations with increasing $|t|$, which are well described by the pQCD model. The spin density matrix element dependent on R ($R = \sigma_L/\sigma_T$) is observed to be constant in the t range studied, indicating that the b-slopes for σ_L and σ_T in ρ proton dissociation are similar. This is suggestive that the non-perturbative contribution to σ_T is small. This is further supported by the measurement of elastic ρ electroproduction [9] in which R was found to be constant as a function of W.

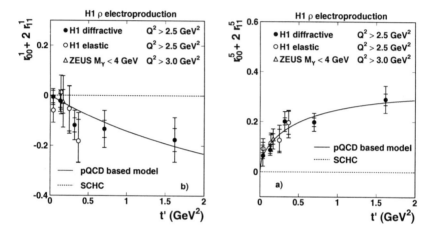

Figure 3. Measurements of (left) $r^1_{00} + 2r^1_{11}$ and (right) $r^5_{00} + 2r^5_{11}$ as a function of t.

9. Vector Meson Production at Large t

The diffractive photoproduction of vector mesons with large negative momentum transfer squared t at the proton vertex is a powerful means to probe the parton dynamics of the diffractive exchange. The variable t provides a relevant scale to investigate the application of pQCD. At sufficiently low values of Bjorken x (i.e. large values of the centre-of-mass energy W), the gluon ladder is expected to include contributions from BFKL evolution [16], as well as from standard DGLAP evolution [17].

Perturbative QCD models for the photoproduction of J/ψ mesons have been developed in the leading logarithmic approximation using either BFKL [18,19,20,21] or DGLAP [22] evolution. In the pQCD models, a non-relativistic approximation [23] for the J/ψ wavefunction is used in which the longitudinal momentum of the vector meson is shared equally between the quark and the anti-quark. In this approximation, the vector meson retains the helicity of the photon such that s-channel helicity conservation (SCHC) is satisfied [24].

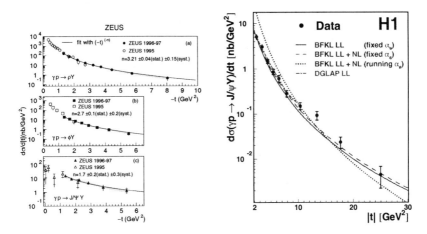

Figure 4. The cross section $d\sigma/dt$ for (left) ρ, ϕ and J/ψ photoproduction from ZEUS and (right) J/ψ photoproduction from H1.

In figure 4, the large $|t|$ cross sections for ρ, ϕ and J/ψ [25,26] are shown. The data are described with a power law $d\sigma/dt \propto |t|^{-n}$, although the value of n in each case increases with the starting value of $|t|$ used in the fit. In figure 4 (right) the data are compared with the predictions from pQCD calculations in the BFKL leading logarithmic approximation [21] (solid curve), including non-leading corrections with fixed α_s [21] (dashed curve) and including non-leading corrections with running α_s [21] (dotted curve). The t dependence and normalisation of the data are well described by the BFKL LL approximation. The inclusion of NL corrections with a fixed strong coupling α_s leads to only a small difference with respect to the LL prediction. However, with a running α_s the t dependence becomes steeper and the prediction is unable to describe the data across the whole t range. The data are also well described by calculations in the DGLAP LL approximation [22] (dashed-dotted curve) in the region of validity for the model $|t| < M^2_{J/\psi}$.

In figure 5, the cross section is plotted as a function of W for different intervals of t, and also as a function of t in two different W regions. As can be seen from the plots, the W dependence is constant with t, and, the t dependence is constant with W. The LL BFKL model predicts a

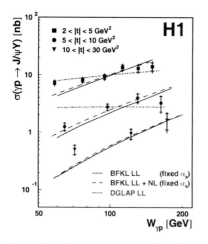 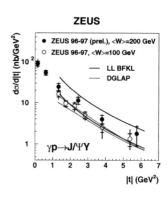

Figure 5. The cross section for J/ψ photoproduction (left) as a function of W in t-intervals (right) as a function of t at two different values of W.

stronger rise with W than is observed in the data, whereas the LL DGLAP model gives a flatter W dependence than the data. The cross section in W may be used to extract the pomeron trajectory assuming a linear form $\alpha_{\mathbb{P}}(t) = \alpha_{\mathbb{P}}(0) + \alpha'_{\mathbb{P}} t$, a fit to the three $\alpha_{\mathbb{P}}$ values yields a slope of $\alpha'_{\mathbb{P}} = -0.0135 \pm 0.0074$ (stat.) ± 0.0051 (syst.) GeV^{-2} with an intercept of $\alpha_{\mathbb{P}}(0) = 1.167 \pm 0.048$ (stat.) ± 0.024 (syst.). The value of the slope parameter α' is lower than that observed for the elastic photoproduction of J/ψ mesons at low $|t|$ [7]. It is also significantly different from the observations at low $|t|$ in hadron-hadron scattering.

To obtain information about the helicity structure of the interaction, the spin density matrix elements are extracted and shown in figure 6 as a function of $|t|$. In contrast to the ρ^0 meson, the measured spin density matrix elements of the J/ψ meson are all compatible with zero, within experimental errors, and are thus compatible with SCHC. The J/ψ results are therefore consistent with the longitudinal momentum of the photon being shared symmetrically between the heavy quarks. Hence, the approximations made in the pQCD models [18,19,20,22,21] for the J/ψ wavefunction are satisfactory for the present data.

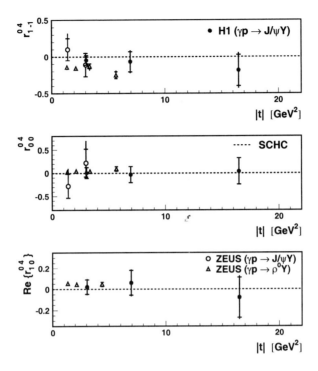

Figure 6. The three spin density matrix elements (top) r^{04}_{1-1}, (middle) r^{04}_{00} and (bottom) $\text{Re}\{r^{04}_{10}\}$ for the J/ψ as a function of $|t|$. The dashed line shows the expectation from SCHC. The results for the photoproduction of J/ψ and ρ^0 mesons are also shown.

10. Summary

The exclusive production of vector mesons at HERA allows the dynamics of vacuum-exchange processes to be studied. The transition from soft to hard behaviour is observed with increasing Q^2 or M^2_{VM}. The analysis of helicity structure provides unique insight into the structure of the vector meson wavefunctions. The production of vector mesons at high $|t|$ shows a hard behaviour in $|t|$ and W suggesting that processes beyond DGLAP may be necessary to describe the data. In general, pQCD has been able to provide qualitative descriptions of the experimental data and the second phase of HERA running will continue to improve our understanding of the strong interaction.

References

1. A. Donnachie and P. V. Landshoff, Nucl. Phys. B **231** (1984) 189.
2. S. J. Brodsky et al, Phys. Rev **50** (1994) 3134 [hep-ex/9402283].
3. J. C. Collins, L. Frankfurt and M. Strikman, Phys. Rev **56** (1997) 2982.
4. A. D. Martin, M. G. Ryskin and T. Teubner, Phys. Rev **55** (1997) 4329.
5. J. C. Collins, L. Frankfurt M. McDermott and M. Strikman, JHEP **103** (2001) 45.
6. L. Frankfurt W. Koepf and M. Strikman, Phys. Rev D **57** (1998) 512.
7. ZEUS Collaboration, S. Chekanov et al., Eur. Phys. J. C **24** (2002) 345.
8. H1 Collaboration, abstract 108, paper to EPS Conference 2003, Aachen.
9. ZEUS Collaboration, abstract 594, paper to EPS Conference 2001, Budapest.
10. H1 Collaboration, abstract 092, paper to EPS Conference 2003, Aachen.
11. ZEUS Collaboration, abstract 813, paper to ICHEP Conference 2002, Amsterdam.
12. K. Schilling and G. Wolf, Nucl. Phys. B **61** (1973) 381.
13. H1 Collab., C. Adloff et al., Eur.Phys.J. C13 (2000) 371;
    H1 Collab., C. Adloff et al., Phys. Lett. B483 (2000) 360.
14. D.Yu. Ivanov, R. Kirshner, Phys. Rev **D58** (1998) 114026 [hep-ph/9907324].
15. H1 Collab., C. Adloff et al., Phys. Lett. B539 (2002) 25.
16. E. A. Kuraev, L. N. Lipatov and V. S. Fadin, Sov. Phys. JETP **44** (1976) 443 [Zh. Eksp. Teor. Fiz. **71** (1976) 840];
    I. I. Balitsky and L. N. Lipatov, Sov. J. Nucl. Phys. **28** (1978) 822 [Yad. Fiz. **28** (1978) 1597].
17. V. N. Gribov and L. N. Lipatov, Yad. Fiz. **15** (1972) 781 and 1218 [Sov. J. Nucl. Phys. **15** (1972) 438 and 675];
    G. Altarelli and G. Parisi, Nucl. Phys. B **126** (1977) 298.
18. J. R. Forshaw and M. G. Ryskin, Z. Phys. C **68** (1995) 137.
19. J. Bartels, J. R. Forshaw, H. Lotter and M. Wüsthoff, Phys. Lett. B **375** (1996) 301.
20. J. R. Forshaw and G. Poludniowski, Eur. Phys. J. C **26** (2003) 411.
21. R. Enberg, L. Motyka and G. Poludniowski, Eur. Phys. J. C **26** (2002) 219.
22. E. Gotsman, E. Levin, U. Maor and E. Naftali, Phys. Lett. B **532** (2002) 37.
23. M. G. Ryskin, Z. Phys. C **57** (1993) 89.
24. E. V. Kuraev, N. N. Nikolaev and B. G. Zakharov, JETP Lett. **68** (1998) 696 [Pisma Zh. Eksp. Teor. Fiz. **68** (1998) 667].
25. ZEUS Collaboration, S. Chekanov et al., Eur. Phys. J. C **26** (2003) 389.
26. H1 Collab., A. Aktas et al., Phys. Lett. B **568** (2003) 205.

5
Elastic and Diffractive *ep* Scattering

DIFFRACTIVE STRUCTURE FUNCTIONS AND QCD FITS

S. LEVONIAN
DESY,
Notkestrasse 85,
22607 Hamburg, Germany
E-mail: levonian@mail.desy.de

Recent progress is reviewed in our understanding of inclusive diffraction in the deeply inelastic regime at HERA. New precision measurements are now available from both H1 and ZEUS collaborations. A new set of diffractive parton distribution functions, determined from the H1 data, is presented. It can be used to test factorization properties of diffractive scattering in different reactions.

1. Introduction

Significant progress has been achieved over the last decade in understanding the nature of *diffractive phenomena* at high energies. This is to a large extent due to the electron-proton collider HERA, which is an ideal place to study hard diffraction in deep-inelastic scattering (DIS). Already the term 'deep-inelastic diffraction' itself sounds as a paradox and goes against our intuition, based upon the experience of hadronic interactions in which diffraction was known to be a predominantly soft peripheral process. HERA for the first time offers the possibility for the partonic structure of colour singlet exchange to be probed.

At the hadronic level diffraction is best described in the framework of Regge formalism[1,2] as a t-channel exchange of a leading trajectory with the vacuum quantum numbers, named *Pomeron*. The specific interest in diffraction as 'physics of the Pomeron' is related to the fact that Pomeron exchange asymptotically dominates over all other contributions to the scattering amplitude, and thus represents the essence of strong interactions in the high energy limit. Translated into modern partonic language this reveals that colourless exchange is important in the low-x regime (which is the high energy limit of QCD) where gluons are expected to dominate. Since the diffractive cross section, being proportional to the gluon density squared, rises with energy faster than the total cross section, unitarity correction effects, e.g. in the form of gluon saturation, are expected to be first seen in diffraction. Hence an interesting question is: can this be observed already at HERA?

Another aspect in which diffraction may play a key rôle is in the interplay between soft and hard processes. Since both short distance and long distance physics contribute to diffractive DIS, HERA has a high potential to provide new insight into non-perturbative QCD phenomena, as well as to bridge the soft (Regge) and hard (pQCD) domains.

1.1. Kinematics of Diffractive DIS at HERA

Figure 1 sketches the generic diffractive process in electron-proton scattering. In addition to the usual DIS kinematic variables: photon virtuality, $Q^2 = -q^2$, Bjorken-x, $x = -q^2/2P \cdot q$ and inelasticity $y = Q^2/sx$ this process is characterised by the 4-momentum squared transferred at the proton vertex, $t = (P-p_Y)^2$ and the longitudinal momentum fraction of the colour singlet exchange relative to the incoming proton, $x_{I\!P} = q \cdot (P - p_Y)/q \cdot P$. Here P and q are the 4-momenta of the interacting proton and photon respectively, and s is the ep centre of mass energy squared. The ratio of Bjorken-x to $x_{I\!P}$ defines another diffractive variable, $\beta = x/x_{I\!P}$, which can be interpreted as the longitudinal momentum fraction of the exchange that is carried by the struck quark.

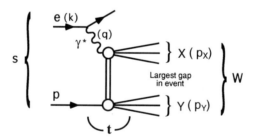

Figure 1. The generic diffractive process $ep \to eXY$ at HERA, in which the interaction is mediated by colour singlet exchange between the photon and proton vertecies, leading to a large rapidity interval separating final state systems X and Y.

After integrating over M_Y, which is not well measured at HERA, a fourfold-differential diffractive reduced cross section, $\sigma_r^{D(4)}$, can be defined through

$$\frac{d^4\sigma^D}{dx_{I\!P}\, dt\, d\beta\, dQ^2} = \frac{4\pi\alpha^2}{\beta Q^4}\left(1 - y + \frac{y^2}{2}\right)\sigma_r^{D(4)}(x_{I\!P}, t, \beta, Q^2), \quad (1)$$

which is related to the diffractive structure functions F_2^D and F_L^D by

$$\sigma_r^D = F_2^D - \frac{y^2}{1+(1-y)^2} F_L^D. \tag{2}$$

To a very good approximation $\sigma_r^D = F_2^D$ everywhere except possibly at the highest values of y. Furthermore, if the outgoing proton is not detected, the measurements integrate over t: $\sigma_r^{D(3)} = \int \sigma_r^{D(4)} dt$.

1.2. Understanding Diffraction in DIS

The variety of phenomenological models for diffractive DIS can be classified into three main categories.

1. Regge motivated models with a factorizable Pomeron[3]. The internal partonic structure of such an object can be resolved in DIS. Regge factorization means that diffractive cross sections can be factorized into a Pomeron flux and its structure function. More generally, sub-leading reggeon contributions, which are essential at high values of $x_{I\!P}$, should also be taken into account:

$$F_2^{D(4)}(x_{I\!P}, t, \beta, Q^2) = f_{I\!P}(x_{I\!P}, t) F_2^{I\!P}(\beta, Q^2) + f_{I\!R}(x_{I\!P}, t) F_2^{I\!R}(\beta, Q^2) \tag{3}$$

with flux factors defined as

$$f_i(x_{I\!P}, t) = x_{I\!P}^{1-2\alpha_i(t)} e^{bt}, \tag{4}$$

where $\alpha_i(t) = \alpha_i(0) + \alpha_i' t$ ($i = I\!P, I\!R$) are Pomeron and reggeon trajectories respectively with intercept $\alpha_i(0)$ and slope α_i'.

2. Non-Pomeron mechanisms generating large rapidity gap events. For example, in the soft colour interaction model[4] diffraction occurs through the normal boson-gluon fusion with additional emission of soft gluons, neutralizing the colour flow. As a consequence the model predicts leading gluon behaviour in which one gluon takes most of the momentum of the colour singlet exchange.

3. Colour dipole approach[5] to DIS in which the interaction is viewed in the proton rest frame. This is especially convenient at low x, in which case the virtual photon chooses a specific Fock state to fluctuate into long before arriving at the proton target ($c\tau \approx 1/xM_p \simeq 1000$ fm at HERA). Hence the resulting cross section can be represented as a simple convolution of the photon wave function $|\gamma\rangle = \alpha_1 |q\bar{q}\rangle + \alpha_2 |q\bar{q}g\rangle + ...$ with the colour dipole cross section ($\sigma_{(q\bar{q})p}$, or $\sigma_{(q\bar{q}g)p}$). The models of this class[6] differ in the way they parameterise dipole cross sections. Their common strong feature is related to the fact, that once fitted to the inclusive DIS data, the

same dipole cross sections with only one additional free parameter, related to the t-dependence, should also describe diffractive phenomena in DIS.

Alternatively, one can employ a model independent approach using the concept of *diffractive parton distributions*, f_i^D. It is based on the rigorous proof[7] that QCD hard scattering factorization is valid for diffractive DIS, and hence the diffractive cross section (1) at fixed values of $x_{I\!P}$ and t can be expressed as a convolution of universal partonic cross sections $\hat{\sigma}^{\gamma^* i}$ with f_i^D:

$$\sigma_r^D(4) \propto \sum_i \hat{\sigma}^{\gamma^* i}(x, Q^2) \otimes f_i^D(x, Q^2; x_{I\!P}, t). \qquad (5)$$

The partonic cross sections $\hat{\sigma}^{\gamma^* i}$ are the same as in inclusive DIS, and f_i^D, which are not known from first principles, should obey the DGLAP evolution equations. Hence the standard framework of NLO QCD can be applied to diffractive DIS in a similar manner as to inclusive DIS.

2. Experimental Techniques and data Samples

Several experimental techniques are used to measure diffractive processes at HERA. They have different advantages and drawbacks and are complementary to each other.

The clearest way is to detect the scattered proton in the forward direction using so called *Roman pot* technique - several detectors inserted in the beam pipe which together with machine magnets form a proton spectrometer. Such a method was exploited by both H1 and ZEUS experiments. It provides a diffractive data sample free of proton dissociative admixture and in addition enables to measure t, which in other methods is not directly measured and has to be integrated over. The method however suffers from low statistics, due to limited acceptance and the dependence on beam conditions, which do not allow full luminosity to be collected.

To obtain high statistics sample one can use characteristic properties of the hadronic final state, as shown in Fig. 1. H1 and ZEUS follow slightly different approaches trying to fully utilize their detector capabilities.

The method iof H1 is based on the requirement of a large rapidity gap separating the leading baryon system Y from the photon dissociation system X. The rapidity gap is positively identified by the absence of activity in the pseudorapidity interval $3.2 < \eta < 7.5$, covered by various detector components. In this approach non-diffractive background is small and can be estimated using Monte Carlo simulation.

The extraction of the diffractive contribution in ZEUS is performed using the so called M_X method, which is based on the fact that diffractive and non-diffractive final states have very different $\ln M_X^2$ distributions. The non-diffractive contribution, exhibiting an exponential fall-off towards lower M_X values, is extrapolated underneath the diffractive plateau and statistically subtracted from the total M_X distribution.

Diffractive samples selected with the latter two methods are dominated by the single dissociative process $ep \to eXp$ with a small admixture of double dissociation in which the low mass proton dissociation system M_Y escapes in the beam hole undetected. Hence the measured cross sections are corrected to the region $M_Y < 1.6$ GeV and $|t| < 1$ GeV^2 in case of H1 and to $M_Y < 2.3$ GeV in case of ZEUS.

Both collaborations have released recently several new measurements with a 2- to 5-fold increase in statistics and a significant extension in the covered phase space as compared to previously published data[8].

New measurements of the diffractive reduced cross section in the range $1.5 < Q^2 < 120$ GeV^2 by the H1 collaboration is presented in Fig. 2. Two data samples[9,10] based on the rapidity gap technique are in a good agreement with each other and with a third measurement[11], covering $2 < Q^2 < 50$ GeV^2, in which the leading proton is detected in the forward spectrometer.

Inclusive diffractive cross sections as measured by the ZEUS collaboration[12] are shown in Fig. 3. With the exception of the lowest M_X bin a strong rise with energy, W, is observed. The data[13] in the transition region between DIS and quasi-real photon-proton interactions (see Fig. 4) exhibit a behaviour very similar to that of the total photon-proton cross section[14]. Namely, the diffractive cross section falls rapidly with Q^2 at high virtualities, while for $Q^2 \to 0$ the cross section dependence on Q^2 flattens off. This behaviour is broadly described by the colour dipole model[15] in which the virtual photon fluctuates into $q\bar{q}$ and $q\bar{q}g$ states.

3. General Properties of Inclusive Diffraction at HERA

To compare the dynamics of diffractive DIS with those of inclusive DIS the ratio of the diffractive to the total inclusive cross sections is studied as a function of energy and Q^2. Fig. 5 represents just two of several such measurements. The ratio is observed to be everywhere remarkably flat, both in Q^2 and in W, except at high β (low M_X). The latter can be readily explained by simple kinematics (suppression of available phase space

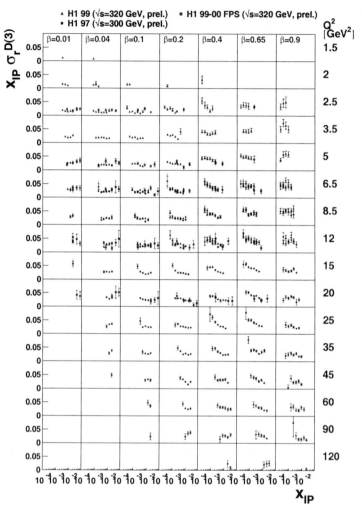

Figure 2. The diffractive reduced cross section as measured in three different samples and with different techniques is shown as a function of $x_{I\!P}$ in bins of β and Q^2.

for gluon radiation close to the kinematic limit), or by the onset of higher twist effects. It is fair to note that such behaviour is naturally expected in models of the 2-nd class (see Sec. 1.2), in which rapidity gap formation is due to purely probabilistic mechanisms.

Approximately flat ratio as a function of W implies the same energy dependence for diffractive and inclusive $\gamma^* p$ cross section, which is not expected neither in the Regge picture, nor in a simple two-gluon exchange

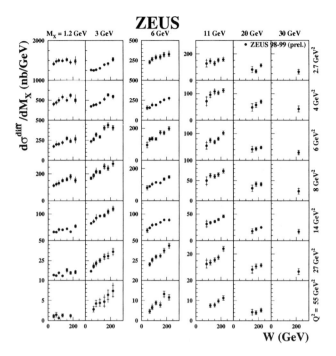

Figure 3. The differential cross section $d\sigma^{\text{diff}}_{\gamma^*p \to XN}/dM_X$, $M_N < 2.3$ GeV, as a function of W in bins of M_X and Q^2.

QCD model for diffractive DIS.

To quantify the difference between the energy dependence of σ^D and σ^{tot} one can compare the values of the Pomeron intercept, $\alpha_{I\!P}(0)$, extracted using Regge motivated fits from diffractive and inclusive DIS data. In the inclusive case, the proton structure function can be parameterised at low $x < 0.01$ as $F_2(x, Q^2) = cx^{-\lambda(Q^2)}$ with $\lambda = \alpha_{I\!P}(0) - 1$. In diffractive DIS $\alpha_{I\!P}(0)$ can be obtained, after correcting for the finite values of t, from the W-, or equivalently from the $x_{I\!P}$-dependence of the diffractive cross section.

The results are summarized in Fig. 6. $\alpha_{I\!P}(0)$ is observed to rise with Q^2, although the error bars are still large in case of diffraction. In contrast to photoproduction, where the same universal 'soft' Pomeron trajectory describes both inclusive and diffractive data, in the DIS regime the Pomeron intercept extracted in diffraction is approximately half of that obtained in inclusive data. This is consistent with the above conclusion that the energy dependence is very similar in diffractive and inclusive DIS. This surprising result is not so easy to explain, unless some sort of unitarity corrections are

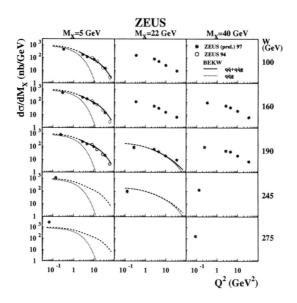

Figure 4. Q^2 dependence of the diffractive cross section $d\sigma^D_{\gamma^*p}/dM_X$ at different values of W and M_X. The solid (dashed) lines represent results of a specific dipole model fit[16] (and its extrapolation). The $q\bar{q}g$ contribution is shown by the dotted curves.

assumed to play a significant rôle in diffraction at HERA. Alternatively, it could point to the possibility that different proportions of soft and hard physics contribute to diffractive processes as compared to fully inclusive ones, suggesting that Q^2 is not completely equivalent or the only relevant scale in those two cases.

4. NLO QCD Fit to H1 Data and Diffractive PDFs

H1 has performed[16] a new NLO DGLAP QCD fit to the medium Q^2 data sample[10] from which diffractive parton densities are obtained. To simplify the fit Regge factorization (3) was assumed, implying that the shape of pdf's is independent on $x_{I\!P}$. At the present level of precision this assumption is justified by the data, as can be seen in Fig. 7, in which the β and Q^2 dependences are compared to the fit results.

Both $I\!P$ and $I\!R$ contributions are included in the fit. The Pomeron structure is parameterised by a light flavour singlet, Σ, and a gluon, g, distribution, while for the reggeon term the *pion* pdf[17] is used. The latter choice does not significantly influence the results, as the sub-leading contribution is negligible at $x_{I\!P} < 0.01$. Diffractive parton densities (*dpdf's*) are

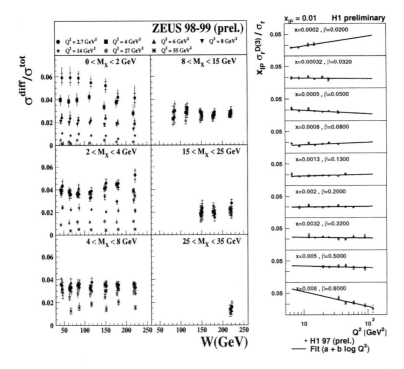

Figure 5. Ratio of diffractive to inclusive DIS cross sections as a function of W (left) and Q^2 (right).

parameterised at a starting scale $Q_0^2 = 3$ GeV^2 and evolved to higher Q^2 using NLO DGLAP equations. 313 data points from the region $Q^2 > 6.5$ GeV^2 and $M_X > 2$ GeV are used to constrain 7 free parameters (3 for each of the distributions, i.e. for the singlet and the gluon, and one for the normalisation of the sub-leading exchange at high $x_{I\!P}$). It should be emphasised that for the first time in diffraction the experimental and model uncertainties are fully propagated to obtain error bands for resulting dpdf's.

The result of the fit is shown in Fig. 8. Whereas the singlet part is well constrained, there is a substantial uncertainty in the gluon distribution at large fractional momenta z (or β), mainly due to model assumptions. The diffractive pdf's remain large up to large z and are dominated by gluons. In total $75 \pm 15\%$ of the exchanged momentum is carried by gluons. Fig. 9 presents the β and Q^2 dependences of the diffractive reduced cross section at two selected values of $x_{I\!P}$. Even extrapolated beyond the region used in the fit, the new pdf's provide a very good description of new H1 data[18]

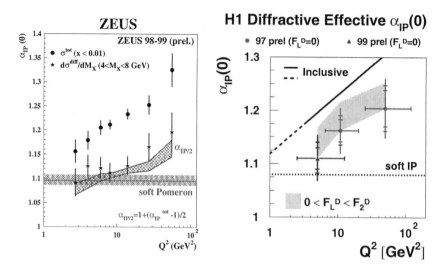

Figure 6. The intercept of the Pomeron trajectory, $\alpha_{I\!P}(0)$, as a function of Q^2 in diffractive and inclusive DIS, as determined by ZEUZ(left) and H1 (right). The cross hatched band represents 'half' of the W-rise of the total γ^*p cross section, $\alpha_{I\!P/2} = 1 + (\alpha_{I\!P}^{tot}(0) - 1)/2$.

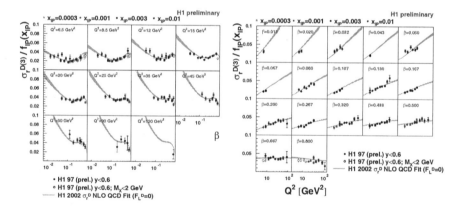

Figure 7. β (left) and Q^2 (right) dependence of the reduced cross section scaled at each $x_{I\!P}$ by the values assumed for the t-integrated Pomeron flux in the QCD fits.

at high Q^2. One can also see that at high $x_{I\!P}$ the reggeon contribution becomes important. It is worth mentioning that new precise low Q^2 data[9] have a large potential to further constrain dpdf's.

Another possible test involves comparisons with diffractive final states[19] at HERA and TEVATRON.

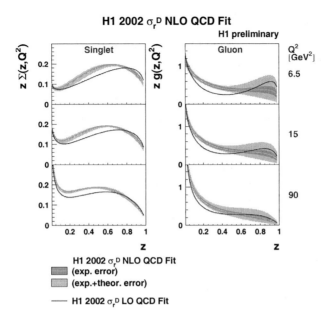

Figure 8. Diffractive parton densities obtained from the QCD fits and normalized such that the 'Pomeron flux' is unity at $x_{I\!P} = 0.003$.

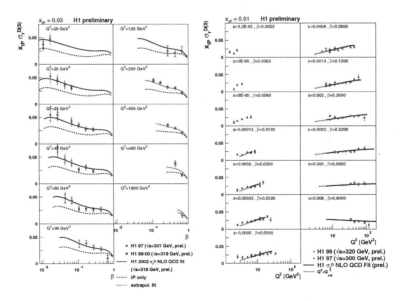

Figure 9. Diffractive reduced cross section at fixed $x_{I\!P}$ as a function of β in bins of Q^2 (left) and as a function of Q^2 in bins of β (right). Also shown is the prediction for $x_{I\!P}\sigma_r^{D(3)}$ at $\sqrt{s} = 319$ GeV from the NLO QCD fit performed to the medium Q^2 data.

5. Summary

New precision measurements of inclusive diffractive DIS are now available from HERA. These data are extremely valuable to further constrain QCD motivated phenomenological models and to distinguish between different approaches to diffractive phenomena in the DIS regime. A new generation of diffractive parton distributions together with their uncertainties have been determined from H1 data in NLO QCD framework. They can be used to test factorization properties of diffraction in different reactions. The energy dependence of the diffractive cross section has no simple explanation at the moment. It may indicate that the onset of unitarity corrections (or saturation effects) is already seen in diffraction at HERA. Alternatively, it could be a manifestation of a complicated interplay between soft and hard phenomena in diffractive DIS. In spite of the recent progress a complete understanding of the nature of colour singlet exchange remains a major challenge in QCD.

References

1. T. Regge, *Nuovo Cimento* **14**, 951 (1959); *ibid.*, **18**, 947 (1960); G. Chew and S. Frautschi, *Phys. Rev. Lett.*, **7**, 394 (1961); G. Chew, S. Frautschi and S. Mandelstam, *Phys. Rev.*, **126**, 1202 (1962).
2. V.N. Gribov, *Sov. Phys. JETP*, **26**, 414 (1968).
3. G. Ingelman and P. Schlein, *Phys. Lett.*, **B152**, 256, (1985).
4. A. Edin, G. Ingelman and J. Rathsman, *Phys. Lett.*, **B366**, 371 (1996); J. Rathsman, *Phys. Lett.*, **B452**, 364 (1999).
5. N.N. Nikolaev and B.G. Zakharov, *Z. Phys.*, **C 49**, 607 (1991); *ibid.*, **C 53**, 331 (1992); A.H. Mueller, *Nucl. Phys.*, **B 415**, 373 (1994).
6. W. Buchmüller, T. Gehrmann and A. Hebecker, *Nucl. Phys.*, **B 537**, 477 (1999); K. Golec-Biernat and M. Wüsthoff, *Phys. Rev.*,**D 59**, 014017 (1999); J. Bartels, K. Golec-Biernat and H. Kowalski, *Phys. Rev.*,**D 66**, 014001 (2002).
7. J. Collins, *Phys. Rev.*, **D 57**, 3051 (1998); err. - *ibid.* **D 61**, 019902 (2000).
8. ZEUS Coll., M. Derrick et al., *Z. Phys.* **C 70**, 391 (1966); H1 Coll., C. Adloff et al., *Z. Phys.* **C 76**, 613 (1997).
9. H1 Coll., paper 88 subm. to EPS 2003.
10. H1 Coll., paper 808 subm. to EPS 2001.
11. H1 Coll., paper 984 subm. to ICHEP 2002.
12. ZEUS Coll., paper 821 subm. to ICHEP 2002; paper 538 subm. to EPS 2003.
13. ZEUS Coll., paper 566 subm. to EPS 2001; paper 540 subm. to EPS 2003.
14. ZEUS Coll., J. Breitweg et al., *Phys. Lett.*, **B487**, 53 (2000).
15. J. Bartels et al., *Eur.Phys. J.*, **C7**, 443 (1999).
16. H1 Coll., paper 89 subm. to EPS 2003.
17. J. Owens, *Phys. Rev.*, **D 30**, 913 (1984)
18. H1 Coll., paper 90 subm. to EPS 2003.
19. N. Vlasov, these proceedings.

DIFFRACTIVE JET AND CHARM PRODUCTION

N. VLASOV

Freiburg University
Physikal Institute
Hermann-Herder st. 3, 79104 Freiburg, Germany
E-mail: vlasov@mail.desy.de

The predictions of diffractive models are compared with H1 measurements of jet production cross sections in diffractive Deep Inelastic Scattering (DIS) and photoproduction and with H1 and ZEUS measurements of $D^{*\pm}(2010)$ meson production in diffractive DIS.

1. Introduction

The diffractive process at HERA, characterized by a large rapidity gap in the distribution of the final-state hadrons, is considered to proceed through the exchange of an object carrying the quantum numbers of the vacuum, known as the Pomeron ($I\!P$). In the resolved-Pomeron model[1], the exchanged Pomeron acts as a source of partons, one of which interacts with the virtual photon. In an alternative view, the diffractive process at HERA can be described by the dissociation of the virtual photon into a $q\bar{q}$ or $q\bar{q}g$ state which interacts with the proton by the exchange of two gluons or, more generally, a gluon ladder with the quantum numbers of the vacuum [2,3,9].

Jet and charm production in diffractive Deep Inelastic Scattering (DIS), which have also been measured by the H1 and ZEUS collaborations [5,6,7], allow quantitative tests of the models due to the sensitivity of jet and charm production to gluon-initiated processes [8]. Final states containing heavy quarks or high transverse momentum jets are of particular interest, since the additional hard scales ensures the applicability of perturbative QCD.

Measurements of inclusive diffractive DIS at HERA have been used to extract the diffractive parton distributions (diffractive pdfs) of the proton. If QCD factorisation holds in diffractive DIS, these diffractive pdfs are universal and may be used to predict the cross sections for exclusive hard diffractive DIS processes such as jet and heavy flavour production.

2. Models of Diffractive Charm and Jet Production

2.1. Resolved Pomeron Model

In the resolved-Pomeron model, proposed by Ingelman and Schlein [1], the exchanged Pomeron is assumed to be a object with a partonic structure. The diffractive cross section factorises into a Pomeron flux factor, describing the probability to find a Pomeron in the proton, the Pomeron's parton density functions which specify the probability to find a given parton in the Pomeron and the interaction cross section with that parton. Within this model, open charm and jet production are produced in diffractive DIS via the boson-gluon-fusion (BGF) process, where the virtual photon interacts with a gluon from the Pomeron. The HERA measurements of the inclusive diffractive differential cross sections were found to be consistent with the resolved-Pomeron model with a Pomeron structure dominated by gluons [9,10].

A combined fit of the Pomeron parton densities to the H1 and ZEUS inclusive diffractive DIS measurements and to the ZEUS data on diffractive dijet photoproduction has been made by Alvero et al. (ACTW) [11]. The resulting diffractive parton distributions have been compared to recent ZEUS measurements of D^* meson producton in diffractive DIS [6]. Leading order (LO) and next-to-leading order (NLO) DGLAP QCD fits have been performed by the H1 Collaboration to determine diffractive parton distributions [10] which were then used for comparisons with recent H1 measurements of D^* meson [7] and dijet [5] production in diffraction. The fit results have been interfaced to the program HVQDIS [12] to calculate cross sections for diffractive charm production in DIS [13] and to the program DISENT [14] to calculate cross sections for diffractive dijet production in DIS, both leading and next-to-leading order.

2.2. Two-gluon-exchange Models

The two-gluon-exchange models consider fluctuations of the virtual photon into $q\bar{q}$ or $q\bar{q}g$ colour dipoles that interact with the proton via colour-singlet exchange, the simplest form of which is a pair of gluons [15,16,17]. The virtual-photon fluctuations into $c\bar{c}$ and $c\bar{c}g$ states can lead to diffractive open-charm production. At high $x_{I\!P}$ values, quark exchanges are expected to become significant, implying that the two-gluon-exchange calculations are expected to be valid only at low $x_{I\!P}$ values ($x_{I\!P} < 0.01$). In the calculations [3,18,19,20], used here, the cross section for two-gluon exchange is related to the square of the unintegrated gluon distribution of the proton which depends on the gluon transverse momentum, k_T, relative to the proton direction. In the "saturation" model [20,21], the calculation of the $q\bar{q}g$ cross section is performed under the assumption of strong k_T ordering of the final-state partons, which corresponds to $k_T^{(g)} \ll k_T^{(q,\bar{q})}$. The parameters of the model were tuned to describe the total photon-proton cross section measured at HERA. Alternatively, in the model of Bartels et al. [3,18,19], configurations without strong k_T ordering are included in the $q\bar{q}g$ cross-section calculation and the minimum value for the final-state-gluon transverse momentum, $k_{T,g}^{cut}$, is a free parameter which has been tuned to describe the H1 dijet cross sections [5]. The sum of the $q\bar{q}$ and $q\bar{q}g$ contributions in the saturation model and the model of Bartels et al. are hereafter also referred to as SATRAP and BJLW, respectively.

3. Diffractive Jets and Charm in DIS

In this section the recent H1 and ZEUS measurements of dijets and D^* meson production in diffractive DIS are compared with the different models.

In diffractive DIS, a photon with virtuality Q^2 emitted from the beam electron interacts with the proton, which loses only a small fraction $x_{I\!\!P}$ of its incident momentum and stays intact (or dissociates into a small mass system Y). The longitudinal momentum fraction of the parton entering the hard scattering process relative to the diffractive exchange is labelled $z_{I\!\!P}$ (equivalent to β at LO). In the hard scattering process, a pair of high transverse momentum (p_T) jets or heavy quarks is produced. The photon-proton centre-of-mass energy is W, which is related to the inelasticity y via $ys = Q^2 + W^2$, where s is the ep centre-of-mass energy squared. The invariant mass of the diffractively produced system X is M_X, and the invariant mass of the two partons emerging from the hard sub-process is given by $\sqrt{\hat{s}} = M_{12}$.

3.1. Diffractive Dijets in DIS

The H1 Collaboration has measured differential cross sections for the diffractive production of dijets in DIS [5,22]. The jets are found using the CDF cone algorithm with cone radius $R = 1$. The kinematic range of the measurement is $4 < Q^2 < 80$ GeV^2, $0.1 < y < 0.7$, $x_{I\!\!P} < 0.05$, $|t| < 1.0$ GeV^2 and $M_Y < 1.6$ GeV. The transverse momenta and pseudorapidities of the two jets, which are searched for in the hadronic centre-of-mass frame, are required to satisfy $p_{T,jet}^* > 4$ GeV and $-3 < \eta_{jet}^* < 0$, respectively.

To facilitate a comparison of the cross sections with NLO calculations, the data were corrected to the kinematic region where the transverse momentum of the first (second) jet is $p^*_{T,jet} > 5(4)$ GeV, as the next-to-leading order dijets cross sections are not reliable in regions of phase space where the two jets have the same transverse momentum [22].

To calculate diffractive dijet cross sections to NLO in QCD, the DISENT program was used. Since the calculations refer to partonic jets, whereas the measurements refer to hadronic jets, the NLO calculations have been corrected for the hadronisation effects.

In Fig. 1, the differential cross sections for dijet production in diffractive DIS are presented as functions of $z_{I\!P}^{jets}$, Q^2, W, $log_{10}x_{I\!P}$ and the average pseudorapidity $\langle \eta \rangle_{lab}^{jets}$ of the jets. The data are compared with leading and next-to-leading order calculations using DISENT interfaced to the H1 diffractive pdf's [10]. The uncertainty on the NLO calculations was estimated by varying the renormalisation scale $\mu_R^2 = p_T^2$ by factors of 1/4 and 4 and the hadronisation corrections. The NLO calculation provides within the theoretical and experimental uncertainties a reasonable description of the shapes and normalisations of the measured cross sections.

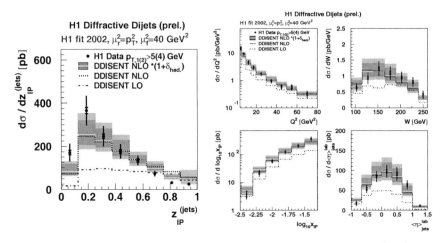

Figure 1. Differential cross sections for dijet production in diffractive DIS (dots) as a function of $z_{I\!P}^{jets}$, Q^2, W, $log_{10}x_{I\!P}$ and $\langle \eta \rangle_{lab}^{jets}$. The data are compared with predictions from NLO calculations.

In Fig. 2, the differential dijet cross sections in the region $x_{I\!P} < 0.01$ are compared with the predictions of the Saturation, BJLW and Resolved Pomeron (Pomeron parametrisation 'h1 fit2' [7]) models. The $P_{T,rem}^{(I\!P)}$ variable measures the transverse momentum of all hadronic final state particles in the Pomeron hemisphere of the $\gamma^* I\!\!P$ centre-of-mass frame which do not belong to the two highest p_T^* jets.

The saturation model is able to reproduce the shapes of the measured cross sections, although the overall predicted dijet rate is too low by a factor of about 2. In the BJLW model, the contribution from $q\bar{q}$ states is negligible and the dominant contribution comes from $q\bar{q}g$ final states. The normalisation of the BJLW model for $q\bar{q}g$ production can be controlled by tuning the lower limit on the transverse momentum of the gluon, $p_{T,g}^{cut}$, in the calculations. If this limit is set to 1.5 GeV, the total cross section for dijet production with $x_{I\!\!P} < 0.01$ is in agreement with the model [5].

In Fig. 2, the Resolved Pomeron model gives a good description of all observables, including the $P_{T,rem}^{(I\!\!P)}$ distribution. The good description of the $P_{T,rem}^{(I\!\!P)}$ distribution by both the Resolved Pomeron and the BJLW models indicates that the present data are not able to discriminate between models with a soft "remnant" and those with a third high-p_T parton.

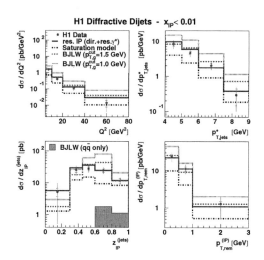

Figure 2. Differential cross sections for dijet production in diffractive DIS (dots), shown as functions of Q^2, $p_{T,jets}^*$, $z_{I\!\!P}^{jets}$ and $P_{T,rem}^{(I\!\!P)}$. The data are compared with different models.

3.2. Diffractive D^* in DIS

The ZEUS Collaboration has measured charm production in diffractive DIS [6]. Figure 3 shows the differential cross sections measured in the kinematic range $1.5 < Q^2 < 200$ GeV^2, $0.02 < y < 0.7$, $\beta < 0.8$, $p_T(D^*) > 1.5$ GeV and $|\eta(D^*)| < 1.5$ as functions of β, $\log(\beta)$, $\log(Q^2)$ and W for $x_{I\!P} < 0.035$ and as functions of $p_T(D^*)$, $\eta(D^*)$, $\log(M_X^2)$ and β for $x_{I\!P} < 0.01$. The data are compared with the ACTW NLO predictions, calculated with the gluon-dominated fit B, the SATRAP and the BJLW predictions (only for $x_{I\!P} < 0.01$).

The two-gluon-exchange BJLW model predictions, obtained with the requirement that $k_{T,g}^{\rm cut} = 1.5$ GeV, describe the differential cross sections in the range $x_{I\!P} < 0.01$ both in shape and normalisation. The saturation model (SATRAP) predictions reproduce the shapes and the normalisations of the differential cross sections measured in both $x_{I\!P}$ ranges. The ACTW NLO predictions, obtained with the gluon-dominated fit B, describe the data reasonably well in both $x_{I\!P}$ ranges.

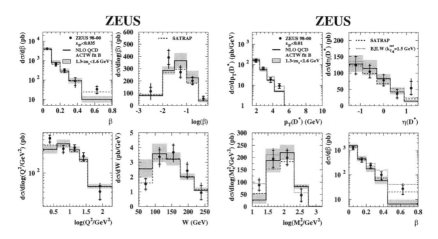

Figure 3. Differential cross sections for diffractive D^* production (dots) compared with different models (see text). The inner error bars indicate the statistical uncertainties, while the outer ones correspond to the statistical and systematic uncertainties added in quadrature. The shaded area shows the effect of varying the charm-quark mass in the ACTW NLO prediction.

Fig. 4 shows the quantity $x_{I\!P} F_2^{D(3),c\bar{c}}$ measured by the ZEUS Collaboration [6] where $F_2^{D(3),c\bar{c}}$ is the open-charm contribution to the diffractive proton structure function. It is shown as a function of $\log(\beta)$ for different values of Q^2 and $x_{I\!P}$. In all cases, $x_{I\!P} F_2^{D(3),c\bar{c}}$ rises as β decreases. The curves show the predictions obtained from the ACTW NLO calculations based on fits B, D and SG [11]. The fit B prediction generally agrees with the data. The fit D (SG) prediction overestimates (underestimates) the measured $x_{I\!P} F_2^{D(3),c\bar{c}}$ at low β.

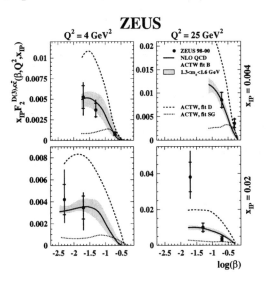

Figure 4. The measured charm contribution to the diffractive structure function of the proton multiplied by $x_{I\!P}$, $x_{I\!P} F_2^{D(3),c\bar{c}}$, as a function of β for different values of Q^2 and $x_{I\!P}$ (dots). The inner error bars indicate the statistical uncertainties, while the outer ones correspond to the statistical and systematic uncertainties added in quadrature. The shaded area shows the effect of varying the charm-quark mass in the ACTW NLO prediction.

Differential cross sections for D^* production in diffractive DIS have also been measured by H1[7,?]. The kinematic range of the data corresponds to $2 < Q^2 < 100$ GeV^2, $0.05 < y < 0.7$, $x_{I\!P} < 0.04$, $|t| < 1$ GeV^2, $M_Y < 1.6$ GeV, $p_T(D^*) > 2$ GeV and $|\eta(D^*)| < 1.5$.

Figure 5 shows the differential cross sections as functions of $\log(Q^2)$, $p_T(D^*)$, $\eta(D^*)$ and $x_{I\!P}$. The data are compared with the leading and next-to-leading order calculations from the diffractive version of HVQDIS, interfaced to the H1 diffractive pdf's [10]. Within the experimental and theoretical uncertainties, good agreement is observed between the data and NLO calculations.

3.3. Dijets in Diffractive Photoproduction

As shown in Sec. 3.1, diffractive dijet production in DIS is in good agreement with the QCD fits to inclusive diffractive data. However, applying

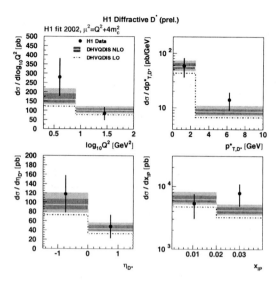

Figure 5. Differential cross sections for diffractive D^* production (dots) compared with NLO calculations. The error bars of the data correspond to the sum of statistical and systematic uncertainties, added in quadrature. The inner error band represents the renormalisation scale uncertainity, while the outer band also includes the uncertainty in the hadronization corrections, added lineary.

the diffractive parton densities to predict diffractive jet production at the Tevatron, leads to an overestimation of the observed rate by one order of magnitude [23]. This discrepancy has been attributed to the presence of the additional beam hadron remnant in $p\bar{p}$ collisions which leads to secondary interactions and a break-down of factorisation. The suppression cannot be calculated perturbatively and has been parametrised in various ways [24,25,26].

The transition from DIS to hadron-hadron scattering has been studied with the H1 detector in the photoproduction regime, where the beam lepton emits a quasi-real photon which interacts with the proton [27]. Processes in which the photon is first resolved into partons which then initiate the hard scattering resemble hadron-hadron scattering. Via the resolved photon process in hard photoproduction, hadronic final states are accessible, which are present in the equivalent $p\bar{p}$ collisions but not in DIS.

Dijet events are identified using the inclusive k_T cluster algorithm [27]. Fig. 6 shows the differential cross sections with respect to $z_{I\!P}^{jets}$ and the estimator x_γ^{jets} of the fractional photon momentum taking part in the hard scattering. The cross sections are given at hadron level and correspond to the kinematic range $Q^2 < 0.01$ GeV^2, $165 < W < 240$ GeV, $x_{I\!P} < 0.03$, $E_T^{jet1} > 5$ GeV and $E_T^{jet2} > 4$ GeV. The H1 2002 fit [10] gives a good description of the diffractive dijet photoproduction. The contribution from the direct photon process ($x_\gamma^{true} = 1$) is indicated by the hatched histogram. The sum of resolved and direct photon processes in the model gives a good description of the data both in normalisation and in shape throughout the x_γ^{jets} range. Hence there is no evidence for any suppression of the diffractive cross section in the region dominated by resolved photons, as might be expected on the basis of data from the Tevatron [23] or phenomenological models [24,25,26].

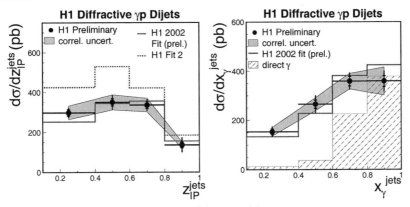

Figure 6. Differential cross sections in $z_{I\!P}^{jets}$ and x_γ^{jets} in diffractive photoproduction (dots). The inner error bars represent the statistical errors and the outer error bars the quadratic sum of the statistical and uncorrelated systematic errors.

References

1. G. Ingelman and P.E. Schlein, *Phys. Lett.* **B152**, 256 (1985).
2. H. Lotter, *Phys. Lett.* **B406**, 171 (1997).
3. J. Bartels, H. Jung and M. Wüsthoff, *Eur. Phys. J.* **C11**, 111 (1999).
4. E.M. Levin et al., *Z. Phys.* **C74**, 671 (1997).
5. H1 Collab., Adloff et al., *Eur. Phys. J.* **C20**, 29 (2001).
6. ZEUS Collab., S. Chekanov et al., *Nucl. Phys.* **B672**, 3-35 (2003).
7. H1 Collab., Adloff et al., *Phys. Lett.* **B520**, 191 (2001).
8. M.F. McDermot and G. Briskin, *Proc. Workshop on Future Physics at HERA*, G. Ingelman, A. De Roeck and R. Klanner (eds.), Vol.2, p.691, Hamburg, Germany, DESY (1996).
9. H1 Collab., C. Adloff et al., *Z. Phys.* **C76**, 613 (1997).
10. H1 Collaboration, *Measurement and NLO DGLAP QCD Interpretation of Diffractive Deep-Inelastic Scattering at HERA*, paper **980** submitted to the

31st Intl. Conference on High Energy Physics, ICHEP2002, Amsterdam (2002).
(http://www-h1.desy.de/h1/www/publications/htmlsplit/H1prelim-02-012.long.html)
11. L. Alvero et al., *Phys. Rev.* **D59**, 074022 (1999).
12. B.W. Harris and J. Smith, *Phys. Rev.* **D57**, 2806 (1998).
13. L. Alvero, J.C. Collins and J.J. Whitmore, [hep-ph/9806340].
14. S. Catani, M.H. Seymour, *Nucl. Phys.* **B485**, 29 (1997).
15. F.E. Low, *Phys. Rev.* **D12**, 163 (1975).
16. S. Nussinov, *Phys. Rev. Lett.* **34**, 1286 (1975).
17. S. Nussinov, *Phys. Rev.* **D14**, 246 (1976).
18. J. Bartels, H. Lotter and M. Wüsthoff, *Phys. Lett.* **B379**, 239 (1996).
19. J. Bartels, H. Jung and A. Kyrieleis, *Eur. Phys. J.* **C24**, 555 (2002).
20. K. Golec-Biernat and M. Wüsthoff, *Phys. Rev.* **D59**, 014017 (1999).
21. H. Kowalski, *Proc. Workshop on New Trends in HERA Physics*, G. Grindhammer, B.A. Kniehl and G. Kramer (eds.), pp. 361-380 (1999), available on http://www-library.desy.de/conf/ringberg99.html
22. H1 Collaboration, *Comparison at NLO between Predictions from QCD Fits to F_2^D and Diffractive Final State Observables at HERA*, paper **113** submitted to the Intl. Europhysics Conference on High Energy Physics, EPS2003, Aachen (2003).
(http://www-h1.desy.de/h1/www/publications/htmlsplit/H1prelim-03-015.long.html)
23. CDF Collaboration, T. Affolder et al., *Phys. Rev. Lett.* **84**, 5043 (2000).
24. E. Gotsman, E. Levin and U. Maor, *Phys. Lett.* **B438**, 229 (1998).
25. B. Cox, J. Forshaw and L. Lönnblad, [hep-ph/9908464].
26. A. Kaidalov, V. Khoze, A. Martin and M. Ryskin, [hep-ph/0306134].
27. H1 Collaboration, *Dijets in Diffractive Photoproduction at HERA*, paper **087** submitted to the Intl. Europhysics Conference on High Energy Physics, EPS2003, Aachen (2003).
(http://www-h1.desy.de/h1/www/publications/htmlsplit/H1prelim-02-113.long.html)

BEYOND BFKL

R. B. PESCHANSKI

CEA/DSM/SPhT, Unité de recherche associée au CNRS,
CE-Saclay, F-91191 Gif-sur-Yvette Cedex, France;
Email: pesch@spht.saclay.cea.fr

The Balitsky-Fadin-Kuraev-Lipatov (BFKL) evolution equation is known to be "unstable" with respect to fluctuations in gluon virtuality, transverse momentum and energy requiring to go beyond the leading order BFKL. Still, these instabilities point to fruitful improvements of our deep understanding of QCD. Recent applications to next-leading order and to saturation problems are outlined.

1. "Instabilities" of the BFKL Equation

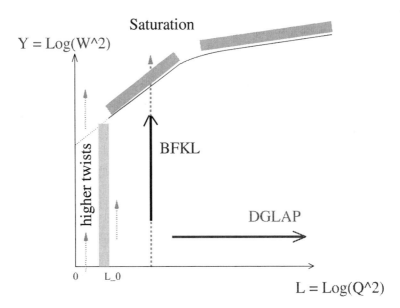

Figure 1. *The QCD evolution landscape.* The BFKL evolution and its limitations. **i)** to the left: non-perturbative region; **ii)** to the right: DGLAP evolution; **iii)** to the top: saturation region.

The Balitsky-Fadin-Kuraev-Lipatov (BFKL) evolution equation has already a venerable past[1]. It appears as a key tool in many recent works on small-x physics (in the broad sense). It is interesting to notice that its limitations themselves are the seeds of interesting fields of research. Let us discuss limitations which can be associated with the idea of "instabilities".

- **i)** *Instability towards the non-perturbative regime*
  It is well known that the perturbative "gluon ladder" contributing to the BFKL cross-section is characterized by a "cigar-shape" structure[2] of the transverse momenta. Hence, it is difficult to avoid an excursion inside the near-by non-perturbative region
  (Fig. (1), to the left).
- **ii)** *Instability towards the renormalization group regime*
  Calculations of the next-leading BFKL kernel[3] has proven that the inclusion of next-leading logs gives a (too) strong correction to the leading log result. After a resummation motivated by the suppression of spurious singularities[4,5], the results show that the resummed NLO-BFKL kernels are very similar, *e.g.* "attracted" towards the Dokshitzer Gribov Lipatov Altarelli Parisi (DGLAP) evolution[6], at least for the structure functions
  (Fig. (1), to the right).
- **iii)** *Instability towards the high density (saturation) regime*
  The BFKL evolution implies a densification of gluons and sea quarks, while they keep in average the same size. It is thus natural to expect[7] a modification of the evolution equation by non-linear contributions in the gluon density. Recently, the corresponding theoretical framework has been settled[8,9], and is based on an extension of the BFKL kernel acting on non-linear terms. It leads to a transition to the saturation regime (Fig. (1), to the top).

The main subjects of my talk will concern contributions[a] to point (**ii**), with a discussion of the phenomenological relevance of (resummed) NLO-BFKL kernels and point (**iii**), with a discussion of traveling wave solutions of non-linear QCD equations, as being deeply related to geometric scaling and the transition to saturation.

---

[a]We shall leave the point (**i**) outside of the scope of the present conference, despite some recent developments related to the AdS/CFT correspondence[10] and the "BFKL treatment" of the 4-supersymmetrical gauge field theory[11].

2. "Instability towards DGLAP"

The "instability" of the BFKL equation's solution w.r.t. the renormalization group evolution is well-known[12]. Indeed, the first correction to the leading-log $1/x$ approximation of the BFKL kernel[3] is large enough to apparently endanger the whole BFKL approach. It was soon realized that a large part of the problem was due to the appearance of singularities which contradict the renormalization group properties. Hence requiring an harmonization between the next leading log BFKL calculations and the renormalization group requirements through higher orders' resummation leads[4,5] to a possible way out of the problem.

Let us focus[13] on the impact of these developments on the proton structure functions and recall the parametrization of the proton structure functions in the (LO) BFKL approximation[14]:

$$F_i = \int \frac{d\gamma}{2i\pi} \left(\frac{Q^2}{Q_0^2}\right)^\gamma x_{Bj}^{-\frac{\alpha_s N_c}{\pi}\chi_{LO}(\gamma)} h_i(\gamma)\, \eta(\gamma)\,, \tag{1}$$

where F_i denotes respectively F_T, F_L, G (resp. transverse, longitudinal and gluon) structure functions and h_i are the known perturbative couplings to the photon ($h_G = 1$ for the gluon structure function), usually called "impact factors"[15]. χ_{LO} is the the LO-BFKL kernel, α_s the (fixed) coupling constant and $\eta(\gamma)$ an (unknown but factorizable[15]) non-perturbative coupling to the proton. Mellin-transforming (1) in $j - 1 \equiv \omega$ space, one easily finds

$$\tilde{F}_i = \int \frac{d\gamma}{2i\pi} \left(\frac{Q^2}{Q_0^2}\right)^\gamma \frac{1}{\omega - \frac{\alpha_s N_c}{\pi}\chi_{LO}(\gamma)} h_i(\gamma)\, \eta(\gamma)\,, \tag{2}$$

and, taking the pole contribution, one has the important relation

$$\omega = \frac{\alpha_s N_c}{\pi}\chi_{LO}(\gamma_k(\omega))\,. \tag{3}$$

Let us try and find the equivalent relation at NLO. At (resummed) next-to-leading order, one can similarly write[b]

$$\tilde{F}_i = \int \frac{d\gamma}{2i\pi} \left(\frac{Q^2}{\Lambda_{QCD}^2}\right)^\gamma e^{-\frac{X(\gamma,\omega)}{b\omega}} h_i(\gamma,\omega)\, \eta(\gamma,\omega)\,, \tag{4}$$

---

[b]Eq.(4) is already an approximation of the (still) unknown complete (resummed) NLO-BFKL formula, since the photon and proton impact factors are not yet known at NLO. However, one expects Eq.(4) to be a phenomenologically valid approximation containing the information on the NLO kernel.

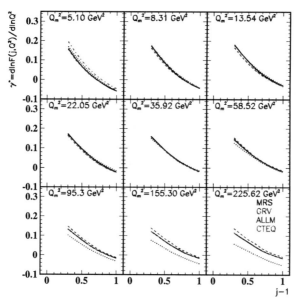

Figure 2. $\bar{\gamma}(\omega, Q^2)$: The "effective" F_2 anomalous dimension. $\bar{\gamma}(\omega, Q^2)$ has been evaluated from four known different parametrizations. They are all compatible in the range $.3 < \omega < 1, 5 < Q^2 < 100\ GeV^2$, where we restrict our analysis.

where, by construction

$$\frac{\partial}{\partial \gamma} X(\gamma, \omega) \equiv \chi_{NLO}(\gamma, \omega) \ . \qquad (5)$$

The function $X(\gamma, \omega)$ appears in the solution of the Green function derived[c] from the renormalization-group improved small-x_{Bj} equation[4], $\chi_{NLO}(\gamma, \omega)$ is a resummed NLO-BFKL kernel and

$$\frac{N_c}{\pi} \alpha_s(Q^2) = \left[b \ln \left(Q^2 / \Lambda_{QCD}^2 \right) \right]^{-1} \ , \qquad (6)$$

---

[c]The second variable of $X(\gamma, \omega)$ in (5) corresponds to the choice of a reference scale $\mu \to \omega \equiv j - 1$ dictated by the treatment of the Green function fluctuations near the saddle-point[4].

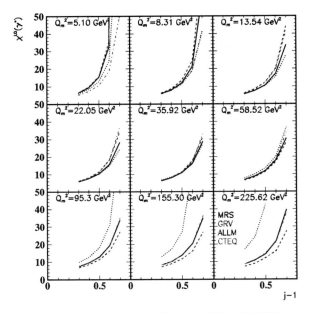

Figure 3. *Test of Relation ii): LO-BFKL.*

with $b = 11/12 - N_f/18$.

At large enough Q^2/Λ^2_{QCD}, one can use the saddle-point appoximation in γ to evaluate (4). Assuming that the perturbative impact factors and the non-perturbative function η do not vary much[d], the saddle-point condition reads

$$\omega \sim \frac{\chi_{NLO}(\bar\gamma,\omega)}{b \ln\left(Q^2/\Lambda^2_{QCD}\right)} = \frac{N_c\, \alpha_s(Q^2)}{\pi}\chi_{NLO}(\bar\gamma,\omega) , \qquad (7)$$

where $\bar\gamma \equiv \bar\gamma(\omega, Q^2)$ is the saddle-point value.

---

[d]We do not take into account modifications *e.g.* coming from powers of γ in the prefactors which may shift the saddle point[4]. We thus assume a smoothness of the structure function integrand around the saddle-point in agreement with the phenomenology[13].

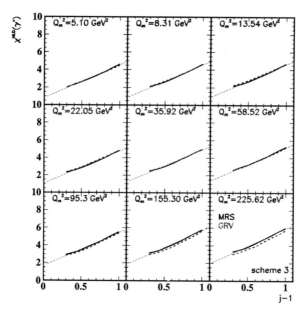

Figure 4. *Test of Relation ii): NLO-BFKL.*

Inserting the saddle point defined by (7) in formula (4), one obtains a set of conditions to be fulfilled at (resummed) NLO level as follows:

i) The Mellin transform $\tilde{F}_2 \equiv \tilde{F}_T + \tilde{F}_L$ defines:

$$\frac{\partial}{\partial \ln(Q^2)} \ln \tilde{F}_2(\omega, Q^2) \sim \bar{\gamma}(\omega, Q^2) \,. \tag{8}$$

ii) $\bar{\gamma}$ verifies

$$\chi_{NLO}(\bar{\gamma}) \equiv \frac{\pi \, \omega}{\alpha_s(Q^2) N_c} \,, \tag{9}$$

where χ_{NLO} is a resummed NLO-BFKL kernel.

iii) The gluon structure function (one may also choose the obervable

F_L) verifies, *via* Mellin transform:

$$\ln(\tilde{G}(\omega,Q^2)) = \ln\left(\tilde{F}_2(\omega,Q^2)\right) - \bar{\gamma}\,\ln\left[h_T(\bar{\gamma}) + h_L(\bar{\gamma})\right] . \qquad (10)$$

We test[13] the properties **i)-iii)** using NLO kernels proposed in[4], and compared with the LO kernel condition (3).

As an example an extraction of an "effective" F_2 anomalous dimension **i)**, see Fig. (2), is performed[13] using different parametrisations in a range of ω verifying the stability with respect to cuts on unknown (smallest) or irrelevant (large) x_{Bj}. The comparison of the property **ii)** to the LO BFKL kernel is displayed in Fig.(3) and the one with a resummed NLO-BFKL kernels (*cf.* Scheme[4] 3) in Fig.(4). As is clearly seen on the figures the Mellin-transform analysis disfavors the BFKL-LO kernel, while it is qualitatively compatible with the resummed BFKL-NLO kernel. The remaining discrepancies at NLO could be attributed to finite NNLO corrections to the kernel or to still unknown NLO contributions to the impact factors[16]. A systematic study of the proposed NLO kernels is thus made possible using the method[e].

3. "Saturation instability"

As well-known, the BFKL evolution (even including next-to-leading contributions) leads to a multiplication of partons with non-vanishing size and thus inevitably leads to a dense medium. This may be called the "Saturation Instability" of the BFKL evolution.

The back-reaction of parton saturation on the BFKL equation has been originally[7] described by adding a non-linear damping term. More recently, the evolution equation to saturation has been theoretically derived in the case of scattering on a "large nucleus", *e.g.* when the development of the parton cascade is tested by uncorrelated probes[8,9].

In the following we will focus on the solutions of the Balitsky-Kovchegov (BK) equations[9] where one consider the evolution within the QCD dipole Hilbert space[17]. To be specific let us consider $N(Y,x_{01})$, the dipole forward scattering amplitude and define

$$\mathbb{N}(Y,k) = \int_0^\infty \frac{dx_{01}}{x_{01}} J_0(kx_{01})\, N(Y,x_{01}) . \qquad (11)$$

Within suitable approximations (large N_c, summation of fan diagrams,

---

[e]Similarly, relation **iii)** can be looked at using the gluon structure function parametriza-

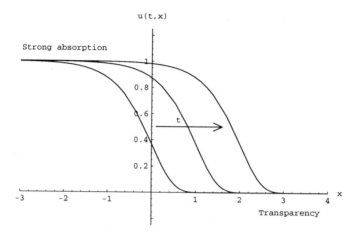

Figure 5. *Typical traveling wave solution.* The function $u(t,x)$ is represented for three different times. The wave front connecting the regions $u = 1$ and $u = 0$ travels from the left to the right as t increases. That illustrates how the "strong absorption" or saturated phase region invades the "transparency" region.

spatial homogeneity) and starting from the Balitsky-Kovchegov equation[9], it has been shown that this quantity obeys the nonlinear evolution equation

$$\partial_Y \mathbb{N} = \bar{\alpha}\chi(-\partial_L)\mathbb{N} - \bar{\alpha}\mathbb{N}^2 , \qquad (12)$$

where $\bar{\alpha} = \alpha_s N_c/\pi$, $\chi(\gamma) = 2\psi(1) - \psi(\gamma) - \psi(1-\gamma)$ is the characteristic function of the BFKL kernel[1], $L = \log(k^2/k_0^2)$ and k_0 is some fixed low momentum scale. It is well-known that the BFKL kernel can be expanded to second order around $\gamma = \frac{1}{2}$, if one sticks to the kinematical regime $8\bar{\alpha}Y \gg L$. We expect this commonly used approximation to remain valid for the full nonlinear equation. The latter boils down to a parabolic nonlinear partial derivative equation:

$$\partial_Y \mathbb{N} = \bar{\alpha}\bar{\chi}(-\partial_L)\mathbb{N} - \bar{\alpha}\mathbb{N}^2 , \qquad (13)$$

with

$$\bar{\chi}(-\partial_L) = \chi(\tfrac{1}{2}) + \frac{\chi''(\tfrac{1}{2})}{2}(\partial_L + \tfrac{1}{2})^2 . \qquad (14)$$

The key point of our recent approach[18] is to remark that the structure of Eq.(13) is identical (for fixed α) to a mathematical and physical archetype of

---

tions. However assumptions on the perturbative make the conclusions more qualitative or indicating some discrepancies to be solved at NLO.

non-linear evolution equation for which useful tools can be applied, namely the Fisher and Kolmogorov-Petrovsky-Piscounov (KPP) equation[19] for a function u:

$$\partial_t u(t,x) = \partial_x^2 u(t,x) + u(t,x)(1 - u(t,x)), \quad (15)$$

which is directly related[18] to ℕ. The equation can be generalized to many physical situations, including running α.

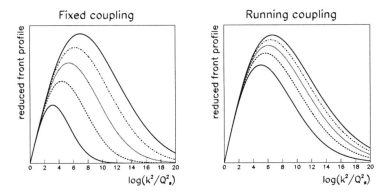

Figure 6. *Evolution of the reduced front profile.* Fixed coupling: left; Running coupling: right. The reduced front profile $(k^2/Q_s^2)^{\gamma_c}\, \mathbb{N}(k/Q_s(Y), Y)$ is plotted against $\log(k^2/Q_s^2)$ for different rapidities. The various lines correspond to rapidities from 2 (lower curves, full line) up to 10 (upper curves). Note the similarity of the wave fronts, but the quicker time evolution (in $\sqrt{t}$) for fixed coupling, by contrast with the slow time evolution (in $t^{1/3}$) for the running coupling case.

Our main results are the following. The well-known geometric scaling property[20] is obtained for the solution of the non-linear equation (13) for the gluon amplitude at large energy. In our notation, the geometric scaling property reads

$$\mathbb{N}(Y,k) = \mathbb{N}\left(\frac{k^2}{Q_s^2(Y)}\right), \quad (16)$$

where $Q_s^2(Y)$ is the saturation scale. We prove that geometric scaling is directly related to the existence of traveling wave solutions of the KPP equation[19] at large times. This means that there exists a function of one variable w such that

$$u(t,x) \underset{t\to +\infty}{\sim} w(x - m(t)) \quad (17)$$

uniformly in x. Such a solution is depicted on Fig.(5). The function $m(t)$ depends on the initial condition. For the QCD case[18], one has to consider

$$m(t) = 2t - \tfrac{3}{2}\log t + \mathbb{O}(1) \ . \tag{18}$$

When transcribed in the appropriate physical variables, this mathematical result implies directly the known geometric scaling properties[18]. It is interesting to note how the traveling wave solutions provide a particularly striking mathematical realization of an "instability" as depicted in Fig.(5), when a stable fixed point (strong absorption) "invades" an unstable one (transparency).

This mathematical analysis can be extended[18] to the study of the transition towards geometrical scaling , *i.e.* the formation of the front wave as a function of time, both for fixed and running α, see Fig.(6).

4. Conclusion

In the present contribution, we have discussed some aspects of the "instabilities" of the BFKL equations, *i.e.*:

- **i)** *Instability towards the non-perturbative regime*
- **ii)** *Instability towards the renormalization group regime*
- **iii)** *Instability towards the high density (saturation) regime*

At first sight, these instabilities could have appeared as drawbacks of the whole approach. On the very contrary, we have seen that the extensions of BFKL equation raised up by the treatment of "instabilities" appear to be the building blocks of most interesting recent developments towards a better understanding of QCD dynamics. As an example, I chose to present some personal recent contributions to this discussion, which are far from giving an idea of the whole extent of the works[f] which attack the problem nowadays.

As a brief outlook, let us mention:

About Point (**i**), not discussed here, let us mention the formal but informative discussion on the $N = 4$ supersymmetric QCD field theory and the AdS/CFT correspondence[11].

Point (**ii**): It is the subject of a developing activity which will allow to master the rather high technicality of the BFKL-NLO calculations and

---

[f]I mentioned quite a few of them in the reference list but I want to apologize for the authors and studies which I may have forgotten in this necessarily shortened review.

thus to penetrate the subtle aspects of the compatibility between BFKL and DGLAP evolution equations.

Point (**iii**): Saturation with both its phenomenological and theoretical aspects will certainly retain the attention of the Particle Physics community. The challenge here is the quest for a new phase of intense QCD fields and the understanding of its dynamical properties.

Acknowledgments

I want to warmly thank my collaborators in the work which has been discussed here: Stéphane Munier, Christophe Royon, Laurent Schoeffel and many colleagues with whom I had fructuous discussions, including those taking place in the charming decor of Ringberg Castle, in front of the Bavarian Alps.

References

1. L.N.Lipatov, *Sov.J.Nucl.Phys.* **23** (1976) 642; V.S.Fadin, E.A.Kuraev and L.N.Lipatov, *Phys. Lett.* **B60** (1975) 50; E.A.Kuraev, L.N.Lipatov and V.S.Fadin, *Sov.Phys.JETP* **44** (1976) 45, **45** (1977) 199; I.I.Balitsky and L.N.Lipatov, *Sov.J.Nucl.Phys.* **28** (1978) 822.
2. J.Bartels, H.Lotter and M. Vogt, *Phys. Lett.* **B373** (1996) 210.
3. V.S. Fadin and L.N. Lipatov, *Phys. Lett.* **B429** (1998) 107; M.Ciafaloni, *Phys. Lett.* **B429** (1998) 363; M. Ciafaloni and G. Camici, *Phys. Lett.* **B430** (1998) 349.
4. G.P. Salam, *JHEP 9807* (1998) 019; M. Ciafaloni, D. Colferai and G.P. Salam, *Phys.Rev.* **D60** (1999) 114036 , *JHEP 9910* (1999) 017;
   M. Ciafaloni, D. Colferai, G.P. Salam and A.M. Stasto, *Phys. Lett.* **B541** (2002) 314.
5. S.J. Brodsky, V.S. Fadin, V.T. Kim, L.N. Lipatov and G.B. Pivovarov, *JETP Lett.* **70** (1999) 105.
6. G.Altarelli and G.Parisi, *Nucl. Phys.* **B126** (1977) 298; V.N.Gribov and L.N.Lipatov, *Sov.J.Nucl.Phys.* (1972) 438 and 675; Yu.L.Dokshitzer, *Sov.Phys. JETP* **46** (1977) 641. For a review, see *e.g.* Yu.L. Dokshitzer, V.A. Khoze, A.H. Mueller, S.I. Troyan, *Basics of Perturbative QCD* .
7. L.V. Gribov, E.M. Levin and M.G. Ryskin, *Phys.Rep.* **100** (1983) 1; A.H. Mueller and J. Qiu, *Nucl. Phys.* **B268** (1986) 427.
8. L. McLerran and R. Venugopalan, *Phys. Rev.* **D49**, 2233 (1994); *ibid.*, 3352 (1994); *ibid.*, **D50**, 2225 (1994); A. Kovner, L. McLerran and H. Weigert, *Phys. Rev.* **D52**, 6231 (1995) ; *ibid.*, 3809 (1995); R. Venugopalan, *Acta Phys.Polon.* **B30**, 3731 (1999); E. Iancu, A. Leonidov and L. McLerran, *Nucl. Phys.* **A692**, 583 (2001); *idem*, *Phys. Lett.* **B510**, 133 (2001); E. Iancu and L. McLerran, *Phys. Lett.* **B510**, 145 (2001); E. Ferreiro, E. Iancu, A. Leonidov

and L. McLerran, *Nucl. Phys.* **A703**, 489 (2002); H. Weigert, *Nucl. Phys.* **A703**, 823 (2002).
9. I. Balitsky, *Nucl. Phys.* **B463**, 99 (1996); Y. V. Kovchegov, *Phys. Rev.* **D60**, 034008 (1999); *ibid.*, **D61**, 074018 (2000).
10. J. Maldacena, *Adv.Theor.Math.Phys.* **2** (1998) 231;
S.S. Gubser, I.R. Klebanov and A.M. Polyakov, *Phys. Lett.* **B428** (1998) 105;
E. Witten, *Adv. Theor. Math. Phys.* **2** (1998) 253; O. Aharony, S.S. Gubser, J. Maldacena, H. Ooguri and Y. Oz, *Phys. Rep.* **323** (2000) 183.
11. R.A. Janik and R. Peschanski, *Nucl. Phys.* **B565** (2000) 193, *Nucl. Phys.***B625** (2002) 279; R.A. Janik, *Phys. Lett.* **B500** (2001) 118; J. Polchinski and M.J. Strassler, *Phys. Rev. Lett.* **88** (2002) 031601; A.V. Kotikov, L.N. Lipatov and V.N. Velizhanin *Phys. Lett.* **B557** (2003) 114.
12. See, for instance, Kopeliovitch B.Z., Peschanski R.: *Working Group report on Diffraction Highlights of the Theory*, Proceedings of the "6th International Workshop on Deep Inelastic Scattering and QCD (DIS 98) COREMANS G., ROOSEN R., eds. pp. 858-862, (World Scientific, 1998) Brussels, Belgium, 1998.
13. R.Peschanski, C.Royon and L.Schoeffel, to appear. See a preliminary discussion in R.Peschanski, *Acta Phys. Pol.* **34** (2003) 3001; L.Schoeffel, contribution to the '11th International Workshop on Deep Inelastic Scattering and QCD (DIS 03), Proceedings to appear.
14. H Navelet, R.Peschanski, Ch. Royon and S.Wallon, *Phys. Lett.* **B385** (1996) 357; S.Munier and R.Peschanski, *Nucl. Phys.* **B524** (1998) 377.
15. S. Catani, M. Ciafaloni and F. Hautmann, *Nucl.Phys.* **B366** (1991) 135; S. Collins and R.K. Ellis, *Nucl.Phys.* **B360** (1991) 3; E.M. Levin, M.G. Ryskin, Yu.M. Shabelski and A.G. Shuvaev, *Sov.J.Nucl.Phys.* **B53** (1991) 657.
16. J. Bartels, D. Colferai, S. Gieseke and A. Kyrieleis, *Phys.Rev.* **D66** (2002) 094017, and references therein.
17. A.H. Mueller, *Nucl.Phys.* **B415** (1994) 373; A.H. Mueller and B. Patel, *Nucl.Phys.* **B425** (1994) 471; A.H. Mueller, *Nucl.Phys.* **B437** (1995) 107.
18. S. Munier and R. Peschanski, *"Geometric scaling as traveling waves"*, to appear in Phys. Rev. Lett., [arXiv:hep-ph/0309177]; *"Traveling wave fronts and the transition to saturation"*, [arXiv: hep-ph/0310357].
19. R. A. Fisher, Ann. Eugenics **7** (1937) 355; A. Kolmogorov, I. Petrovsky, and N. Piscounov, *Moscou Univ. Bull. Math.* **A1** (1937) 1. For Mathematical properties : M. Bramson, *Memoirs of the American Mathematical Society* **44** (1983) 285. For Physical Applications, see B. Derrida and H. Spohn, E. Brunet and B. Derrida, *Phys. Rev.* **E56**(1997) 2597. For a recent general review, Wim van Saarloos, *Phys. Rep.* **386** (2003) 29.
20. A. M. Staśto, K. Golec-Biernat, and J. Kwiecinski, *Phys. Rev. Lett.* **86** (2001) 596.

POMERON PHYSICS AND QCD

O. NACHTMANN

Institut für Theoretische Physik
Universität Heidelberg
Philosophenweg 16, D-69120 Heidelberg, Germany
E-mail: O.Nachtmann@thphys.uni-heidelberg.de

We review some theoretical ideas concerning diffractive processes. We discuss the Regge Ansatz for the pomeron and the two pomeron model. Then we present the results obtained from nonperturbative QCD for high energy scattering. There we can extract from elastic scattering data the parameters describing the QCD vacuum, in particular the string tension.

1. Introduction

In this contribution to the Ringberg workshop 2003 we shall discuss a topic of hadron physics: diffractive processes at high energies. We speak of a diffractive scattering process if one or more large rapidity gaps occur in the final state. Examples are the reactions (see Fig. 1)

$$a + b \to X + Y,$$
$$a + b \to X + Z + Y, \tag{1}$$

where a, b are the incoming particles and X, Y, Z can be particles or groups of particles. Large rapidity gaps are required between X and Y and X, Y and Z. The object exchanged across the rapidity gaps is called the pomeron (see for instance Ref. 1,2 for the history and many refences).

We can distinguish three classes of reactions involving pomerons.

- Soft reactions, where no large momentum transfers occur and no short distance physics is involved. Examples are the following elastic scattering processes near the forward direction:

$$p + p \to p + p \tag{2}$$
$$\pi + p \to \pi + p \tag{3}$$
$$\gamma + p \to \gamma + p. \tag{4}$$

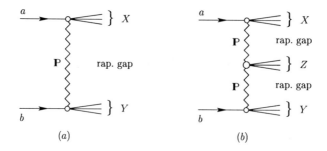

Figure 1. Processes with (a) one and (b) two large rapidity gaps.

In (4) we consider real photons. In all these cases the objects scattering are relatively large, with diameters of order 0.5 to 1.0 fm.

- Semi-hard reactions, where both large and small objects and/or momentum transfers are involved. Examples are

$$p + p \to p + H + p, \qquad (5)$$
$$p + p \to p + 2 \text{ high } p_T \text{ jets } + p \qquad (6)$$
$$\gamma^* + p \to \gamma^* + p. \qquad (7)$$

Here H denotes the Higgs particle. The virtual photon γ^* in (7) is supposed to have high Q^2, such that it is a small object probing the large proton. The imaginary part of the amplitude of forward virtual Compton scattering (7) is, of course, directly related to the structure functions of deep inelastic scattering. The diffractive central production of high p_T jets (6) was first discussed in Ref. 3 where also the important concept of the partonic structure of the pomeron was introduced.

- Hard reactions, where all participating objects are small and ideally all momentum transfers should be large. The prime example for this is $\gamma^* - \gamma^*$ scattering with highly virtual photons

$$\gamma^* + \gamma^* \to \gamma^* + \gamma^*. \qquad (8)$$

Here the observable quantity is the imaginary part of the forward scattering amplitude, that is the total cross section for $\gamma^*\gamma^*$ going to hadrons

$$\gamma^* + \gamma^* \to X. \qquad (9)$$

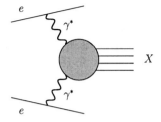

Figure 2. The two photon process in e^+e^- collisions giving hadrons X.

This can be measured for instance at e^+e^- colliders, see Fig. 2.

2. Pomerons: What Are They?

The main question is now: what happens in a diffractive collision at high energy, that is when a pomeron is exchanged? Let us discuss this first for the prototype of such collisions, elastic hadron-hadron scattering: two hadrons come in, they interact, two hadrons go out (Fig. 3a). What happens in between? This question was, of course, already studied before the advent of QCD. As an example of a pre-QCD approach let us discuss Regge theory which is based on general analyticity and crossing properties of scattering

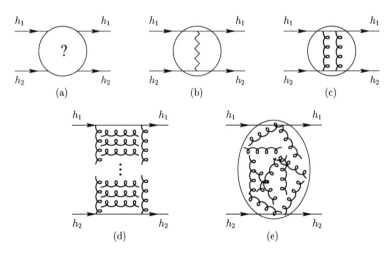

Figure 3. Hadron-hadron scattering: (a) what happens?, (b) phenomenological pomeron and (c) two gluon exchange, (d) exchange of a reggeised gluon ladder and (e) in the fluctuating vacuum gluon field.

amplitudes. The Regge approach was developed in the nineteensixties and the answer provided then to our question was: an object is exchanged, the pomeron (Fig. 3b), which corresponds to the rightmost singularity in the complex angular momentum plane. But to this date Regge theory alone cannot predict if the pomeron is a single Regge pole, consists of two Regge poles, is maybe a Regge cut, and so on. Nevertheless, the assumption that the pomeron seen in elastic hadron-hadron scattering is a simple Regge pole works extremely well phenomenologically.

We discuss as an example the Donnachie-Landshoff (DL) Ansatz [4] for the soft pomeron in Regge theory. The DL pomeron is assumed to be effectively a simple Regge pole. Consider as an example $p - p$ elastic scattering

$$p(p_1) + p(p_2) \to p(p_3) + p(p_4), \tag{10}$$
$$s = (p_1 + p_2)^2, \; t = (p_1 - p_3)^2, \tag{11}$$

where s and t are the c.m. energy squared and the momentum transfer squared. The corresponding diagram is as in Fig. 3b, setting $h_1 = h_2 = p$. In the DL Ansatz the wavy line in Fig. 3b can be interpreted as an effective pomeron propagator given by

$$(-is\alpha'_\mathbb{P})^{\alpha_\mathbb{P}(t)-1}. \tag{12}$$

Here $\alpha_\mathbb{P}(t)$ is the pomeron trajectory which is assumed to be linear in t.

$$\alpha_\mathbb{P}(t) = \alpha_\mathbb{P}(0) + \alpha'_\mathbb{P} t \tag{13}$$

with $\alpha_\mathbb{P}(0) = 1 + \epsilon_1$ the pomeron intercept and $\alpha'_\mathbb{P}$ the slope of the pomeron trajectory. The $pp\mathbb{P}$ vertex, or Regge residue factor, is assumed to have the form

$$-i3\beta_\mathbb{P} F_1(t)\gamma^\mu \tag{14}$$

where $F_1(t)$ is the isoscalar Dirac electromagnetic form factor of the nucleons and $\beta_\mathbb{P}$ the $\mathbb{P}$-quark coupling constant. A good representation of the form factor is given by the dipole formula

$$F_1(t) = \frac{4m_p^2 - 2.79t}{(4m_p^2 - t)(1 - t/m_D^2)^2},$$
$$m_D^2 = 0.71 \text{ GeV}^2. \tag{15}$$

Putting everything together we find from these rules the following expression for the scattering amplitude at large s

$$\langle p(p_3, s_3), p(p_4, s_4)|T|p(p_1, s_1), p(p_2, s_2)\rangle$$
$$\sim 2s\left(3\beta_\mathbb{P} F_1(t)\right)^2 \delta_{s_3 s_1} \delta_{s_4 s_2} i(-is\alpha'_\mathbb{P})^{\alpha_\mathbb{P}(t)-1} \tag{16}$$

Here $s_1, \ldots, s_4$ are the spin indices. This formula (16) describes the data surprisingly well and from the fits one gets the following values for the DL-pomeron parameters:

$$\epsilon_1 = 0.0808, \quad \alpha'_{\mathbb{P}} = 0.25 \text{ GeV}^{-2}, \quad \beta_{\mathbb{P}} = 1.87 \text{ GeV}^{-1}. \tag{17}$$

Examples of the fits are shown in Figs. 4 and 5 where in addition to the pomeron also non-leading Regge exchanges are taken into account. Note in

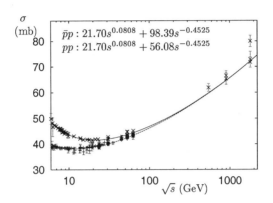

Figure 4. Total pp and $p\bar{p}$ cross sections as function of $\sqrt{s}$ (from Ref. 1).

Fig. 4 that the Tevatron measurements of $\sigma_{tot}(p\bar{p})$ at $\sqrt{s} \simeq 1.8$ TeV give two incompatible values. A resolution of this longstanding discrepancy would be very welcome. The simple DL Ansatz failed, however, when applied to semi-hard diffractive reactions. A second pomeron had to be introduced [5] in order to describe the structure functions of deep inelastic lepton-nucleon scattering, that is the total cross section of a highly virtual photon being absorbed by a proton. This second pomeron, the hard pomeron, was found to have an intercept much higher than the soft one:

$$\alpha_{\mathbb{P}}(0)|_{\text{hard}} = 1 + \epsilon_0, \quad \epsilon_0 \cong 0.44. \tag{18}$$

This, of course, spoiled the simplicity of the original DL-Regge picture for the pomeron and raised a number of questions. Is the hard pomeron also present in pp scattering? Then it should show up at higher energies and maybe the LHC experiments will be able to tell. The biggest question for a theorist is, of course, how to understand all this in the framework of the theory of strong interactions, that is quantum chromodynamics, QCD.

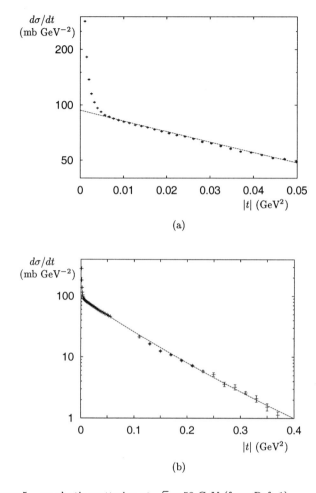

Figure 5. pp elastic scattering at $\sqrt{s} = 53$ GeV (from Ref. 1).

Clearly, the advent of QCD in the early nineteenseventies changed the whole scene in hadronic physics radically. We have since then a quantum field theory where in principle all hadronic phenomena should be derivable from the fundamental Lagrangian. The question arose how to describe the pomeron in QCD. The first answers were given in Ref. 6. These authors modeled the pomeron as a perturbative two-gluon exchange (Fig. 3c). However, this approach has problems. Gluons in QCD are massless particles. Exchanging one of them leads in perturbation theory to a long-range Coulomb potential $\propto 1/r$ and a singularity in $d\sigma/dt$ at $t = 0$. Of course, in elastic hadron-hadron scattering one must exchange at least two gluons due to colour conservation. But this still gives a long-range potential falling with some power of $1/r$ and singularities in derivatives of $d\sigma/dt$ at $t = 0$. From general principles we know the absence of massless hadrons to imply that $d\sigma/dt$ is not singular at $t = 0$. This is, of course, consistent with experiment. The two-gluon exchange is the starting point of the BFKL approach [7,8] where the interaction of the gluons is taken into account. Instead of two gluons, one or even several gluon ladders, where the gluons themselves are "reggeised", are exchanged (Fig. 3d). Even if this is usually called the perturbative pomeron of QCD, one should keep in mind that this pomeron represents a very sophisticated summation of an infinite number of diagrams in perturbation theory. For each diagram only the leading term for high energies is kept. There is no guaranty that the summation of the leading terms really gives the leading term of the full theory. Also, this approach as it is based on perturbation theory does not generate a finite length scale for the potential and thus leads to singularities in derivatives of $d\sigma/dt$ at $t = 0$. But the BFKL pomeron should be important for hard diffractive reactions.

For hadron-hadron scattering the total cross section and the parameters describing $d\sigma/dt$, for instance the slope parameter b,

$$b(s) = \frac{\partial}{\partial t} \ln \frac{d\sigma}{dt}(s,t)|_{t=0}, \qquad (19)$$

are dimensioned parameters. This fact suggested to some authors to consider the soft pomeron as a nonperturbative object in QCD, see Refs. 9-12,1 and references therein. We shall treat this point of view in Sec. 3. In this approach hadrons scatter at small $|t|$ since the quarks in the hadrons feel the nonperturbative fluctuations of the gluon fields in the vacuum (Fig. 3e). These fluctuations are assumed to be of Gaussian nature in the stochastic vacuum model [13]. We will show that the application of this model to high energy scattering allows us to extract from high energy data on $d\sigma/dt$ values for the vacuum parameters, the gluon condensate, the correlation length, the non-abelian parameter and related to them the string tension. As we shall see, the results compare well to lattice calculations and other information on these parameters.

3. Soft Hadron Reactions

In this section we will outline a microscopic approach towards hadron-hadron diffractive scattering (see Refs. 10-12). Consider as an example elastic scattering of two hadrons h_1, h_2

$$h_1 + h_2 \to h_1 + h_2 \qquad (20)$$

at high energies and small momentum transfer. We will look at reaction (20) from the point of view of an observer living in the "femto-universe", that is we imagine having a microscope with resolution much better than 1 fm for observing what happens during the collision. Of course, we should choose an appropriate resolution for our microscope. If we choose the resolution much too good, we will see too many details of the internal structure of the hadrons which are irrelevant for the reaction considered and we will miss the essential features. The same is true if the resolution is too poor. In [10] we used a series of simple arguments based on the uncertainty relation to estimate this appropriate resolution. Let $t = 0$ be the nominal collision time of the hadrons in (20) in the c.m. system. This is the time when the hadrons h_1 and h_2 have maximal spatial overlap. Let furthermore be $t_0/2$ the time when, in an inelastic collision, the first produced hadrons appear. We estimate $t_0 \approx 2$ fm from the Lund model of particle production [14]. Then the appropriate resolution, that is the cutoff in transverse parton momenta k_T of the hadronic wave functions to be chosen for describing reaction (20) in an economical way is

$$k_T^2 \leq \sqrt{s}/(2t_0) \qquad (21)$$

where $\sqrt{s}$ is the c.m. energy. Modes with higher k_T can be assumed to be integrated out. With this we could argue that over the time interval

$$-\frac{1}{2}t_0 \leq t \leq \frac{1}{2}t_0 \qquad (22)$$

the following should hold:

(a) The parton state of the hadrons does not change qualitatively, that is parton annihilation and production processes can be neglected for this time.
(b) Partons travel in essence on straight lightlike world lines.
(c) The partons undergo "soft" elastic scattering.

The strategy is now to study first soft parton-parton scattering in the femto-universe. There, the relevant interaction turns out to be mediated by the gluonic vacuum fluctuations. It is common belief that these have a highly nonperturbative character. In this way the nonperturbative QCD vacuum structure enters the picture for high energy soft hadronic reactions. Once we have solved the problem of parton-parton scattering we have to fold the partonic S-matrix with the hadronic wave functions of the appropriate resolution (21) to get the hadronic S-matrix elements.

Here we can only give an outline of the various steps in this program. Let us start by considering meson-meson scattering:

$$M_1(P_1) + M_2(P_2) \to M_1(P_3) + M_2(P_4). \tag{23}$$

We represent mesons as $q\bar{q}$ dipoles. We use standard techniques of quantum field theory, the LSZ reduction formula and the functional integral. This allows us in a first step to represent the amplitude for the scattering of a $q\bar{q}$ colour dipole on another $q\bar{q}$ dipole at high energies in terms of a correlation function of two lightlike Wegner-Wilson loops (Fig. 6). The second step

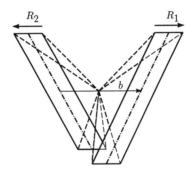

Figure 6. Lightlike Wegner-Wilson loops in Minowski space. The sides of the loops are given by the paths of quarks and antiquarks of the colour dipoles of sizes $\vec{x}_T = \vec{R}_1$ and $\vec{y}_T = \vec{R}_2$. The impact parameter is b (from Ref. 1).

is to fold with the meson wave functions which describe the distribution of the colour dipoles in the mesons. The final formula reads

$$S_{fi} = \delta_{fi} + i(2\pi)^4 \delta(P_3 + P_4 - P_1 - P_2) T_{fi},$$

$$T_{fi} \equiv \langle M_1(P_3), M_2(P_4)|T|M_1(P_1), M_2(P_2)\rangle \tag{24}$$

$$= -2is \int d^2 b_T d^2 x_T d^2 y_T e^{i\vec{q}_T \cdot \vec{b}_T} w_1(\vec{x}_T) w_2(\vec{y}_T)$$

$$\times \langle \mathcal{W}_+(\tfrac{1}{2}\vec{b}_T, \vec{x}_T) \mathcal{W}_-(-\tfrac{1}{2}\vec{b}_T, \vec{y}_T) - 1 \rangle_G.$$

Here $s = (P_1 + P_2)^2$ is the c.m. energy squared and $\vec{q}_T$ is the momentum transfer, which is purely transverse in the high energy limit:

$$\vec{q}_T = (\vec{P}_1 - \vec{P}_3)_T \tag{25}$$

and $\bar{b}_T$ is the impact parameter. Furthermore $\vec{x}_T, \vec{y}_T$ are the transverse sizes of the colour dipoles and $w_{1,2}$ describe the distribution of the colour dipoles in the mesons $M_{1,2}$. The symbol $\langle\ \rangle_G$ denotes the functional integration over the gluonic degrees of freedom and $\mathcal{W}$ are the Wegner-Wilson loops.

The task is now to evaluate the functional integral $\langle\rangle_G$ in (24). Surely we do not want to make a perturbative expansion there, remembering our argument of Sec. 2. Instead, we will turn to the stochastic vacuum model (SVM) introduced in Ref. 13 which provides a method to calculate in a certain approximation functional integrals in the nonperturbative domain of QCD. In the SVM the QCD vacuum is described by three parameters

- G_2, the gluon condensate introduced in Ref. 15,
- κ, the nonabelian parameter,
- a, the vacuum correlation length.

The SVM relates these parameters to the string tension

$$\sigma = \frac{32\pi\kappa G_2 a^2}{81}. \tag{26}$$

The SVM has been applied extensively in low energy hadron physics with considerable success, see Ref. 16 for a review. The QCD vacuum structure has also been investigated in lattice calculations [17,18]. In these one finds the values for the vacuum parameters shown in the second column of Table 1 where the value for the string tension is the input parameter. In the third column of the table the vacuum parameters of the lattice results are used as input and the resulting string tension is calculated from the SVM relation (26). The outcome, $\sqrt{\sigma} = 415$ MeV, is very satisfactory and gives confidence that the SVM catches some true features of the QCD vacuum.

Now we come back to high energy scattering. To evaluate the correlation function of the two lightlike Wegner-Wilson loops in (24) the SVM was continued from Euclidean space, where it was originally formulated, to Minkowski space [11]. This opened the possibility for many applications, see for instance, Ref. 19. Here I can just cite some results as examples.

The differential cross section for proton-proton elastic scattering for $|t| \lesssim 1$ GeV^2 was calculated in Ref. 20. The proton was considered as a quark-diquark system. Thus the simple formula (24) for meson-meson scattering could be taken over with the diquarks replacing the antiquarks. The dipoles in the proton were assumed to have a Gaussian distribution where the parameters can be related [21] to those of the electromagnetic form factors. The data could then be described reasonably well with the

Table 1. Summary of the vacuum parameters G_2: the gluon condensate, κ: the nonabelian parameter, a: the correlation length, and of the string tension σ, as determined with different methods. Quantities which are underlined are input values. In the second column we list the result of the lattice calculations [18], where σ is taken as input. In the third column the result of the string tension in the SVM is listed, where the vacuum parameters G_2, κ, a are input, see (26). The fourth column lists the results for σ, κ, a obtained in Ref. 20 from the data on high energy proton-proton elastic scattering. Here the value of G_2 is calculated using the SVM relation (26). Errors of the numbers can be estimated to be around 10 %.

parameter	lattice calculation quenched	SVM static pot.	high energy scattering
$\sqrt{\sigma}$/MeV	<u>420</u>	415	435
$G_2^{1/4}$/MeV	486	<u>486</u>	529
κ	0.89	<u>0.89</u>	0.74
a/fm	0.33	<u>0.33</u>	0.32

values for the vaccum parameters listed in column 4 of the table, see Fig. 7. Note that the data extends over many orders of magnitude. It was found in Ref. 20 that the shape of this t-distribution was quite sensitive to the vacuum parameters, in particular to κ.

We see from the table that the vacuum parameters extracted from high-energy scattering are very well compatible with the values obtained from the lattice calculations. In particular, the value for the square root of the string tension $\sqrt{\sigma} = 435$ MeV is well in the range obtained from the study of heavy charmonium states. There one finds $\sqrt{\sigma} \simeq 420 - 440$ MeV. In our opinion this gives support to the idea that high energy hadron-hadron scattering is dominated by nonperturbative QCD effects.

But maybe pp is special. Let us therefore look at one other result of this approach. In Fig. 8 we show the pomeron contributions to various hadron-hadron total cross sections at $\sqrt{s} = 20$ GeV as function of the hadron size parameter R_h divided by R_p. The curve, normalised to the pp point, gives the predicted functional dependence. Again, the results are quite satisfactory.

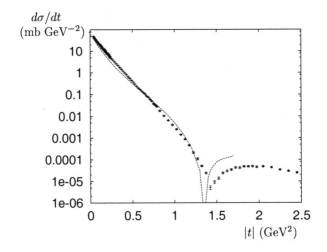

Figure 7. Results from [20] for the elastic differential cross section for proton-proton scattering at $\sqrt{s} = 23$ GeV using the matrix cumulant expansion method (dashed line) compared to the experimental data (from Ref. 1).

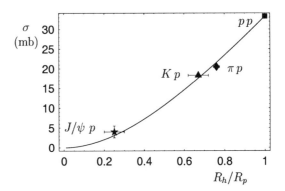

Figure 8. The pomeron contribution at $\sqrt{s} = 20$ GeV to hadron-proton total cross sections as function of the hadron size (from Ref. 1).

4. Conclusions

Here we list some points which we consider important in the study of the pomeron in QCD.

- The phenomenological description in terms of the Regge pole Ansatz works surprisingly well. The challenge is to understand / derive this from QCD.

- The soft pomeron and QCD: We have developed a calculational framework based on quantum field theoretic methods. We suggest that the soft pomeron is related to nonperturbative features of QCD, in particular to the QCD vacuum structure. The vacuum parameters like the string tension and the gluon condensate can be quantitatively extracted from high energy scattering data.
- Semi-hard diffractive phenomena, in particular the structure functions of deep inelastic lepton-nucleon scattering are well described for instance by the two pomeron model of Donnachie and Landshoff. Is this directly related to the usual DGLAP evolution as suggested in Ref. 22? What is the role of the BFKL, that is the perturbative, pomeron there? Is there saturation? Why does the dipole picture (see Refs. 23,24,1 and the references therein) work so well? Is there a chance to make nonperturbative calculations for the structure functions at small x from first principles? Some steps in this direction have been presented in Refs. 25,26. For hard pomeron phenomena like the total cross section for $\gamma^* + \gamma^* \to$ hadrons presumably the BFKL approach should work. But there are problems of large higher order corrections [27]. The experimental data [28] from the LEP experiments does not give evidence for the BFKL pomeron, so far.
- What and where is the odderon, the $C = -1$ partner of the pomeron? See Refs. 1,29 for a discussion of this question. Experimental efforts [30] to look for the odderon should be continued and extended.

Acknowledgements

The author is grateful to the organisers of the Ringberg Workshop on HERA Physics 2003 for inviting him to give a talk there. The atmosphere there was stimulating and enjoyable. Many of the results discussed in this contribution could not have been presented without the privilege the author had in being able to collaborate with many colleagues. Special thanks in this respect are due to A. Donnachie, H. G. Dosch and P. V. Landshoff. Finally, we thank C. Ewerz for discussions, for reading the manuscript and for help with the figures.

References

1. A. Donnachie, H. G. Dosch, P. V. Landshoff and O. Nachtmann, *Pomeron Physics and QCD*, Cambridge University Press, 2002.
2. J. R. Forshaw and D. A. Ross, *Quantum Chromodynamics and the Pomeron*, Cambridge University Press, 1997.
3. G. Ingelman and P. Schlein, Phys. Lett. **B 152**, 256 (1985).
4. A. Donnachie and P. V. Landshoff, Nucl. Phys. **B244**, 322 (1984); Nucl. Phys. **B267**, 690 (1986); Phys. Lett. **B185**, 403 (1987).
5. A. Donnachie and P. V. Landshoff, Phys. Lett. **B437**, 408 (1998); Phys. Lett. **B518**, 63 (2001).
6. F. E. Low, Phys. Rev. **D12**, 163 (1975);
   S. Nussinov, Phys. Rev. Lett. **34**, 1286 (1975);
   J. F. Gunion, D. E. Soper, Phys. Rev. **D15**, 2617 (1977).
7. E. A. Kuraev, L. N. Lipatov, V. S. Fadin, Sov. Phys. J.E.T.P. **44**, 443 (1976), **45**, 199 (1977);
   L. N. Lipatov, Sov. J. Nucl. Phys. **23**, 338 (1976);
   Ya. Ya. Balitskii, L. N. Lipatov, Sov. J. Nucl. Phys. **28**, 822 (1978).
8. L. N. Lipatov, *Pomeron in Quantum Chromodynamics*, in "Perturbative Quantum Chromodynamics" (A. H. Mueller, Ed.), World Scientific, Singapore, 1989.
9. P. V. Landshoff, O. Nachtmann, Z. Phys. **C35**, 405 (1987).
10. O. Nachtmann, Ann. Phys. **209**, 436 (1991).
11. H. G. Dosch, E. Ferreira, A. Krämer, Phys. Rev. **D50**, 1992 (1994).
12. O. Nachtmann, *High Energy Collisions and Nonperturbative QCD* in "Perturbative and Nonperturbative Aspects of Quantum Field Theory", H. Latal, W. Schweiger (Eds.), Springer-Verlag, Berlin, Heidelberg (1997).
13. H. G. Dosch: Phys. Lett. **B190**, 177 (1987);
    H. G. Dosch, Yu. A. Simonov, Phys. Lett. **B205**, 339 (1988);
    Yu. A. Simonov, Nucl. Phys. **B307**, 512 (1988).
14. B. Andersson, G. Gustafson, G. Ingelman, T. Sjöstrand, Phys. Rep. **C97**, 33 (1983).
15. M. A. Shifman, A. I. Vainshtein, V. I. Zakharov, Nucl. Phys. **B147**, 385, 448, 519 (1979).
16. H. G. Dosch, *Nonperturbative Methods in Quantum Chromodynamics*, Progr. in Part. and Nucl. Phys. **33**, 121 (1994).
17. A. DiGiacomo, H. Panagopoulos, Phys. Lett. **B285**, 133 (1992)
    M. D'Elia, A. DiGiacomo, E. Meggiolaro, Phys. Lett. **B408**, 315 (1997).
18. E. Meggiolaro, Phys. Lett. **B451**, 414 (1999).
19. H. G. Dosch, T. Gousset, G. Kulzinger, H. J. Pirner, Phys. Rev. **D55**, 2602 (1997);
    H. G. Dosch, T. Gousset, H. J. Pirner, Phys. Rev. **D57**, 1666 (1998);
    M. Rueter, H. G. Dosch, Phys. Rev. **D57**, 4097 (1998);
    E. Ferreira, F. Pereira, Phys. **D56**, 179 (1997);
    H. G. Dosch, F. S. Navarra, M. Nielsen, M. Rueter, Phys. Lett. **B466**, 363 (1999);

H. G. Dosch, F. S. Navarra, M. Nielsen, Nucl. Phys. Proc. Suppl. **86**, 539 (2000);
G. Kulzinger, H. G. Dosch, H. J. Pirner, Eur. Phys. J. **C7**, 73 (1999);
A. I. Shoshi, F. D. Steffen, H. J. Pirner, Nucl. Phys. **A709**, 131 (2002).
20. E. Berger, O. Nachtmann, Eur. Phys. J. **C7**, 459 (1999).
21. H. G. Dosch, O. Nachtmann, T. Paulus, and S. Weinstock, Eur. Phys. J. **C21**, 339 (2001).
22. A. Donnachie and P. V. Landshoff, Phys. Lett. **B533**, 277 (2002).
23. N. Nikolaev and B. G. Zakharov, Z. für Physik **C53**, 331 (1992).
24. K. Golec-Biernat and M. Wüsthoff, Phys. Rev. **D59**, 014017 (1999); **D60**, 114023 (1999).
25. A. Hebecker, E. Meggiolaro and O. Nachtmann, Nucl. Phys. **B571**, 26 (2000).
26. O. Nachtmann, Eur. Phys. J. **C26**, 579 (2003).
27. V.S. Fadin and L.N. Lipatov, Phys. Lett. **B429**, 127 (1998);
G. Camici and M. Ciafaloni, Phys. Lett. **B412**, 396 (1997).
28. P. Achard *et al.* [L3 Collaboration], Phys. Lett. B **531**, 39 (2002);
G. Abbiendi *et al.* [OPAL Collaboration], Eur. Phys. J. C **24**, 17 (2002).
29. C. Ewerz, *The Odderon in Quantum Chromodynamics*, hep-ph/0306137.
30. C. Adloff et. al, H1 Coll., Phys. Lett. **B544**, 35 (2002).

THEORY OF DIFFRACTIVE MESON PRODUCTION

L. MOTYKA

Institute of Physics, Jagellonian University,
Reymonta 4, 30-059 Kraków, Poland
E-mail: leszekm@th.if.uj.edu.pl

In this talk some recent developments in the theory of diffractive meson production are reviewed: (1) the issue of end point divergencies and diffractive factorisation; (2) elastic vector meson production in the context of the gluon saturation in the impact parameter plane; (3) diffractive vector meson photoproduction at large momentum transfer; (4) C-even meson photoproduction and the Odderon.

1. Introduction

Diffractive meson production offers unique opportunities to probe the colour singlet exchange in high energy scattering. In HERA experiments one measures properties of the scattering amplitude in a wide range of energy W, momentum transfer t, and the incoming photon virtuality Q^2, for various mesons.[1] The multidimensional set of data is, with no doubts, a challenge for the theory. Standard methods of QCD, like the fixed order perturbative calculus and the hard factorisation, are not always sufficient to get a correct picture, and the use of resummation schemes, like the BFKL formalism[2] or phenomenological models as the saturation model,[3] is necessary. Therefore a lot of effort is directed towards improving the phenomenological models and rooting them in the fundamental theory. The main goal is to obtain a robust global description of the diffractive meson production data, in a simple conceptual framework grounded in QCD.

On the other hand, new measurements may be performed that probe some interesting mechanisms (e.g. the Odderon exchange) which are not fully understood yet. Furthermore, some quantities which are not calculable from first principles may be determined or constrained. The hadron structure is a good example here.

In this talk I am presenting short notes on four selected topics in which substantially new results have recently appeared and further progress is expected in the near future.

2. Diffractive factorisation

The theory of exclusive meson electroproduction is based on the diffractive factorisation theorem.[4] The proof holds for a longitudinally polarized incoming photon that carries a large virtuality. In this case the scattering amplitude is related to parton distributions in the target. To be specific, for the diffraction at high energies the dominant contribution comes from the gluon evaluated at the scale of the order of the photon virtuality or of the meson mass. It should be added, however, that the photon – meson transition requires a non-zero energy-momentum transfer from the proton to the photon. Hence, the operators defining the gluon distribution in the proton are evaluated in non-forward kinematics. and the conventional gluon distribution has to be replaced by its off-diagonal counterpart.[5] The evolution of the off-diagonal partons is known in QCD,[5] and it was applied to construct models of these quantities.

In the case of transversely polarized photons, though, one finds some troublesome pieces in the amplitude which seem to spoil the diffractive factorisation. The hard part of the amplitude at the leading twist gives $1/[z(1-z)]^2$ dependence on the longitudinal momentum fraction z of the quark. The singular behaviour of this part at the end-points, $z \to 0$ and $z \to 1$, is not compensated by the asymptotic form of the meson distribution amplitude $\phi(z) = 6z(1-z)$, and infrared logarithmic divergencies appear in the integrals over z, breaking down the hard factorisation. Speaking in terms of the colour dipole model the photon wave function does not suppress large colour dipoles with an asymmetric sharing of the longitudinal momentum and those dipoles may interact with the target exchanging soft gluons.

An interesting concept was put forward[6] to use the singular behaviour of the hard amplitude to explain some peculiarities of the high-t diffractive meson photoproduction. Namely, the hard physics dictates a $1/|t|^4$ dependence of the amplitude without a helicity flip between the photon and the meson and a $1/|t|^3$ dependence of the helicity flip amplitude. Indeed, the data for ρ and ϕ exhibit approximately $1/|t|^3$ behaviour. Surprisingly, the analysis of angular distribution of the vector meson decay products, points to a strong dominance of the no-flip amplitude for all the range of momentum transfer.[7]

An assumption[6] that the distribution amplitudes do not vanish at $z \to 0$ and $z \to 1$ leads to an appearance of a power divergence at the end-points that replaces the usual logarithmic divergence. The necessary physical infrared cut-off is naturally provided by some soft scale $\sim \Lambda_{QCD}$. Thus, the end-point contributions would have a different scaling with t than it follows from the perturbative analysis. This is how one could explain the observed t-dependence $\sim 1/|t|^3$ of the dominant no-flip amplitude. Another consequence of this scenario is that the helicity need not be conserved for the quark–anti-quark configurations which contain a very slow parton and which are sensitive to non-perturbative effects, e.g. the instantons. Thus, this model could explain the qualitative features of the data for high-t meson production but no quantitative predictions have been provided so far. Although the idea is interesting, the assumptions behind it are controversial. For instance, the invariant mass $M_{q\bar{q}}$ of the end-point configurations at the quark level, given by $M_{q\bar{q}}^2 \simeq (m_q^2 + k_T^2)/[z(1-z)]$, is much larger than the meson mass M_V (with m_q and k_T denoting quark mass and transverse momentum respectively). Thus, such configurations should be strongly suppressed. The reason is that in an exclusive process, the invariant mass at the parton level should be close to the meson mass if the local parton-hadron duality is to hold.

Recently, a thorough analysis of the factorisation at the end-points beyond the leading twist contribution has been carried out.[8] The momentum dependent meson wave function is modelled in the following way. The starting point is the perturbative photon wave function, next its point-like component is removed using the Borel transform. The obtained meson wave function $\Psi_V(z, k_T) \sim \exp\left(-\frac{k_T^2 + m_q^2}{M_V^2 z(1-z)}\right)$ contains contributions from all twists. The resulting helicity structure is the same as the helicity structure of the photon. Note that this wave function gives the asymptotic form of the distribution amplitudes after the integration over the transverse momentum, and that it naturally suppresses the configurations with large invariant mass at the parton level.

This model permits a detailed analysis of the end-point contributions at all twists. An apparent singular behaviour of the integrals at the end-points is found to be only an effect of inappropriate extrapolations at the leading twist. In the complete model of the wave function the end-point divergences do not appear. Instead, at large photon virtuality Q^2 one finds logarithms of Q^2 which can be fully absorbed into the generalized evolution of t-channel $gg \to q\bar{q}$ exchange. Thus the end point contributions are separated into two factorised terms, containing the gluon distribution and the an effective $q\bar{q}$ distribution. The latter term may be neglected at small x and the gluon dominates.[8]

3. Elastic production of vector mesons

It was shown a long time ago[9,10] that elastic production of vector mesons may be used as a probe of the collinear parton densities in the proton, giving direct access to the off-diagonal gluon. In fact, a detailed measurement of the forward peak provides even more information. The t-dependence of the scattering is related to the profile of the scattering amplitude in the impact parameter plane by the Fourier transform. Furthermore, in collisions at high energies the transverse position of the dipole is conserved in the scattering, thus, in the b-representation the S-matrix is diagonal and much easier to determine. There is one difficulty left, however. Experiments measure cross sections, and the information about relative complex phases of the scattering matrix elements is partly lost. A procedure that was recently proposed for S-matrix determination[11] is based on a simplifying assumption that this matrix is real. In this scheme a closed formula was derived for the S-matrix elements in a dipole scattering off the proton, at a given impact parameter. The necessary experimental input was the differential cross section $d\sigma/dt$ for the elastic electroproduction of longitudinally polarised ρ meson.[11] The differential cross section of this process $d\sigma/dt$ is known for a few values of Q^2, in each case covering about a decade in x and the range of $-t < 0.6$ GeV^2. The results obtained were parameterised in a form following from the hypothesis of b-dependent saturation, and the saturation scale was determined as a function of the impact parameter.

In the colour dipole representation the dependence of dipole scattering amplitude on the impact parameter is influenced by the matter (mostly gluons) distribution in the proton. Thus, in principle, the cross section for vector meson production could be used to provide some clues on the three dimensional structure of the proton.

The b-dependent saturation of the elastic colour dipole – proton cross section is a phenomenon which follows directly from the unitarity constraints. A vector meson production process may probe dipole scattering in the kinematical domain where rescattering effects should be important. It opens a possibility to analyse in detail how the amplitude saturates in the impact parameter plane and for increasing collision energy. In the classical saturation model proposed by Golec-Biernat and Wüsthoff (GBW)[3] it is assumed that the nucleon-dipole interaction is confined to a disc in the transverse plane. In a more realistic picture the scattering matrix elements should be characterised by a smooth distribution $S(b)$ in the impact parameter. The large-b behaviour is constrained by the fact that QCD is a theory with a mass gap, thus $S(b)$ should decrease faster than $\exp(-mb)$ for $b \to \infty$. It may be argued that the Gaussian tail $S(b) \sim \exp(-Ab^2)$ accounts better for the confining properties of the QCD vacuum.[12]

Two recent extensions of the GBW model[13,14] use the Gaussian profile $S(b)$ of a single dipole scattering. The unitarity constraints are imposed by the Glauber-Muller prescription at each impact parameter[13] or by inserting[14] the b-dependence into solutions of Balitsky-Kovchegov equation.[15] Both the models were confronted with the data for elastic J/ψ photo- and electroproduction, giving a good fit. For the J/ψ though, the typical dipole size is related to the inverse c-quark mass, and the unitarity corrections are not large.[13] An important goal to achieve in the near future is to provide a global fit of all the data for exclusive vector meson production in the saturation model with the impact parameter dependence. Such an analysis, based on the present and future HERA data, would provide stringent experimental constraints of the dynamics of b-dependent saturation.

4. Diffractive production at large t

Recently, ZEUS[7] and H1[16] measured the differential cross section $d\sigma/dt$ for the ρ^0, ϕ and J/ψ diffractive production over a wide range of t, and also the corresponding spin density matrix elements. From those data estimates of the helicity amplitudes may be obtained. The amplitude without helicity flip dominates for all mesons and for the accessible t range. The double flip and single flip amplitudes have the magnitudes of one fifth and one tenth of the leading amplitude respectively.

High energy diffractive meson production in non-forward direction tests colour singlet exchange at large momentum transfer t. Elastic scattering of a dipole with size r is strongly suppressed for $r^2|t| \gg 1$ because the dipole tends to break up. Also gluon radiation that spoils the rapidity gap is important in this regime. Therefore the momentum transfer provides the hard scale for the problem and the perturbative approach is justified for the photon-meson transition, even for light mesons and the photoproduction limit.

In the colour dipole representation at high energies, the dipole scattering amplitude is diagonal in the quark helicity basis. In the leading twist approximation the quark momenta are collinear and the vector meson helicity is a simple sum of quark helicities. Beyond this approximation, the angular momentum originating from a motion of quarks inside the meson and a possible appearance of additional partons in higher Fock states has to be taken into account and the sum of quark helicities need not match the meson helicity. This explains how the double helicity flip occurs in the high-t photoproduction of ρ^0 and ϕ.[17,18,19]

The lowest order approximation to the C-even colour singlet in QCD is given by two gluons combined into a colour neutral state. This simple picture is not sufficient, though, when the rapidity gap $\hat{y}$ between the meson and the proton remnant is large. Perturbative corrections to the two-gluon exchange contain terms that behave like $(\hat{y}\alpha_s)^n$ at the n-th order of expansion. Since $\hat{y}\alpha_s \sim 1$, the fixed order perturbative expansion gives inaccurate results and a resummation of leading logarithms $(\hat{y}\alpha_s)^n$ (with $\hat{y} \sim \log s$) is necessary. Those contributions are diagrammatically represented as gluon ladders. The resummation is performed by the BFKL equation[2] providing a consistent framework to analyse the non-forward diffractive amplitudes.

Vector meson production data at $|t| > 1$ GeV^2 are dominated by events with a dissociated proton. Thus, in most theoretical analyses all the final states of the proton remnant are included, having a given transverse momentum and the total colour charge equal to zero. Then, the hard colour singlet couples predominantly to individual partons in the nucleon[20] resolving them at the hard scale $\sim |t|$. This approximation improves with increasing $|t|$, and this allows to factorise the cross sections into parton densities in the proton and a partonic cross section for the elastic meson production.

The scattering amplitude A in the BFKL formalism is given by a convolution over the gluon transverse momenta, of the universal gluon $2 \to 2$ Green function G and of two specific impact factors Φ_i describing the coupling of two gluons to the systems at the two sides of the rapidity gap, $A \sim \Phi_1 \otimes G \otimes \Phi_2$. The analytic approach to the BFKL equation is based on its conformal invariance[21] which permits a decomposition of the Green function in the basis of its conformal Eigenfunctions, indexed by a discrete conformal spin n and a continuous variable, ν. This method was successfully applied to the J/ψ diffractive production. At first, only the leading conformal component with $n = 0$ [22,23] was taken into account, and then the calculation was extended to all the conformal spins.[24]

For the production of light vector mesons, the impact factor representing photon-meson transition depends substantially on the details of both the photon and the meson structure. For the photon, one may use the perturbative wave function with a constituent quark mass. An alternative approach is based on the Operator Product Expansion in which only the leading dependence on the dipole vector $\vec{r}$ is retained at $r \to 0$. The dependence on the longitudinal momentum fraction is parameterised by the distribution amplitudes.[25] The approximation may be systematically improved by including higher twist terms and the QCD evolution. The same method may be applied also for the vector mesons.[26]

Recently, a few attempts have been made[17,23,24,18,19,27] to describe the high-t photoproduction data. In those papers important observations about the meson production dynamics are performed, but none of the approaches is completely successful. It was argued[17] that non-perturbative condensates in the QCD vacuum generate a sizeable chiral odd coupling of the photon to the quark anti-quark pair. This component significantly enhances the amplitude without helicity flip, and the discrepancy between the theory and the data decreases. Furthermore, an analysis was carried out[23,24,18,19] that went beyond the two-gluon exchange approximation[17] and the complete BFKL amplitudes for all helicities were determined. The differential cross sections for all the mesons were fitted very well[23,24,19] as well as the energy dependence and spin density matrix for J/ψ production. Although the inclusion of BFKL effects greatly improves the fit of the spin density matrix elements, this description is not satisfactory for light vector mesons. In particular, both the sign and the magnitude of the single helicity flip amplitude come out wrong when the most probable values of wave function parameters are used. The model is capable to provide a reasonable fit of the data with an unrealistically large value of the constituent quark mass in the photon wave function. This may indicate that the model lacks some physical mechanism that suppresses the scattering of large dipoles e.g. the Sudakov form-factor.[19]

Another important idea is to use the meson wave function beyond the fixed twist approximation.[27] In the model, the BFKL evolution is included in the doubly logarithmic approximation but the chiral odd components in the photon are not taken into account. As a consequence, the single helicity flip amplitude is found to dominate in severe contradiction with the data.

The lesson learned from those three approaches may be summarised in the following. The chiral odd components of the photon play an important rôle in the vector meson photoproduction. Calculations based on the BFKL equation[23,24,18,19] give a better picture of all the observables than the two gluon approximation. An improvement is needed in the treatment of the large dipole scattering, perhaps an inclusion of the Sudakov form-factor. Another interesting task is to apply the BFKL formalism to compute the helicity amplitudes at a large photon virtuality and non-zero $|t|$.

5. The Odderon

The Odderon is the leading part of the colour singlet exchange at high energies with odd C-parity, a partner of the C-even Pomeron. The surprising feature of the Odderon is that only very little evidence of this exchange has been found so far.[28] HERA experiments may be able to shed some light on the puzzle.

A simple and robust signal of the Odderon could be provided by a diffractive photoproduction of a C-even meson. In the perturbative QCD, three t-channel gluons is the simplest system which can take the quantum numbers of the Odderon and that contributes at asymptotic energies. The necessary inclusion of the leading logarithmic corrections leads to the BKP evolution equation[29] for the Odderon. The equation yields amplitudes with the intercept equal to one,[30] that is the diffractive cross sections constant with the energy.

Diffractive η_c photoproduction was studied within the perturbative framework. The charmed quark mass sets the scale for the problem, and the $\gamma \to \eta_c$ impact factor may be calculated in perturbative QCD. For the coupling of the Odderon to the proton, one is forced to use a model, and thus, to introduce some uncertainties. The first estimates of the cross section based on the three gluon approximation gave $\sigma(\gamma p \to \eta_c p) \simeq 10$ pb,[31] a signal probably too weak to detect given the low acceptance for η_c. The QCD evolution effects increased the prediction significantly[32] to yield $\sigma(\gamma p \to \eta_c p) \simeq 50$ pb. In addition, the cross section for a similar diffractive process with the proton dissociation was found to be even larger,[33] $\sigma(\gamma p \to \eta_c X) \simeq 60$ pb. Thus, the combined cross section for the diffractive η_c production exceeds earlier estimates by one order of magnitude and may turn out to be large enough for HERA experiments to discover the perturbative Odderon exchange.

In search of the Odderon, other promising observables have been also proposed. Estimates of the diffractive photoproduction cross-section of π^0, f_2, a_2^0 were obtained in the Stochastic Vacuum Model.[34] The H1 collaboration set the upper limits for the corresponding cross sections[35] and, unfortunately, the results contradict the prediction for π^0 production. Besides that, an important idea was developed to use the Pomeron-Odderon interference effects in diffractive processes in the Odderon searches. The charge and spin asymmetries in $\pi^+\pi^-$ diffractive electroproduction were shown to be sensitive to the Odderon amplitude.[36]

6. Conclusions

The theory of diffractive meson production has been developing rapidly since about ten years. It touches some central problems of QCD, transition between the soft and the hard regime of the scattering, the resummation schemes at large energies and the saturation phenomenon. In addition, the analysis of exclusive processes showed importance of the non-forward parton distributions and stimulated studies of their properties and their QCD evolution. Analysis of the high-t photoproduction of vector mesons clearly indicates that the photon and light vector mesons contain sizeable chiral odd components. Diffractive production of C-even mesons may shed some light on the Odderon puzzle. All of this shows how much inspiration originates from this field.

Recent advances of the theory narrow the gap between the first principles of QCD and some successful phenomenological models. The mechanism of diffractive factorisation which is the basis for the perturbative approach is getting better understood. The Balitsky-Kovchegov equation provides a QCD description of the perturbative parton saturation phenomenon within a well defined approximation scheme. Exchange amplitudes of the perturbative Pomeron and Odderon are computed for increasing number of processes at the leading logarithmic accuracy, and estimates of non-leading corrections are also known in some cases. The good news is, that together with the improving theoretical foundations of the phenomenological analysis, the data become better described. This may indicate that the crucial mechanisms that govern the diffractive meson production are correctly identified. Still, the quantitative description of the data is rather crude.

Wealth of HERA data, to be supplemented by the new data from HERA2 will certainly keep stimulating efforts to obtain a more complete QCD description of the diffractive meson production. With improved statistics, the constraints on the theoretical estimates will become more stringent. We will learn more on some important phenomena, like the unitarisation of QCD amplitudes in the impact parameter plane. Finally, there is a lot of room left for measurements of new observables, for surprising results and for new theoretical ideas.

Acknowledgments

I would like to thank the Organizers for creating an excellent, inspiring atmosphere during the Workshop. The author is partially supported by the Polish Committee for Scientific Research (KBN) grant no. 5P03B 14420.

References

1. see for instance P. Thompson, these Proceedings.
2. L. N. Lipatov, Sov. J. Nucl. Phys. **23** (1976) 338; E. A. Kuraev, L. N. Lipatov and V. S. Fadin, Sov. Phys. JETP **44**, 443 (1976); *ibid.* **45** (1977) 199; I. I. Balitsky and L. N. Lipatov, Sov. J. Nucl. Phys. **28** (1978) 822.
3. K. Golec-Biernat and M. Wüsthoff, Phys. Rev. D **59** (1999) 014017; Phys. Rev. D **60** (1999) 114023.
4. J. C. Collins, L. Frankfurt and M. Strikman, Phys. Rev. D **56** (1997) 2982.
5. A.V. Radyushkin, Phys. Lett. B **385** (1996) 333; Phys. Rev. D **56** (1997) 5524.
6. P. Hoyer, J. T. Lenaghan, K. Tuominen and C. Vogt, arXiv:hep-ph/0210124.
7. S. Chekanov et al. [ZEUS Collaboration], Eur. Phys. J. C **26** (2003) 389.
8. A. Ivanov and R. Kirschner, Eur. Phys. J. C **29** (2003) 353.
9. M. G. Ryskin, Z. Phys. C **57** (1993) 89.
10. S. J. Brodsky et al. Phys. Rev. D **50** (1994) 3134.
11. S. Munier, A. M. Staśto and A. H. Mueller, Nucl. Phys. B **603** (2001) 427.
12. L. Motyka, Acta Phys. Polon. B **34** (2003) 3069.
13. H. Kowalski and D. Teaney, arXiv:hep-ph/0304189.
14. E. Gotsman et al. Acta Phys. Polon. B **34** (2003) 3255.
15. I. Balitsky, Nucl. Phys. B **463** (1996) 99; Y. V. Kovchegov, Phys. Rev. D **60** (1999) 034008.
16. C. Adloff et al. [H1 Collaboration], Phys. Lett. B **539** (2002) 25; A. Aktas et al. [H1 Collaboration], Phys. Lett. B **568** (2003) 205.
17. D. Y. Ivanov et al. Phys. Lett. B **478** (2000) 101.
18. R. Enberg et al. JHEP **0309** (2003) 008;
19. G. G. Poludniowski et al. arXiv:hep-ph/0311017.
20. J. Bartels et al. Phys. Lett. B **348** (1995) 589.
21. L. N. Lipatov, Phys. Rept. **286** (1997) 131.
22. J. R. Forshaw and M. G. Ryskin, Z. Phys. C **68** (1995) 137.
23. J. R. Forshaw and G. Poludniowski, Eur. Phys. J. C **26** (2003) 411.
24. R. Enberg, L. Motyka and G. Poludniowski, Eur. Phys. J. C **26** (2002) 219.
25. P. Ball, V. M. Braun and N. Kivel, Nucl. Phys. B **649** (2003) 263.
26. P. Ball et al. Nucl. Phys. B **529** (1998) 323.
27. A. Ivanov and R. Kirschner, arXiv:hep-ph/0311077.
28. C. Ewerz, arXiv:hep-ph/0306137.
29. J. Bartels, Nucl. Phys. B **175** (1980) 365; J. Kwieciński and M. Praszałowicz, Phys. Lett. B **94** (1980) 413.
30. J. Bartels et al. Phys. Lett. B **477** (2000) 178.
31. J. Czyżewski et al. Phys. Lett. B **398** (1997) 400.
32. J. Bartels et al. Eur. Phys. J. C **20** (2001) 323.
33. J. Bartels, M. A. Braun and G. P. Vacca, arXiv:hep-ph/0304160.
34. E. R. Berger et al. Eur. Phys. J. C **9** (1999) 491; Eur. Phys. J. C **14** (2000) 673.
35. C. Adloff et al. [H1 Collaboration], Phys. Lett. B **544** (2002) 35.
36. P. Hagler et al. Phys. Lett. B **535** (2002) 117; Eur. Phys. J. C **26** (2002) 261.

UNINTEGRATED PARTON DENSITIES AND APPLICATIONS

GÖSTA GUSTAFSON

Dept. of Theor. Physics, Lund University,
Sölvegatan 14A, SE-22362 Lund, Sweden
E-mail: gosta@thep.lu.se

Different formalisms for unintegrated parton densities are discussed, and some results and applications are presented.

1. Introduction

Calculations based on *collinear factorization* work very well in many applications, including e.g.:
  - DIS at large Q^2.
  - High $p_\perp$ jets in $p\bar{p}$ collisions.
  - Inclusive observables.

In these cases DGLAP evolution works, and $k_\perp$-ordered chains up to a hard subcollision or a highly virtual photon dominate the parton evolution. The unintegrated parton density, $\mathcal{F}(x, k_\perp^2)$, then satisfies the relations

$$F(x, Q^2) = \int^{Q^2} \frac{dk_\perp^2}{k_\perp^2} \mathcal{F}(x, k_\perp^2); \quad \mathcal{F}(x, k_\perp^2) = \left. \frac{\partial F(x, Q^2)}{\partial \ln Q^2} \right|_{Q^2 = k_\perp^2}. \quad (1)$$

This formalism has, however, problems for observables which are sensitive to the $k_\perp$ of the "active" quark. Some examples are:
  - Transverse momentum unbalance in 2-jet events.
  - Heavy quark production.
  - Forward jets at small x_{Bj}.

In many of these cases calculations, which allow for one extra gluon, e.g. LO pQCD + parton showers or NLO DGLAP calculations, are able to give a good description of the data, but also these calculations have problems for observables which involve a large rapidity separation. In the following I will discuss different formalisms for $k_\perp$-factorization, non-$k_\perp$-ordered evolution and some applications.

2. Non-$k_\perp$-ordered Evolution and $k_\perp$-Factorization

At small x and limited Q^2 non-$k_\perp$-ordered chains give important contributions. In a formalism based on $k_\perp$-factorization the non-integrated pdfs $\mathcal{F}(x, k_\perp^2, Q^2)$ may depend on 2 scales, the $k_\perp$ of the active parton and the virtuality of the photon or the hard scattering. The second scale is then related to a limiting angle for the emissions, as discussed further below. It is also important to remember that the gluon distribution is not an observable, and depends on the theoretical formalism.

2.1. BFKL

In the BFKL evolution[1], accurate to leading $\log 1/x$, $\mathcal{F}(x, k_\perp^2)$ depends only on a single scale $k_\perp^2$, and satisfies an integral equation with a kernel $K(\mathbf{k}_\perp, \mathbf{k}'_\perp)$:

$$\mathcal{F}(x, k_\perp^2) = \mathcal{F}_0 + \frac{3\alpha_s}{\pi} \int \frac{dz}{z} \int d^2 k'_\perp K(\mathbf{k}_\perp, \mathbf{k}'_\perp) \mathcal{F}(\frac{x}{z}, k'^2_\perp). \qquad (2)$$

We note that the dominant leading log behaviour originates from the $1/z$ pole in the splitting function. This leading order result has problems because the NLO corrections are very large. There are e.g. large effects from energy conservation, as demonstrated e.g. in MC studies by J. Andersen *et al.*[2], which imply that analytic calculations often are unreliable.

The kernel K in Eq. (2) describes the emission of a quasi-real gluon with transverse momentum $\mathbf{q}_\perp$ from a virtual link with momentum $\mathbf{k}'_\perp$, which after the emission gets momentum $\mathbf{k}_\perp = \mathbf{k}'_\perp - \mathbf{q}_\perp$. The kernel has the property that small values of $\mathbf{q}_\perp$ are suppressed. Such emissions are compensated by virtual corrections, and in the BFKL formalism this is taken into account by treating the links as Reggeized gluons.

We can compare this situation with e^+e^--annihilation. Here the total cross section is determined by the lowest order diagram (α_s^0). The lowest order contribution to 3-jet events is $\mathcal{O}(\alpha_s)$, and to this order there are negative contributions to $\sigma_{2\text{jet}}$, such that σ_{tot} is approximately unchanged. In this case the gluon emissions can be treated with Sudakov form factors.

For a link in the BFKL chain the $\mathcal{O}(\alpha_s)$ corrections give a compensation for emissions for which $q_\perp < k_\perp, k'_\perp$. These soft emissions give no contribution to the inclusive cross section, i.e. to F_2. They must, however, be added for exclusive final states, with appropriate Sudakov form factors. The net result of this is that downward steps in $k_\perp$ are suppressed by a factor $k_\perp^2/k_\perp'^2$. In the relevant variable $\ln k_\perp^2$, this corresponds to an exponential suppression allowing downward steps within ~ 1 unit in $\ln k_\perp^2$ [3].

2.2. The CCFM model

An evolution equation which interpolates between BFKL and DGLAP was formulated by Catani, Ciafaloni, Fiorani, and Marchesini[4]. This formalism, the CCFM model, is based on a different separation between initial and final state radiation (denoted ISR and FSR respectively). Some soft emissions are included in the ISR, but this increase is compensated by "non-eikonal" form factors. The ISR ladder satisfies the following constraint:

In the ISR emissions *the colour order agrees with the order in energy* (or order in lightcone momentum $p_+ = p_0 + p_L$). (Thus all emissions which in colour order are followed by a more energetic one are treated as FSR.) With this constraint colour coherence implies that *this ordering also coincides with the ordering in angle, or rapidity.*

This angular ordering implies that the emission angle for the last emission must be known, when a new step is taken in the evolution. This angle therefore appears as a limiting angle in the non-integrated distribution function used in the evolution. Thus the distribution $\mathcal{F}(x, k_\perp^2, \bar{q})$ depends on *two* scales, $k_\perp^2$ and $\bar{q}$, where the limiting angle is specified by $y_{\text{limit}} = \ln(xM_p/\bar{q})$ (in the proton rest frame).

For exclusive final states, the final state radiation has to be added in appropriate kinematical regions. As the initial state radiation is ordered in p_+ but not in p_-, also the final state radiation is unsymmetric with respect to the initial proton and photon directions.

We can compare with the BFKL and DGLAP formalisms, where the parton distributions depend on a single scale. The initial state radiation is strongly ordered either in q_+ (for BFKL), or in $q_\perp$ (for DGLAP). In both cases this ordering also implies a strong ordering in y. CCFM interpolates between the two regions at the cost of a more complicated formalism.

That more emissions are included in the ISR implies that the average step is shorter. Therefore the angular ordering constraint becomes more important and implies a strong dependence upon the scale $\bar{q}$. This feature also implies that it has not been easy to implement the CCFM model in an event generator, but such a program, CASCADE by Hannes Jung[5], is now available. The CCFM model does not include quark links in the parton chains. The original model, and also the first version of the CASCADE program, referred to as JS, also included only the singular term $\propto 1/z$ in the splitting function. To include the non-singular terms is not straight forward, but one possible solution is implemented in a new fit, called J2003[6]. In this fit set 1 includes only the singular terms, while in set 2 also the non-singular terms in the splitting function are included.

2.3. *The Linked Dipole Chain Model*

The Linked Dipole Chain Model, LDC[7], is a reformulation and generalisation of the CCFM model. It is based on a different separation between initial and final state radiation. The ISR is ordered in both q_+ and q_-, and satisfies the constraint $q_{\perp i} > \min(k_{\perp i}, k_{\perp i-1})$. Softer emissions are treated as final state emissions. In this respect the LDC formalism is more similar to BFKL. We note that the ordering in q_+ and q_- also implies an ordering in angle. An important property is also that the parton chain is fully symmetric with respect to the two ends of the chain.

The fact that fewer emissions are treated as ISR implies that a single chain in LDC corresponds to the collective contributions from several possible chains in CCFM. Now it turns out that summing over all possible emissions in CCFM, the non-eikonal form factors exactly cancel. The result is a simple evolution equation in terms of a *single scale* unintegrated density function $\mathcal{F}(x, k_\perp^2)$. In the MC implementation it is, however, also possible to add an angular cut and thus obtain results for a two-scale distribution.

We note that to leading order the LDC and CCFM formalisms give the same result for the integrated structure functions. The parton chains and the unintegrated distributions differ, however, and only after addition of final state emissions in the different relevant kinematic regions do the two formalisms also give the same result for exclusive final parton states (to leading order).

Thus the LDC formalism results in a much simplified evolution equation. Other merits of the formalism include:

- It contains the same chains as in DGLAP for Q^2 large, which makes it easier to interpret the differences between large and small Q^2.

- There is a natural generalization to include subleading terms, e.g. quark links, non-singular terms in the splitting functions, and a running α_s.

- It is suitable for implementation in a MC, and the event generator LDCMC is produced by Lönnblad and Kharraziha[8].

– The MC gives very good fits to experimental F_2 data, and it also agrees well with MRST and CTEQ results for the *integrated* gluon distribution[9].

2.4. *Other formalisms for unintegrated parton densities*

A different formalism, which also interpolates smoothly between BFKL and DGLAP, was formulated by Kwieciński, Martin, and Staśto (KMS)[10]. This formulation is based on a single scale evolution equation. The contributions from $k_\perp^2 > Q^2$ are neglected, and thus the gluon distribution satisfies Eq. (1). This formalism has been further developed by Kimber, Martin and Ryskin (KMR)[11]. In their approach the single scale KMS evolution is used, but an angular constraint is applied for the last step. The result is therefore a density distribution, which depends on two scales. Since the underlying KMS evolution does not include the soft initial state radiation in the CCFM model, it also takes fewer and larger steps, which implies that the dependence on the limiting angle is significantly smaller than for the CCFM formalism.

3. Comparison between results from different formalisms

We here want to compare the parton densities obtained in different formalisms, and also study the effects of the non-singular terms in the splitting functions and of quark links in the evolution chains. Some results are shown in Fig. 1[9,12]. Among the LDC fits the result denoted *standard* contains both quark links and non-singular splitting terms, *gluonic* contains only gluon links, and *leading* only singular terms. (In the fit *gluonic-2*, refered to in the next section, the power of $(1-x)$ in the input gluon density is changed to 7 instead of its value 4 in *gluonic*.) In all cases the input distributions for $k_\perp^2 = Q_0^2$ are adjusted to give good fits to experimental F_2 data. The CCFM result J2003 set 1 contains only gluons and only the $1/z$ pole in the splitting function. We see that the effect of quarks and non-singular terms become larger for larger $k_\perp$. The sensitivity to the scale $\bar{q}$ is illustrated in Fig. 2a, which shows the gluon density $\mathcal{F}(x, k_\perp^2, \bar{q})$ as function of $\bar{q}/k_\perp$ for fixed $k_\perp$. As discussed above, in the CCFM formalism the density is very sensitive to the $\bar{q}$ scale, and varies strongly for $\bar{q}/k_\perp$ between 1 and 2.

It is, however, interesting to note that observable cross sections differ much less than the parton densities, as seen in the next section. The reason is that the distributions differ in particular for $\bar{q} < k_\perp$, while in a hard subcollision the dominant contributions are obtained for $k_\perp^2 \approx \bar{q}^2/4$ [9]. In Fig. 2b we see that the results are indeed not so different for $\bar{q} = 2k_\perp$. In

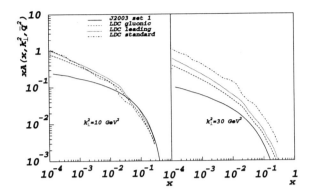

Figure 1. Comparison of different sets of unintegrated gluon densities at scale $\bar{q} = 10$ GeV. For the notations, see the main text.

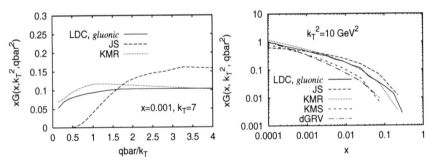

Figure 2. The LDC *gluonic* unintegrated gluon density compared to the results of JS and KMR. *Left*: As function of $\bar{q}/k_\perp$ for fixed x and $k_\perp$. *Right*: As function of x for $k_\perp^2 = 10$ GeV^2 and $\bar{q} = 2k_\perp$. Also shown are here the single scale results from KMS and the derivative of the GRV fit.

this figure also the single scale result from KMS and the derivative of the GRV result (denoted dGRV) are included.

4. Applications

4.1. *Heavy quark production*

Applications of the $k_\perp$-factorization formalism to b quark production at the Tevatron are shown in Fig. 3. We see that for both the LDC and the CCFM models the fits with only the leading term reproduce the data best, while the other fits, which should be expected to be more accurate, although being significantly better than the NLO QCD result, do not give equally good fits. We should note, however, that b production is also well reproduced by collinear factorization plus parton showers, as implemented in the PYTHIA event generator[17][18].

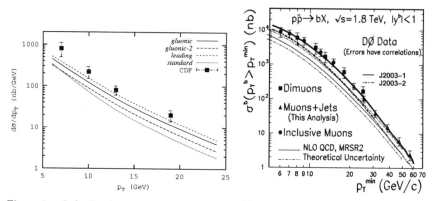

Figure 3. *Left*: Production of b-quarks at CDF[13] compared to results from the LDC model[14]. *Right*: D0 data[15] compared with CCFM results[16]. Here also a NLO QCD result is included. For notations, see the main text.

4.2. *Forward jets*

Results for forward jet production at HERA are shown in Fig. 4. Here we have a comparatively large separation in rapidity, and we see that LO and NLO dijet calculations are far below the data. The CCFM model gives a much better description, but also here we see that the best fit is obtained including only the singular terms in the splitting function. The reason for this is still not understood; is there some dynamical mechanism which somehow compensates the effect of the non-singular terms?

4.3. *Minimum bias and underlying events in pp collisions*

In hadron-hadron collisions collinear factorization works well for calculations of high-$p_\perp$ jets. However, in this formalism the minijet cross section diverges with $\sigma_{jet} \sim 1/p_\perp^4$, which implies that also the total $E_\perp$ diverges. This implies the need for a soft cutoff, and in PYTHIA fits to experimental

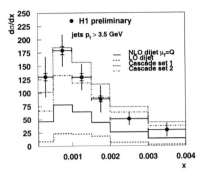 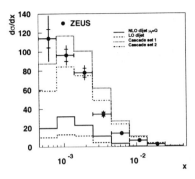

Figure 4. Forward jet data from H1 and ZEUS compared to the CASCADE MC and LO and NLO dijet calculations[19].

data give a cutoff $p_{\perp 0} \sim 2$ GeV. This cutoff is also growing with energy, which makes it difficult to extrapolate safely to the high energies at LHC.

The symmetry between the two ends of the parton chain implies that the LDC formalism also is applicable to hadron-hadron collisions[20]. In the $k_\perp$-*factorization* formalism the off-shell matrix element does not blow up when the exchanged transverse momentum $k_\perp \to 0$, and we have therefore a dynamical cutoff for soft minijets. The cross section for a *chain* in pp collisions (which possibly may contain more than one hard subcollision) can thus be obtained from the fit to DIS data. An important point is here that the result is insensitive to the soft cutoff, Q_0, in the evolution. DIS data can be fitted with different values for Q_0, if the input distribution $f_0(x, Q_0^2)$ is adjusted accordingly. If Q_0 is increased, the number of hard chains decreases, but at the same time the number of soft chains (for which all emissions have $q_\perp < Q_0$) increases, so that the total number of chains is approximately unchanged.

There are two sources for *multiple interactions*: It is possible to have two hard scatterings in the same chain, and there may be more than one chain in a single event. The LDC model, when applied to pp collisions, can predict the correlations between hard scatterings within one chain, and also the average number of chains in a single event. The experimentally observed "pedestal effect" indicates that the hard subcollisions are highly correlated, so that central collisions have many minijets, while peripheral collisions have fewer minijets. In PYTHIA comparisons with data favour a distribution in the number of subcollisions, which is very close to a geometric distribution[21].

Some preliminary results from the LDC model are shown in fig. 5. Here a geometric distribution is assumed for the number of chains in one event. Fig. 5a shows the number of minijets in the "minimum azimuth region" $60° < \phi < 120°$ at $\sqrt{s} = 1.8$ TeV. The two LDC curves are obtained for soft cut-off values 0.99 and 1.3 GeV, showing the insensitivity to this cut-off. The two PYTHIA curves correspond to default parameter values, and parameters tuned to CDF data[22]. We note that the LDC result agrees very well with the tuned PYTHIA result. Fig. 5b shows corresponding results for LHC. Also here the two curves correspond to different cut-off values, and for comparison the result for 1.8 TeV is also indicated. We see that the activity increases by a little more than a factor of 2 between the two energies.

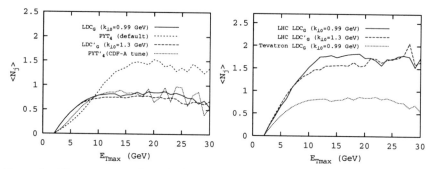

Figure 5. The average number of minijets in the "minimum azimuth region" for $|\eta| < 2.5$ vs. $E_\perp$ for the hardest jet. Left: For $\sqrt{s} = 1.8$ TeV. Right: For 14 TeV.

The symmetry of the formalism implies that the chains join at one end at the same rate as they multiply at the other. The chain cross section grows like s^λ, and therefore the average chain multiplicity satisfies $<n_{\text{chain}}> \propto s^\lambda/\sigma_{\text{tot}}$. Thus the results also may have implications for unitarization, saturation and diffraction. Work in these areas is in progress.

5. Conclusions

Our main conclusions can be summarized as follows:

- Unintegrated parton densities are not observables, and their properties depend strongly on the definitions.

- Different formalisms give similar results for $\mathcal{F}(x, k_\perp^2, \bar{q}^2 = 4k_\perp^2)$. This also implies that the predictions for observable quantities often are similar.
- The rôle of the non-singular terms in the splitting functions is still a problem.
- Observables without a large rapidity separation are often well described by higher order matrix elements plus DGLAP evolution.
- There is a close relation between DIS and high energy pp collisions. The properties of the underlying event and minimum bias events can be predicted from DIS data.

Acknowledgements

I am indebted to Leif Lönnblad and Hannes Jung for help in the preparation of this talk.

References

1. E.A. Kuraev, L.N. Lipatov, V.S. Fadin, *Sov. Phys. JETP* **45** (1977) 199; Ya.Ya. Balitsky, L.N. Lipatov, *Sov. J. Nucl. Phys.* **28** (1978) 822.
2. J. Andersen, J. Stirling, *JHEP* **02** (2003) 018.
3. G. Gustafson, G. Miu, *Eur. Phys. J.* **C23** (2002) 267.
4. M. Ciafaloni, *Nucl. Phys.* **B269** (1988) 49; S. Catani, F. Fiorani, G. Marchesini, *Nucl. Phys.* **B336** (1990) 18.
5. H. Jung, G. Salam, *Eur. Phys. J.* **C19** (2001) 351; H. Jung, *Comp. Phys. Comm.* **143** (2002) 100.
6. M. Hansson, H. Jung, hep-ph/0309009.
7. B. Andersson, G. Gustafson, J. Samuelsson, Nucl. Phys. **B467** (1996) 443; B. Andersson, G. Gustafson, H. Kharraziha, *Phys. Rev.* **D57** (1998) 5543.
8. H. Kharraziha, L. Lönnblad, *JHEP* **03** (1998) 006.
9. G. Gustafson, G. Miu, L. Lönnblad, *JHEP* **0209** (2002) 005.
10. J. Kwieciński, A.D. Martin, A. Staśto, *Phys. Rev.* **D56** (1997) 3991.
11. M.A. Kimber, A.D. Martin, M.G. Ryskin, *Phys. Rev.* **D63** (2001) 114027.
12. H. Jung, private communication.
13. D. Acosta et al. (CDF), *Phys. Rev.* **D66** (2002) 032002.
14. A.V. Lipatov, L. Lönnblad, N.P. Zotov, hep-ph/0309207.
15. B. Abbott et al. (D0), *Phys. Rev. Lett.* **85** (2000) 5068.
16. H. Jung, Contribution to the International Multiparticle Symposium, Cracow, Sept. 2003.
17. E. Norrbin, T. Sjöstrand, *Eur. Phys. J.* **C17** (2000) 137.
18. R.D. Field, *Phys. Rev.* **D65** (2002) 094006.
19. H. Jung, for H1 and ZEUS collaborations, *Nucl. Phys. B - Proceedings supplement* **117** (2002) 352.
20. G. Gustafson, G. Miu, *Eur. Phys. J.* **C23** (2002) 267; G. Gustafson, G. Miu, L. Lönnblad, *Phys. Rev.* **D67** (2003) 034020.

21. T. Sjöstrand et al., *Comp. Phys. Comm.* **135** (2001) 238;
    T. Sjöstrand, M. van Zijl, *Phys. Rev.* **D36** (1987) 2019.
22. R.D. Field, private comm. at the workshop MC@LHC, CERN, July 2003.

INSTANTON-DRIVEN SATURATION AT SMALL X

F. SCHREMPP

Deutsches Elektronen Synchrotron (DESY),
Notkestrasse 85,
D-22607 Hamburg, Germany
E-mail: fridger.schrempp@desy.de

A. UTERMANN

Institut für Theoretische Physik der Universität Heidelberg,
Philosophenweg 16,
D-69120 Heidelberg, Germany
E-mail: A.Utermann@thphys.uni-heidelberg.de

We report on the interesting possibility of instanton-driven gluon saturation in lepton-nucleon scattering at small Bjorken-x. Our results crucially involve non-perturbative information from high-quality lattice simulations. The conspicuous, intrinsic instanton size scale $\langle\rho\rangle \approx 0.5$ fm, as known from the lattice, turns out to determine the saturation scale. A central result is the identification of the "colour glass condensate" with the QCD-sphaleron state.

1. Motivation

1.1. *Saturation in the Parton Picture*

One of the most important observations from HERA is the strong rise of the gluon distribution at small Bjorken-x [1]. On the one hand, this rise is predicted by the DGLAP evolution equations [2] at high Q^2 and thus supports QCD [3]. On the other hand, an undamped rise will eventually violate unitarity. The reason for the latter problem is known to be buried in the linear nature of the DGLAP- and the BFKL-equations [4]: For decreasing Bjorken-x, the number of partons in the proton rises, while their effective size $\sim 1/Q$ increases with decreasing Q^2. At some characteristic scale $Q^2 \approx Q_s^2(x)$, the gluons in the proton start to overlap and so the linear approximation is no longer applicable; non-linear corrections to the linear evolution equations [5] arise and become significant, potentially taming the growth of the gluon distribution towards a "saturating" behaviour.

1.2. Instantons and Saturation?

eP-scattering at small Bjorken-x and decreasing Q^2 uncovers a novel regime of QCD, where the coupling α_s is (still) small, but the parton densities are so large that conventional perturbation theory ceases to be applicable. Much interest has recently been generated through association of the saturation phenomenon with a multiparticle quantum state of high occupation numbers, the "Colour Glass Condensate" that correspondingly, can be viewed [6] as a strong *classical* colour field $\propto 1/\sqrt{\alpha_s}$.

Being extended non-perturbative and topologically non-trivial fluctuations of the gluon field, instantons [7] (I) are naturally very interesting in the context of saturation, since

- classical *non-perturbative* colour fields are physically appropriate;
- the functional form of the instanton gauge fields is explicitly known and their strength is $A_\mu^{(I)} \propto \frac{1}{\sqrt{\alpha_s}}$ as needed;
- an identification of the "Colour Glass Condensate" with the QCD-sphaleron state appears very suggestive[8,9] (c.f. below and Sec 3.2).

Two arguments in favour of instanton-driven saturation are particularly worth emphasizing.

We know already from I-perturbation theory that the instanton contribution tends to strongly increase towards the softer regime [10,11,12]. The mechanism for the decreasing instanton suppression with increasing energy is known since a long time [13,14]: Feeding increasing energy into the scattering process makes the picture shift from one of tunneling between adjacent vacua ($E \approx 0$) to that of the actual creation of the sphaleron-like, coherent multi-gluon configuration [15] on top of the potential barrier of height [10] $E = m_{\rm sph} \propto \frac{1}{\alpha_s \rho_{\rm eff.}}$.

A crucial aspect concerns the I-size ρ. On the one hand it is just a collective coordinate to be integrated over in any observable, with the I-size distribution $D(\rho) = d\,n_I/d^4z d\rho$ as universal weight. On the other hand, according to lattice data, $D(\rho)$ turns out to be *sharply* peaked (Fig. 1 (left)) around $\langle\rho\rangle \approx 0.5$ fm. Hence instantons represent truly non-perturbative gluons that bring in naturally an intrinsic size scale $\langle\rho\rangle$ of hadronic dimension. As we shall see, $\langle\rho\rangle$ actually determines the saturation scale [19,8,9].

Presumably, it is also reflected in the conspicuous *geometrization* of soft QCD at high energies [20,19,8]. For related approaches associating instantons with high-energy (diffractive) scattering, see [21, 22, 23, 14]. Instantons in the context of small-x saturation have also been studied recently by Shuryak

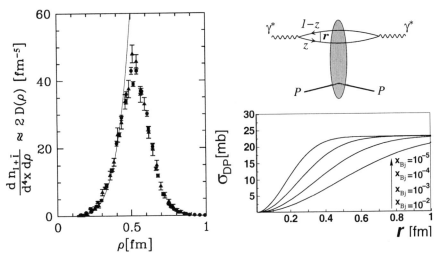

Figure 1. (Left) UKQCD lattice data[16,17,18] of the $(I+\bar{I})$-size distribution for quenched QCD ($n_f = 0$). Both the sharply defined I-size scale $\langle \rho \rangle \approx 0.5$ fm and the parameter-free agreement with I-perturbation theory[17,18] (solid line) for $\rho \lesssim 0.35$ fm are apparent. (Right) $\gamma^* p$ scattering at small x. The photon fluctuates into a $q\bar{q}$ dipole interacting with the proton (top). Generic behaviour of the dipole-proton cross section, as taken from the GBW-model [28] (bottom).

and Zahed [24], with conclusions differing in part from those of our preceding work [20,19,8,9]. Their main emphasis rests on Wilson loop scattering, and lattice information was not used in their approach.

In this paper we shall report about our results on the interesting possibility of instanton-driven saturation at small Bjorken-x. They have been obtained by exploiting crucial non-perturbative information from high-quality lattice simulations [16,17].

2. Setting the Stage

The investigation of saturation becomes most transparent in the familiar colour-dipole picture [25] (cf. Fig. 1 (top right)), notably if analyzed in the so-called dipole frame[26]. In this frame, most of the energy is still carried by the hadron, but the virtual photon is sufficiently energetic, to dissociate before scattering into a $q\bar{q}$-pair (a *colour dipole*), which then scatters off the hadron. Since the latter is Lorentz-contracted, the dipole sees it as a colour source of transverse extent, living (essentially) on the light cone. This colour field is created by the constituents of the well developed hadron wave function and – in view of its high intensity, i.e. large occupation numbers – can be considered as classical. Its strength near saturation is $\mathcal{O}(1/\sqrt{\alpha_s})$. At high energies, the lifetime of the $q\bar{q}$-dipole is much larger than the interaction time between this $q\bar{q}$-pair and the hadron and hence, at small $x_{\rm Bj}$, this gives rise to the familiar factorized expression of the inclusive photon-proton cross sections,

$$\sigma_{L,T}(x_{\rm Bj}, Q^2) = \int_0^1 dz \int d^2\mathbf{r}\, |\Psi_{L,T}(z,r)|^2\, \sigma_{\rm DP}(r,\ldots). \quad (1)$$

Here, $|\Psi_{L,T}(z,r)|^2$ denotes the modulus squared of the (light-cone) wave function of the virtual photon, calculable in pQCD, and $\sigma_{\rm DP}(r,\ldots)$ is the $q\bar{q}$-dipole-nucleon cross section. The variables in Eq. (1) are the transverse ($q\bar{q}$)-size $\mathbf{r}$ and the photon's longitudinal momentum fraction z carried by the quark. The dipole cross section is expected to include in general the main non-perturbative contributions. For small r, one finds within pQCD [25,27] that $\sigma_{\rm DP}$ vanishes with the area πr^2 of the $q\bar{q}$-dipole. Besides this phenomenon of "colour transparency" for small $r = |\mathbf{r}|$, the dipole cross section is expected to saturate towards a constant, once the $q\bar{q}$-separation r exceeds a certain saturation scale r_s. While there is no direct proof of the saturation phenomenon, successful models incorporating saturation do exist [28,29] and describe the data efficiently.

Let us outline more precisely the strategy we shall pursue:

The guiding question is: Can background instantons of size $\sim \langle\rho\rangle$ give rise to a saturating, geometrical form for the dipole cross section,

$$\sigma_{\rm DP}^{(I)}(r,\ldots) \stackrel{r \gtrsim \langle\rho\rangle}{\sim} \pi\langle\rho\rangle^2 \quad (2)$$

We have obtained answers from two alternative approaches [20,19,8,9]:

(1) *From I-perturbation theory to saturation:* Here, we start from the large Q^2 regime and appropriate cuts such that I-perturbation theory is strictly valid. The corresponding known results on I-induced DIS processes[30] are then transformed into the colour-dipole picture. With the crucial help of lattice results, the $q\bar{q}$-dipole size r is next carefully increased towards hadronic dimensions. Thanks to the lattice input, IR divergencies are removed and the original cuts are no longer necessary.

(2) *Wilson-loop scattering in an I-background:* As a second, complementary approach we have considered the semi-classical, non-abelian eikonal approximation. It results in the identification of the $q\bar{q}$-dipole with a Wilson loop, scattering in the non-perturbative

colour field of the proton. The field $A_\mu^{(I)} \propto \frac{1}{\sqrt{\alpha_s}}$ due to background instantons is studied as a concrete example, leading to analytically calculable results in qualitative agreement with the first approach.

3. From I-Perturbation Theory to Saturation

3.1. *The Simplest Process:* $\gamma^* + g \overset{(I)}{\to} q_R + \bar{q}_R$

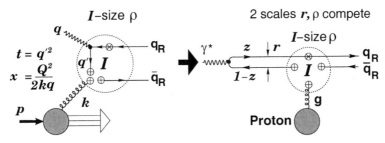

Figure 2. Transcription of the simplest I-induced process into the colour-dipole picture

Let us briefly consider first the simplest I-induced process, $\gamma^* g \Rightarrow q_R \bar{q}_R$, with one flavour and no final-state gluons (Fig. 2 (left)). More details may be found in [8]. Already this simplest case illustrates transparently that in the presence of a background instanton, the dipole cross section indeed saturates with a saturation scale of the order of the average I-size $\langle \rho \rangle$.

We start by recalling the results for the total $\gamma^* N$ cross section within I-perturbation theory from [30],

$$\sigma_{L,T}(x_{Bj}, Q^2) = \int_{x_{Bj}}^{1} \frac{dx}{x} \left(\frac{x_{Bj}}{x}\right) G\left(\frac{x_{Bj}}{x}, \mu^2\right) \int dt \frac{d\hat{\sigma}_{L,T}^{\gamma^* g}(x, t, Q^2)}{dt}; \quad (3)$$

$$\frac{d\hat{\sigma}_L^{\gamma^* g}}{dt} = \frac{\pi^7}{2} \frac{e_q^2}{Q^2} \frac{\alpha_{em}}{\alpha_s} \left[x(1-x)\sqrt{tu} \frac{\mathcal{R}(\sqrt{-t}) - \mathcal{R}(Q)}{t + Q^2} - (t \leftrightarrow u)\right]^2 \quad (4)$$

with a similar expression for $d\hat{\sigma}_T^{\gamma^* g}/dt$. Here, $G(x_{Bj}, \mu^2)$ denotes the gluon density and L, T refers to longitudinal and transverse photons, respectively.

Note that Eqs. (3), (4) involve the "master" integral $\mathcal{R}(\mathcal{Q})$ with dimension of a length,

$$\mathcal{R}(\mathcal{Q}) = \int_0^\infty d\rho \, D(\rho) \rho^5 (\mathcal{Q}\rho) K_1(\mathcal{Q}\rho). \quad (5)$$

In usual I-perturbation theory, the ρ-dependence of the I-size distribution $D(\rho)$ in Eq.(5) is known [31] for sufficiently small ρ,

$$D(\rho) \approx D_{I-\text{pert}}(\rho) \propto \rho^{6-\frac{2}{3}n_f}, \qquad (6)$$

the strong power law increase of which is well known to generically lead to (unphysical) IR-divergencies from large-size instantons. However, in DIS for sufficiently large virtualities $\mathcal{Q}$, the crucial factor $(\mathcal{Q}\rho) K_1(\mathcal{Q}\rho) \sim e^{-\mathcal{Q}\rho}$ in Eq.(5) exponentially suppresses large size instantons and I-perturbation theory holds, as shown first in [30].

The effective size $\mathcal{R}(\mathcal{Q})$ in Eq.(5) correspondingly plays a central rôle in the context of a continuation of our I-perturbative results to smaller $\mathcal{Q}$. Here, crucial lattice information enters. We recall that the I-size distribution $D_{\text{lattice}}(\rho)$, as *measured* on the lattice [16,17,18], is strongly peaked around an average I-size $\langle\rho\rangle \approx 0.5$ fm (cf. Fig. 1 (left)), while being in excellent, parameter-free agreement [17,18] with I-perturbation theory for $\rho \lesssim 0.35$ fm (cf. Fig. 1 (left)).

Our general strategy is thus to generally identify $D(\rho) = D_{\text{lattice}}(\rho)$ in Eq.(5), whence

$$\mathcal{R}(0) = \int_0^\infty d\rho\, D_{\text{lattice}}(\rho) \rho^5 \approx 0.3 \text{ fm} \qquad (7)$$

becomes finite and a $\mathcal{Q}^2$ cut is no longer necessary.

By means of an appropriate change of variables and a subsequent 2d-Fourier transformation, Eqs. (3, 4) may indeed be cast [8] into a colour-dipole form (1), e.g. (with $\hat{Q} = \sqrt{z(1-z)}\, Q$)

$$\left(|\Psi_L|^2 \sigma_{\text{DP}}\right)^{(I)} \approx |\Psi_L^{\text{pQCD}}(z,r)|^2 \frac{1}{\alpha_s} x_{\text{Bj}} G(x_{\text{Bj}}, \mu^2) \frac{\pi^8}{12} \qquad (8)$$

$$\times \left\{ \int_0^\infty d\rho D(\rho)\, \rho^5 \left(\frac{-\frac{d}{dr^2}\left(2r^2 \frac{K_1(\hat{Q}\sqrt{r^2+\rho^2/z})}{\hat{Q}\sqrt{r^2+\rho^2/z}}\right)}{K_0(\hat{Q}r)} - (z \leftrightarrow 1-z) \right) \right\}^2.$$

The strong peaking of $D_{\text{lattice}}(\rho)$ around $\rho \approx \langle\rho\rangle$, implies

$$\left(|\Psi_{L,T}|^2 \sigma_{\text{DP}}\right)^{(I)} \Rightarrow \begin{cases} \mathcal{O}(1) \text{ but exponentially small;} & r \to 0, \\ |\Psi_{L,T}^{\text{pQCD}}|^2 \frac{1}{\alpha_s} x_{\text{Bj}} G(x_{\text{Bj}}, \mu^2) \frac{\pi^8}{12} \mathcal{R}(0)^2; & \frac{r}{\langle\rho\rangle} \gtrsim 1. \end{cases} \qquad (9)$$

Hence, the association of the intrinsic instanton scale $\langle\rho\rangle$ with the saturation scale r_s becomes apparent from Eqs. (8, 9): $\sigma_{\text{DP}}^{(I)}(r,\ldots)$ rises strongly

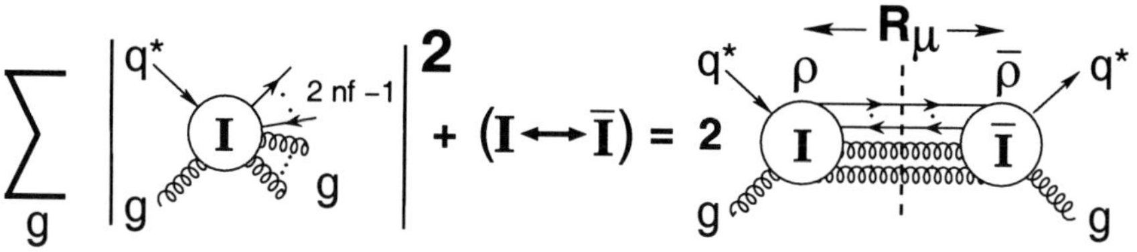

Figure 3. Optical theorem for the I-induced q^*g- subprocess. The incoming, virtual q^* originates from photon dissociation, $\gamma \to \bar{q} + q^*$.

as function of r around $r_s \approx \langle \rho \rangle$, and indeed *saturates* for $r/\langle \rho \rangle > 1$ towards a *constant geometrical limit*, proportional to the area $\pi \mathcal{R}(0)^2 = \pi \left(\int_0^\infty d\rho\, D_{\text{lattice}}(\rho)\, \rho^5 \right)^2$, subtended by the instanton. Since $\mathcal{R}(0)$ would be divergent within I-perturbation theory, the information about $D(\rho)$ from the lattice (Fig. 1 (left)) is crucial for the finiteness of the result.

3.2. The Realistic Process: $\gamma^* + g \xrightarrow{(I)} n_f\, (q_R + \bar{q}_R) + gluons$

On the one hand, the inclusion of an arbitrary number of final-state gluons and $n_f > 1$ light flavours causes a significant complication. On the other hand, it is due to the inclusion of final-state gluons that the identification of the QCD-sphaleron state with the colour glass condensate has emerged [8,9], with the qualitative "saturation" features of the preceding subsection remaining unchanged. In view of the limited space, let us therefore focus our main attention in this section to the emerging sphaleron interpretation of the colour glass condensate.

Most of the I-dynamics resides in I-induced q^*g-subprocesses like

$$q_L^*(q') + g(p) \xrightarrow{(I)} n_f\, q_R + (n_f - 1)\, \bar{q}_R + \text{gluons}, \qquad (10)$$

with an incoming off-mass-shell quark q^* originating from photon dissociation, $\gamma \to \bar{q} + q^*$. The important kinematical variables are the total I-subprocess energy $E = \sqrt{(q'+p)^2}$ and the quark virtuality $Q'^2 = -q'^2$.

It is most convenient to account for the arbitrarily many final-state gluons by means of the so-called "$I\bar{I}$-valley method" [32]. It allows to achieve via the optical theorem (cf. Fig. 3) an elegant summation over the final-state gluons in form of an exponentiation, with the effect of the gluons residing entirely in the $I\bar{I}$-valley interaction $-1 \leq \Omega_{\text{valley}}^{I\bar{I}}(\frac{R^2}{\rho\bar{\rho}} + \frac{\rho}{\bar{\rho}} + \frac{\bar{\rho}}{\rho}; U) \leq 0$, between I's and $\bar{I}$'s. The new collective coordinate R_μ denotes the $I\bar{I}$-distance 4-vector (cf. Fig. 3), while the matrix U characterizes the $I\bar{I}$ relative colour orientation. Most importantly, the functional form of $\Omega_{\text{valley}}^{I\bar{I}}$ is analytically known [33,34] and the limit of I-perturbation theory is attained for $\sqrt{R^2} \gg \sqrt{\rho\bar{\rho}}$.

The strategy we shall apply is identical to the one for the "simplest process" in the previous Sec. 3.1: Starting point is the general form of the I-induced $\gamma^* N$ cross section, this time obtained by means of the $I\bar{I}$-valley method [11]. By exploiting the optical theorem (cf. Fig. 3), the total $q^* g$-cross section is most efficiently evaluated from the imaginary part of the forward elastic amplitude induced by the $I\bar{I}$-valley background. The next step is again a variable and 2d Fourier transformation into the colour-dipole picture like before.

The dipole cross section $\tilde{\sigma}_{DP}^{(I),\text{gluons}}(\mathbf{l}^2, x_{Bj}, \ldots)$ before the final 2d-Fourier transformation[a] $\mathbf{l} \leftrightarrow \mathbf{r}$ to the dipole size $\mathbf{r}$ arises simply as an energy integral over the I-induced total $q^* g$ cross section from [11],

$$\tilde{\sigma}_{DP}^{(I),\text{gluons}} \approx \frac{x_{Bj}}{2} G(x_{Bj}, \mu^2) \int_0^{E_{\max}} \frac{dE}{E} \left[\frac{E^4}{(E^2 + Q'^2) Q'^2} \sigma_{q^* g}^{(I)}(E, \mathbf{l}^2, \ldots) \right], \quad (11)$$

involving in turn integrations over the $I\bar{I}$-collective coordinates $\rho, \bar{\rho}, U$ and the $I\bar{I}$-distance R_μ.

In the softer regime of interest for saturation, we again substitute $D(\rho) = D_{\text{lattice}}(\rho)$, which enforces $\rho \approx \bar{\rho} \approx \langle \rho \rangle$ in the respective $\rho, \bar{\rho}$-integrals, while the integral over the $I\bar{I}$-distance R is dominated by a *saddle point*,

$$\frac{R}{\langle \rho \rangle} \approx \text{function}\left(\frac{E}{m_{\text{sph}}}\right); \quad m_{\text{sph}} \approx \frac{3\pi}{4} \frac{1}{\alpha_s \langle \rho \rangle} = \mathcal{O}(\text{few GeV}). \quad (12)$$

At this point, the mass m_{sph} of the QCD-sphaleron [10,35], i.e the barrier height separating neighboring topologically inequivalent vacua, enters as the scale for the I-subprocess energy E. The saddle-point dominance of the R-integration implies a one-to-one relation,

$$\frac{R}{\langle \rho \rangle} \Leftrightarrow \frac{E}{m_{\text{sph}}}; \quad \text{with } R = \langle \rho \rangle \Leftrightarrow E \approx m_{\text{sph}}. \quad (13)$$

Our careful continuation to the saturation regime now involves, in addition to the I-size distribution $D_{\text{lattice}}(\rho)$, crucial lattice information about the second basic building block of the I-calculus, the $I\bar{I}$-interaction $\Omega^{I\bar{I}}$. The relevant lattice observable is the distribution of the $I\bar{I}$-distance [17,8]

---

[a]The 2-dimensional vector $\mathbf{l}$ denotes the transverse momentum of the quark with four-momentum q'.

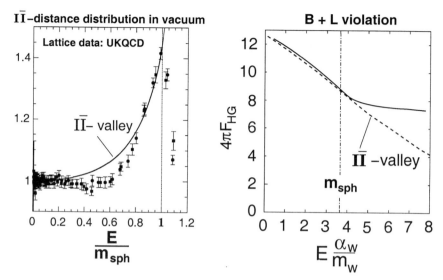

Figure 4. (Left) UKQCD lattice data [16,17] of the (normalized) $I\bar{I}$-distance distribution and the corresponding $I\bar{I}$-valley prediction [8] displayed versus energy in units of the QCD sphaleron mass $m_{\rm sph}$. The lattice data provide the first direct evidence that the $I\bar{I}$-valley approach is adequate right up to $E \approx m_{\rm sph}$, where the dominant contribution to the scattering process arises. Beyond this point a marked disagreement rapidly develops. (Right) The same trend for electroweak $B + L$-violation is apparent from a completely independent semiclassical, numerical simulation of the suppression exponent for two-particle collisions ('Holy Grail' function) $F_{\rm HG}(E)$. [36,37]

$R = \sqrt{R_\mu^2}$, essentially providing information on $\left\langle \exp[-\frac{4\pi}{\alpha_s}\Omega^{I\bar{I}}]\right\rangle_{U,\rho,\bar{\rho}}$ in Euclidean space. Due to the crucial saddle-point relation (12), (13), we may replace the original variable $R/\langle\rho\rangle$ by $E/m_{\rm sph}$. A comparison of the respective $I\bar{I}$-valley predictions with the UKQCD lattice data [16,17,8] versus $E/m_{\rm sph}$ is displayed in Fig. 4 (left). It reveals the important result that the $I\bar{I}$-valley approximation is quite reliable up to $E \approx m_{\rm sph}$. Beyond this point a marked disagreement rapidly develops: While the lattice data show a *sharp peak* at $E \approx m_{\rm sph}$, the valley prediction continues to rise indefinitely for $E \gtrsim m_{\rm sph}$! It is most remarkable that an extensive recent and completely independent semiclassical numerical simulation [36] shows precisely the same trend for electroweak $B + L$-violation, as displayed in Fig. 4 (right). Also here, there is an amazing agreement with the valley approximation [37] up to the electroweak sphaleron mass and a rapid disagreement developing beyond. It is again at hand to identify $\Omega^{I\bar{I}} = \Omega^{I\bar{I}}_{\rm lattice}$ for $E \gtrsim m_{\rm sph}$. Then, on account of Eq. (11), the integral over the I-subprocess energy spectrum in the dipole cross section appears to be dominated by the sphaleron configuration at $E \approx m_{\text{sph}}$ The feature of saturation analogously to the "simplest process" in Sec. 3.1 then implies the announced identification of the colour glass condensate with the QCD-sphaleron state.

4. Wilson-Loop Scattering in an *I*-Background

Let us next turn to our second approach within the colour-dipole picture, which is still in progress [9,38]. It is complementary to our previous strategy of extending the known results of *I*-perturbation theory towards the saturation regime by means of non-perturbative lattice information. $q\bar{q}$-dipole scattering will be described as the scattering of Wilson loops.

We work within the semiclassical, non-abelian eikonal approximation that is appropriate for the scattering of partons at high energies ($s \gg -t$ or small x_{Bj}) from a soft colour field A_μ [39]. The basic approximation is that the soft interaction of the partons with the colour field does not change their direction appreciably, such that they just pick up a non-abelian phase factor during the scattering. Each phase factor is given by a path-ordered integral calculated along the classical path of the respective parton,

$$W(\mathbf{x}) = \text{P} \exp\left\{-i g_s \int_{-\infty}^{+\infty} d\lambda \, q^\mu A_\mu(\lambda q + x_\perp)\right\}, \text{ with } x_\perp \cdot q = 0. \quad (14)$$

This so-called Wilson line depends on the 2-dimensional vector **x** describing the distance to the proton-photon plane. Correspondingly, $q\bar{q}$-dipoles lead to colourless, gauge invariant Wilson loops:

$$\mathcal{W}(\mathbf{r}, \mathbf{b}; A_\mu) = \frac{1}{N_c} \text{tr}\left[W(\mathbf{b} + \mathbf{r}/2) W^\dagger(\mathbf{b} - \mathbf{r}/2)\right]. \quad (15)$$

The Wilson loop (15) depends on the transverse size **r** of the colour dipole and the transverse distance **b** between dipole and colour field A_μ. It is a basic object in the framework of the colour glass condensate language, where the proton is viewed as a source of the classical field A_μ. Averaging over possible field configurations ($\langle \ldots \rangle_{A_\mu}$) and integrating over the impact parameter **b** leads to the total dipole cross section (e.g. [40, 6]),

$$\sigma_{\text{DP}}(r, \ldots) = 2 \int d^2\mathbf{b} \, \langle 1 - \mathcal{W}(\mathbf{r}, \mathbf{b}; A_\mu)\rangle_{A_\mu}. \quad (16)$$

In general, the meaning of the colour glass condensate in the context of the dipole cross section (16) (cf. also [41]) is that of an effective theory, leading to non-linear evolution equations [40,42] for the respective scattering amplitude.

As a first concrete testing ground for the impact of instantons within this framework, let us identify the classical field A_μ in the dipole cross section (16) with the known instanton field $A_\mu^{(I)}$. The functional integration $\langle\ldots\rangle_{A_\mu}$ over the field configurations A_μ is then to be understood as an integration over the I-collective coordinates, i.e.

$$A_\mu(x) \to A_\mu^{(I)}(x,\rho,x_0); \quad \langle\ldots\rangle_{A_\mu} \to \mathcal{D}A_\mu^{(I)} \to d^4x_0 \, d\rho \, D_{\text{lattice}}(\rho). \quad (17)$$

In a first step, one has to calculate the Wilson loop in the I-background, which can be performed analytically. Subsequently, one has to integrate over the collective coordinates. Finally, we get a dipole cross section depending on the dipole size r and the size $\langle\rho\rangle$ of the instanton in the vacuum,

$$\sigma_{\text{DP}}^{(I)}(r,\ldots) \propto \langle\rho\rangle^2 f\left(\frac{r}{\langle\rho\rangle}\right). \quad (18)$$

Like in our first approach (Sec. 3), this dipole cross section turns out to saturate towards a constant limit proportional to $\langle\rho\rangle^2$ for $r \gtrsim \langle\rho\rangle$.

For a more realistic estimate, it is important to notice that one has to take an $I\bar{I}$-configuration (like the valley field [38]) in the total cross section (16). For an estimate of the elastic part of the dipole cross section, one can take the single I-gauge field and square the resulting dipole scattering amplitude. This elastic contribution $\sigma_{\text{DP}}^{(I)}(r)/\sigma_{\text{DP}}^{(I)}(\infty)$, normalized to one for $r \to \infty$, is displayed in Fig. 5 (left) as function of $r/\langle\rho\rangle$. The importance of $\langle\rho\rangle$ in the approach to saturation becomes again apparent. In Fig. 5 (right), the corresponding impact parameter profile for $r = \langle\rho\rangle, \infty$ is displayed. This simplest estimate of the dipole cross section in an I-background can certainly not describe the proton in an adequate way, notably due to the lack of non-trivial proton kinematics. Hence it is not surprising that the resulting dipole cross section and hence the saturation scale comes out x_{Bj}-independent in this case. Nevertheless, this calculation illustrates once more the close connection between an 'extended' classical colour background field of size $\langle\rho\rangle$ and the saturation scale. Taking the instanton solution as an initial condition for the BK-equations [40,42], one could generate the proper x_{Bj}-dependence via the implied evolution.

In order to model the proton more realistically, we have also worked out [38] a generalization to dipole-dipole scattering in an I-background. The formalism used is analogous to [43] within the stochastic vacuum approach [44]. Like in [43], we started with the calculation of the loop-loop contributions to the dipole-dipole scattering amplitude, that are dominant in the large-N_c limit. In this case, the trace in Eq. (15) is taken separately

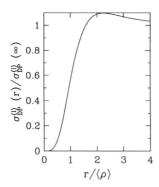 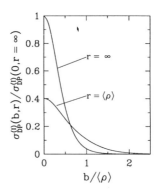

Figure 5. (Left) The elastic contribution to the I-induced dipole cross section (Right) The corresponding impact parameter profile.

for both dipoles. This leads to consistent results and we observe again a saturation of the resulting dipole cross section. In [22, 23, 24] it was pointed out, however, that the dominant contribution in the high-energy limit comes from contribution with a colour exchange between the dipoles. The impact of this contribution for the dipole cross section is presently under investigation [38].

References

1. C. Adloff et al. [H1 Collaboration], Eur. Phys. J. **C21**, 33 (2001);
   S. Chekanov et al. [ZEUS Collaboration], Eur. Phys. J. **C21**, 443 (2001).
2. V.N. Gribov and L.N. Lipatov, Sov. J. Nucl. Phys.**15**, 438 (1972);
   L.N. Lipatov, Sov. J. Nucl. Phys.**20**, 94 (1975);
   G. Altarelli and G. Parisi, Nucl. Phys. **B126**, 298 (1977);
   Y.L. Dokshitzer, Sov. Phys. JETP**46**, 641 (1977).
3. A. De Rujula et al., Phys. Rev. **D10**, 1649 (1974).
4. L.N. Lipatov, Sov. J. Nucl. Phys.**23**, 338 (1976);
   V.S. Fadin, E.A. Kuraev and L.N. Lipatov, Phys. Lett. **B60**, 50 (1975), Sov. Phys. JETP**44**, 443 (1976), Sov. Phys. JETP**45**, 199 (1977);
   I.I. Balitsky and L.N. Lipatov, Sov. J. Nucl. Phys.**28**, 822 (1978).
5. L.V. Gribov, E.M. Levin and M.G. Ryskin, Nucl. Phys. **B188**, 555 (1981);
   L.V. Gribov, E.M. Levin and M.G. Ryskin, Phys. Rept.**100**, 1 (1983).
6. E. Iancu, A. Leonidov and L. D. McLerran, Nucl. Phys. **A692**, 583 (2001);
   E. Ferreiro et al., Nucl. Phys. **A703**, 489 (2002).
7. A. Belavin et al., Phys. Lett. **B59**, 85 (1975).
8. F. Schrempp and A. Utermann, Phys. Lett. **B543**, 197 (2002).
9. F. Schrempp and A. Utermann, Proc. Strong and Electroweak Matter 2002,

Heidelberg, Oct. 2002, ed. M.G. Schmidt, p. 477 [arXiv:hep-ph/0301177].
10. A. Ringwald and F. Schrempp, *Proc. Quarks '94*, ed D.Yu. Grigoriev *et al.* (Singapore: World Scientific) p 170, [arXiv:hep-ph/9411217].
11. A. Ringwald and F. Schrempp, *Phys. Lett.* **B438**, 217 (1998).
12. A. Ringwald and F. Schrempp, *Comput. Phys. Commun.* **132**, 267 (2000).
13. H. Aoyama and H. Goldberg, *Phys. Lett.* **B188**, 506 (1987); A. Ringwald, *Nucl. Phys.* **B330**, 1 (1990); O. Espinosa, *Nucl. Phys.* **B343**, 310 (1990).
14. D.M. Ostrovsky *et al.*, *Phys. Rev.* **D66**, 036004 (2002).
15. F. R. Klinkhamer and N. S. Manton, *Phys. Rev.* **D30**, 2212 (1984).
16. D.A. Smith and M.J. Teper (UKQCD), *Phys. Rev.* **D58**, 014505 (1998).
17. A. Ringwald and F. Schrempp, *Phys. Lett.* **B459**, 249 (1999).
18. A. Ringwald and F. Schrempp, *Phys. Lett.* **B503**, 331 (2001).
19. F. Schrempp and A. Utermann, *Acta Phys. Polon.* **B33**, 3633 (2002).
20. F. Schrempp, *J. Phys.* **G28**, 915 (2002).
21. D.E. Kharzeev, Y.V. Kovchegov and E. Levin, *Nucl. Phys.* **A690**, 621 (2001).
22. E. Shuryak and I. Zahed, *Phys. Rev.* **D62**, 085014 (2000).
23. M. A. Nowak, E. V. Shuryak and I. Zahed, *Phys. Rev.* **D64**, 034008 (2001).
24. E. V. Shuryak and I. Zahed, arXiv:hep-ph/0307103.
25. N. Nikolaev and B.G. Zakharov, *Z. Phys.* **C49**, 607 (1990); *Z. Phys.* **C53**, 331 (1992); A.H. Mueller, *Nucl. Phys.* **B415**, 373 (1994).
26. A.H. Mueller, *Parton Saturation - An Overview*, hep-ph/0111244, in *QCD Perspectives on Hot and Dense Matter*, NATO Science Series, Kluwer, 2002.
27. F. E. Low, *Phys. Rev.* **D12**, 163 (1975); L. Frankfurt, G.A. Miller and M. Strikman, *Phys. Lett.* **B304**, 1 (1993).
28. K. Golec-Biernat and M. Wusthoff, *Phys. Rev.* **D59**, 014017 (1999); K. Golec-Biernat and M. Wusthoff, *Phys. Rev.* **D60**, 114023 (1999).
29. J. Bartels *et al.*, *Phys. Rev.* **D66**, 014001 (2002).
30. S. Moch, A. Ringwald and F. Schrempp, *Nucl. Phys.* **B507**, 134 (1997).
31. G. 't Hooft, *Phys. Rev.* **D14**, 3432 (1976); *Phys. Rev.* **D18**, 2199 (1978) (Erratum).
32. A. Yung, *Nucl. Phys.* **B297**, 47 (1988).
33. V.V. Khoze and A. Ringwald, *Phys. Lett.* **B259**, 106 (1991).
34. J. Verbaarschot, *Nucl. Phys.* **B362**, 33 (1991).
35. D. Diakonov and V. Petrov, *Phys. Rev.* **D50**, 266 (1994).
36. F. Bezrukov, D. Levkov, C. Rebbi, V. Rubakov and P. Tinyakov, *Phys. Rev.* **D68**, 036005 (2003)
37. A. Ringwald, *Phys. Lett.* **B555**, 227 (2003); arXiv:hep-ph/0302112.
38. F. Schrempp and A. Utermann, in preparation.
39. O. Nachtmann, *Annals Phys.* **209**, 436 (1991); J.C. Collins and R.K. Ellis, *Nucl. Phys.* **B360**, 3 (1991).
40. I.I. Balitsky, *Phys. Rev. Lett.* **81**, 2024 (1998), *Phys. Rev.* **D60**, 014020 (1999).
41. W. Buchmüller and A. Hebecker, *Nucl. Phys.* **B476**, 203 (1996).
42. Y.V. Kovchegov, *Phys. Rev.* **D60**, 034008 (1999).
43. A. I. Shoshi, F. D. Steffen and H. J. Pirner, *Nucl. Phys.* **A709**, 131 (2002).
44. A. Krämer and H. G. Dosch, *Phys. Lett.* **B252**, 669 (1990).

6
New Physics at HERA

QCD CORRECTIONS TO SINGLE TOP QUARK AND W BOSON PRODUCTION AT HERA

KAI-PEER O. DIENER

Paul Scherrer Institut,
CH-5232 Villigen,
Switzerland
E-mail: kai.diener@psi.ch

We review theoretical results on QCD corrections to flavour changing neutral current single top as well as single W boson production at HERA. In both cases the inclusion of QCD corrections has a moderate effect on the size of the cross sections but reduces the QCD scale dependence considerably. This writeup summarizes a talk given at the Ringberg Workshop "New Trends in HERA Physics 2003".

1. Introduction

At HERA, 27.5GeV electrons collide with 920GeV protons allowing on the one hand for precise studies of the proton structure and associated non-perturbative aspects of QCD, on the other hand for tests of perturbative QCD and the electroweak (EW) standard model (SM) in general.

We review the production of a single top quark in a FCNC reaction and the production of a single W boson in $e^{\pm}p$ collisions at HERA energies. This work is, thus, concerned with phenomenological aspects of the gauge sector (non-abelian contribution to W boson production) and flavour structure (single top production) of the SM as well as aspects of perturbative QCD. Typical momentum transfers of the reactions we consider define scales at which a perturbative treatment of QCD corrections to the leading-order (LO) EW processes is justified. In fact, the inclusion of such corrections is necessary to obtain theoretical predictions for the relevant observables which are of comparable accuracy as the corresponding experimental measurements at HERA.

In a recent search [1] for events with a high-energy isolated electron or muon and missing transverse momentum the H1 collaboration claimed to find an excess of events with transverse momentum of the hadronic system greater than 25GeV. Within the SM, such events are expected to be mainly due to W boson production with subsequent leptonic decay. It is, thus, important to control the effects of radiative QCD corrections to the LO EW W boson production mechanism within and beyond the SM. A possible explanation of an excess of W boson production events beyond the SM would be the FCNC production of a short-lived top quark decaying into a bottom quark and a W boson.

In section 2, we review a calculation of QCD corrections to FCNC top quark production at HERA in the eikonal approximation as carried out by Belyaev and Kidonakis in [2].

We proceed in section 3 by describing QCD corrections to the W production cross sections at HERA. Section 3.1 is concerned with corrections to the total cross section. We review the corrections to resolved photoproduction as discussed in [3] by Nason, Rückl and Spira. Section 3.2 is devoted to corrections to the W production cross sections differential in the transverse momentum of the W boson p_T and its rapidity y. We discuss next-to-leading order (NLO) QCD corrections to the direct photoproduction part which is the dominant contribution to these distributions. The results of section 3.2 are quoted from [4].

Section 4 contains a short summary and concluding remarks.

2. Soft Gluon Corrections to FCNC Top Production

In [2], FCNC couplings between a top quark t, an up or charm quark q and a photon ($F^{\mu\nu}$ is the corresponding field strength tensor) are introduced through the effective Lagrangian

$$\Delta \mathcal{L}^{\text{eff}} = \frac{1}{\Lambda} \kappa_{tq\gamma} e \bar{t} \sigma_{\mu\nu} q F^{\mu\nu} + \text{h.c.}, \tag{1}$$

where Λ is a dimensionful scale parameter. Clearly, HERA should be quite sensitive to $\kappa_{tu\gamma}$ through the partonic process $e^\pm u \to e^\pm t$.

To reduce the theoretical error in the determination of bounds for $\kappa_{tu\gamma}$, the authors of [2] consider gluonic quantum corrections to the LO EW process. They argue, that since the top quark is very heavy top production at HERA occurs — if at all — mostly close to threshold and that in this regime the eikonal approximation suffices to describe first-order QCD effects.

Fig. 1 is reproduced from [2] and shows the dependence of the Born, one loop and Born plus one loop cross section on the renormalization and factorization scale $\mu = \mu_R = \mu_F$. The parton densities were implemented in the CTEQ5M [5] parameterization and the numerical input parameters

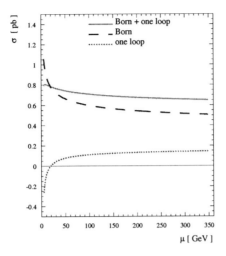

Figure 1. Born, one loop, and Born plus one loop cross section as a function of the QCD scale $\mu = \mu_R = \mu_F$ for FCNC single top quark production at HERA with $m_t = 175$GeV and $\kappa_{tu\gamma} = 0.1$ for $\sqrt{S} = 318$GeV.

were set to

$$\sqrt{S} = 318\text{GeV}, \quad \Lambda = m_t = 175\text{GeV}, \quad \kappa_{tu\gamma} = 0.1.$$

Fig. 1 shows that above $\mu = 10$GeV the inclusion of NLO QCD corrections in the eikonal approximation leads to a significant stabilization of the total cross section with respect to simultaneous variation of the renormalization and factorization scale. It can also be inferred that $\mu = m_t = 175$GeV as central value for the QCD scale is a sensible choice, unlike $\mu = Q = \sqrt{-(p_e - p_{e'})^2}$, which would lead to significant contributions from very low Q values of the order of m_e where the parton density functions are not well-defined.

In Fig. 2, the Born, one loop and Born plus one loop cross sections are plotted as a function of a) the top mass and b) the centre-of-mass energy. The QCD scale μ is set to its central value m_t and $\kappa_{tu\gamma}$ is chosen as above. A strong dependence of the cross sections on the top mass and CM energy is observed. Indeed, a variation of m_t by ±5% leads to roughly 20% uncertainty in the cross section, a 6% increase in CM energy causes an increase of the cross section by 40%.

Analyses of FCNC single top production from an initial state u quark at HERA have been carried out taking into account the results of [2] discussed

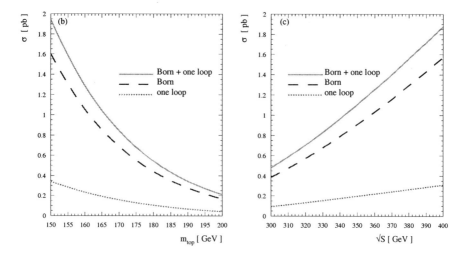

Figure 2. Born, one loop, and Born plus one loop cross section for the FCNC single top quark production at HERA with $\mu = \mu_R = \mu_F = m_t$ and $\kappa_{tu\gamma} = 0.1$ versus (a) the top quark mass with $\sqrt{S} = 318\text{GeV}$ and (b) the CM energy with $m_t = 175\text{GeV}$.

above. The ZEUS collaboration found $\kappa_{tu\gamma} < 0.174$ [6], the H1 collaboration established $\kappa_{tu\gamma} < 0.27$ [7], both at the 95% confidence level.

3. NLO QCD Corrections to W Boson Production

There are three different mechanisms of W boson production in $e^{\pm}p$ collisions:

(1) The deep-inelastic regime defined by the kinematical condition that $Q^2 = -(p_e - p_{e'})^2 > Q^2_{\text{cut}}$ and that the hadronic cross section can be written

$$\sigma_{ep} = \hat{\sigma}_{eq} \otimes f_p^q,$$

with the quark density in the proton f_p^q and "$\otimes$" indicating a convolution.

(2) The photoproduction region with $Q^2 = -(p_e - p_{e'})^2 < Q^2_{\text{cut}}$ where a further distinction is made between

(a) direct photoproduction with an initial state photon instead of the electron in the partonic subprocess, i.e.,

$$\sigma_{ep} = \hat{\sigma}_{\gamma q} \otimes f_p^q \otimes f_e^\gamma,$$

where f_e^γ is given by the Weizsäcker–Williams spectrum and
(b) resolved photoproduction where the initial state photon is split into a $q, \bar{q}$ pair and the photon in the initial state of the partonic subprocess is thus replaced by a (anti) quark:

$$\sigma_{ep} = \hat{\sigma}_{\bar{q}q} \otimes f_p^q \otimes f_e^\gamma \otimes f_\gamma^{\bar{q}}.$$

Which of the production mechanisms is dominant depends to a large extent on whether we consider total or differential cross sections. At lowest order there is no contribution from resolved photoproduction to the p_T distribution of the W as the partonic process in this case is $2 \to 1$ with vanishing p_T in the final state. The resolved contributions to the total cross section are, however, quite significant and we shall thus proceed by reviewing the total and differential W boson production cross sections separately.

3.1. Corrections to the Total Cross Section

In [3], Nason, Rückl and Spira showed that resolved photoproduction is the dominant contribution to the LO total W boson production cross section in $e^\pm p$ collisions. It contributes about 60%, while direct photoproduction and deep-inelastic scattering account for roughly 15% and 25%, respectively. Their analysis of QCD quantum corrections was, thus, confined to the resolved part.

Fig. 3 shows the dependence of the total cross section for $W^\pm$ production with subsequent muonic decay on the QCD renormalization and factorization scale $\mu = \mu_R = \mu_F$. The plots are reproduced from [3] with CTEQ4M [8] and ACFGP [9] parton densities for the proton and the photon, respectively. The calculation was carried out in the $\overline{\text{MS}}$ scheme using an NLO strong coupling constant with $\Lambda_5 = 202$MeV and an angular cut of $5°$ (instead of a fixed Q^2_{cut}) to separate photoproduction from deep inelastic scattering (DIS). At $\mu = M_W$, the QCD corrections enhance the resolved contribution by about 40%. It can be observed that the scale dependence in the sum of direct and resolved contributions is significantly reduced, once the NLO corrections to the resolved part are included. The full curves show the total sum of NLO resolved, LO direct and LO DIS contribution, which is the prediction of [3] for the total $W^\pm$ production cross sections. The residual scale dependence is about 5%. Since the remaining dependence on Q^2_{cut} is of similar size, Nason, Rückl and Spira estimate the total theoretical uncertainty to be less than about 10%.

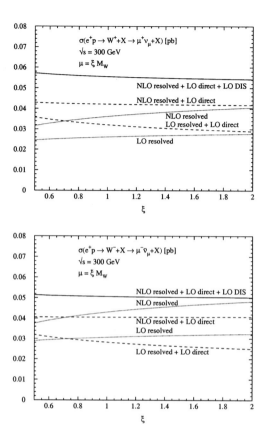

Figure 3. Dependence of the individual contributions to W^+ (up) and W^- (down) production with subsequent muonic decay on the renormalization and factorization scale $\mu = \mu_R = \mu_F = \xi M_W$. The full curves represent the final predictions for the total cross section of $W^\pm$ production in e^+p collisions.

3.2. Corrections to the Differential Cross Section

We proceed with the discussion of QCD quantum corrections to the doubly and singly differential W boson production cross sections $\mathrm{d}^2\sigma/\mathrm{d}y\mathrm{d}p_T$ and $\mathrm{d}\sigma/\mathrm{d}p_T$ as discussed in [4]. As pointed out there, the resolved contribution to the p_T distribution is negligible for $p_T > 15\,\mathrm{GeV}$. Instead, direct photoproduction is the dominant mechanism whereas deep-inelastic scattering is smaller but comparable. Thus, the authors of [4] confined the discussion of QCD corrected differential cross section distributions to the case of direct photoproduction. Events with one jet in the final state in addition to the W boson are distinguished from events with two final state jets by requiring that their cones of size $R = 1$ defined by an inclusive k_T algorithm do not overlap and that both partons have a transverse energy of more than 5GeV. All other configurations (especially with LO kinematics) define one-jet events.

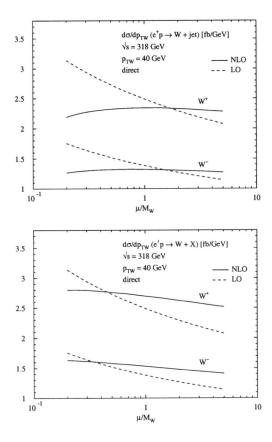

Figure 4. Dependence of the direct contribution to the W production cross section on the renormalization and factorization scale $\mu = \mu_R = \mu_F = \xi M_W$ for $p_T = 40\text{GeV}$. The full curves represent the NLO predictions and the broken curves the LO scale dependences. The upper plot presents $W + 1\text{jet}$ production, the lower one presents the sum of $W + 1\text{jet}$ plus $W + 2\text{jets}$.

To estimate the theoretical uncertainties, the QCD scale dependence of the direct contributions to the processes $e^+ p \to W^\pm + X$ is presented in Fig. 4. The scale dependence is significantly smaller, once the NLO corrections are included. The residual scale dependence is reduced from about 20% down to around 5%. The plots of Fig. 4 clearly indicate that the NLO QCD corrections are small at the central value of the QCD scale m_W. Since the uncertainties of the parton densities are of similar size, the total theoretical uncertainty can be estimated to be less than about 10%.

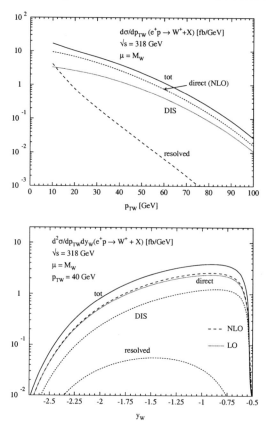

Figure 5. Shown are the transverse momentum distributions (up), and rapidity distribution (down) at $p_T = 40$ GeV of W^+ bosons at HERA. The full curves show the total distributions including the NLO corrections to the direct contribution, while the broken lines exhibit the individual parts at LO and the direct part at NLO.

Fig. 5 depicts the transverse momentum (up) as well as rapidity (down) distribution of W^+ bosons produced at HERA with $\sqrt{S} = 318$ GeV.

The upper plot clearly demonstrates the dominance of the direct photoproduction part, while the DIS part is smaller but still relevant. The resolved photoproduction part can safely be neglected for p_T values larger than about 10–15GeV. The rising total sum at p_T values below about 15GeV signals the breakdown of the purely perturbative results and underlines the necessity of a soft gluon resummation to describe W boson production for smaller transverse momenta which, however, is beyond the scope of [4].

The rapidity distribution of W^+ boson production at HERA for $p_T = 40$GeV is presented in the lower plot of Fig. 5 for the individual production mechanisms. The direct photoproduction part is shown at LO and NLO. As in the case of the transverse momentum distribution, the QCD corrections to the direct contribution amount to less than about 10–15% and are, thus, of moderate size. The figure shows that W bosons are preferably produced at larger rapidity values.

4. Conclusions

We have reviewed FCNC single top quark as well as single W boson production at HERA.

As the authors of [2] could show, the inclusion of lowest-order QCD quantum effect in the eikonal approximation leads to a stabilization of the FCNC single top production cross section with respect to variation of the renormalization and factorization scales. The stabilization is most pronounced for $10\text{GeV} < \mu_R = \mu_F < 75\text{GeV}$. For $\mu_R = \mu_F = m_t$ QCD quantum effects lead to a 25% increase in cross section, rendering reliable control of the radiative corrections an indispensable prerequisite for a precise determination of FCNC couplings.

There are several competing mechanisms to produce W bosons at HERA. The most important contributions to the total cross section stem from resolved photons, whereas the transverse momentum and rapidity distributions (for finite p_T) are dominated by direct photoproduction.

In their analysis of the dominant resolved part the authors of [3] found that QCD corrections enhance the total W production cross section by about 40% while reducing the dependence on the renormalization and factorization scale to 5%. The total remaining theoretical uncertainty is estimated to be smaller than 10%.

Direct photoproduction is the dominant contribution to distributions of the W production cross section in (W) transverse momentum and rapidity. The lowest-order QCD radiative corrections constitute a ± 10–15% effect and reduce the renormalization and factorization scale dependence to about

5%. The total theoretical uncertainty is estimated to be roughly 10%.

We remark that neither anomalous top quark production (with subsequent decay into a W boson and a b quark) nor higher-order QCD corrections to W boson production within the SM can explain the lepton flavour asymmetry in the excess of isolated single lepton events as observed in [1].

Acknowledgments

I would like to thank M. Spira and C. Schwanenberger for the fruitful collaboration and useful discussions. I appreciated the hospitality at Castle Ringberg and I would like to thank the organizers for their excellent work and the invitation to participate in this meeting.

Note Added in Proof

Meanwhile, the soft-gluonic corrections to FCNC single top production at HERA (and the Tevatron) have been evaluated up to next-to-next-to-leading order level [10]. The authors observe a further stabilization of the cross section with respect to variations of the QCD scale in comparison to the NLO order soft gluon-corrected result.

References

1. V. Andreev et al. [H1 Collaboration], Phys. Lett. B **561**, 241 (2003).
2. A. Belyaev and N. Kidonakis, Phys. Rev. D **65**, 037501 (2002).
3. P. Nason, R. Rückl and M. Spira, J. Phys. G **25**, 1434 (1999); M. Spira, arXiv:hep-ph/9905469.
4. K. P. Diener, C. Schwanenberger and M. Spira, Eur. Phys. J. C **25**, 405 (2002); K. P. Diener, C. Schwanenberger and M. Spira, hep-ex/0302040.
5. H. L. Lai et al. [CTEQ Collaboration], Eur. Phys. J. C **12**, 375 (2000).
6. S. Chekanov et al. [ZEUS Collaboration], Phys. Lett. B **559**, 153 (2003).
7. A. Aktas et al. [H1 Collaboration], arXiv:hep-ex/0310032.
8. H. L. Lai et al., Phys. Rev. D **55**, 1280 (1997).
9. P. Aurenche, P. Chiappetta, M. Fontannaz, J. P. Guillet and E. Pilon, Z. Phys. C **56**, 589 (1992).
10. N. Kidonakis and A. Belyaev, arXiv:hep-ph/0310299.

SEARCH FOR BEYOND THE STANDARD MODEL PHYSICS AT HERA

E. GALLO

INFN Firenze,
Via Sansone 1,
50019 Sesto Fiorentino, Italy,
E-mail: gallo@fi.infn.it

The H1 and ZEUS Collaborations have almost completed a rich program of searches for physics beyond the Standard Model using the data collected in the HERA I period. The main highlights, such as leptoquarks and signatures with isolated leptons at high energy, are reviewed.

1. Introduction

The H1 and ZEUS Collaborations have each collected of the order of $110 \div 130$ pb^{-1} of e^+p and 15 pb^{-1} of e^-p collisions data in the HERA I running period. These data have been used for a wide spectrum of searches for physics beyond the Standard Model (SM). In these proceedings some of the highlights are presented, together with some of the prospects for the HERA II running.

2. Leptoquarks

Leptoquarks (LQ) are scalar or vector triplet bosons, carrying both lepton (L) and baryon (B) number, and are foreseen in many extensions of the SM. HERA is the ideal accelerator to look for leptoquarks, as, for large masses, they are produced from the fusion of the electron/positron beam with a valence u, d-quark in the proton. The Buchmüller-Rückl-Wyler model classifies them in 14 different types, according *i.e.* to their quark flavour content (up or down), their fermion number (7 with $F = 3B + L$=0 and 7 with $F = 2$), their spin (7 scalar and 7 vector), and their decay (eq or νq). For masses $M_{LQ} < \sqrt{s} = 300, 318$ GeV, leptoquarks are produced mainly in the s-channel. The cross-section, in the narrow-widthapproximation, is proportional to $\lambda^2 q(x, Q^2)$, where λ is the Yukawa coupling of the leptoquark to the lepton and quark and the quark density $q(x, Q^2)$ is evaluated at Bjorken $x = M_{LQ}^2/s$ and photon virtuality $Q^2 = M_{LQ}^2$. The signature is therefore a narrow resonance decaying into electron+jet or neutrino+jet, with mass peaking at $M_{LQ} \simeq \sqrt{xs}$. The final state is in principle not distinguishable from the SM DIS processes $ep \to eX$ (NC) or $ep \to \nu X$ (CC), however the inelasticity y distribution, which is related to the scattering angle in the $eq(\nu q)$ rest frame, is different for LQs ($d\sigma/dy \sim y$ for scalar, $\sim (1-y)^2$ for vectors) and for DIS ($\sim 1/y^2$). For $M_{LQ} \gg \sqrt{s}$, the effect of the leptoquark exchange can be treated as a contact-interaction term.

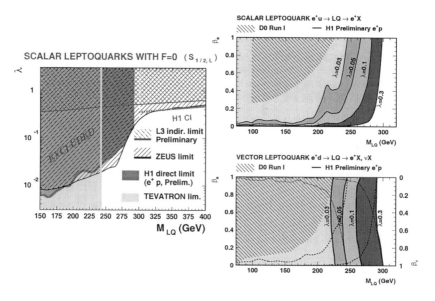

Figure 1. Example of limits on leptoquark production from the HERA I data, compared also to LEP and Tevatron results. The left plot shows the limit of the Yukawa coupling λ as a function of the LQ mass for the $S_{1/2}^L$ type leptoquark. The right plot shows the limit on the branching ratio β_e versus the LQ mass.

Both H1 and ZEUS have performed sophisticated fits of the SM+LQ+interference cross-section to their high Q^2 and high$-x$ data, using the two-dimensional information in the $x-y$ plane[1]. No evidence of a leptoquark signal has been found in the data. H1 has used the neutral current data at high Q^2, while ZEUS has also done a fit to the charged current data, providing strong limits for the 4 types of leptoquark decaying both to eq and to νq. An example of these limits is shown for the $S_{1/2}^L$

leptoquark (produced in e^+p, i.e. $F = 0$) in Fig. 1 (left), where the limits of the coupling λ versus the LQ mass is presented for H1 and ZEUS, in comparison to L3 and the Tevatron combined analysis.

In the mass region between 240 and 300 GeV, HERA set the most stringent limits, excluding couplings in the range of $10^{-2} \div 10^{-1}$. At LEP, the strongest limits come from indirect searches from the reaction $e^+e^- \to q\bar{q}$ and are competitive only in the contact-interaction region for few types of leptoquarks. At Tevatron, leptoquarks are produced in pairs and the cross-section is independent from the Yukawa coupling λ, as shown in the figure. However their limits depend on the branching ratio β_e to electron and quark in the final state, as illustrated in Figure 1 (right). HERA has better sensitivity than D0 for small branching ratios and there is still a window open for discovery for small values of λ and β_e. The search for leptoquarks remain therefore a *must* also for the HERA II searches, which will complement the Tevatron II results in this field.

3. Lepton Flavour Violation

As there is now clear evidence for oscillations in the neutrino sector, it is natural to look for flavour violation interactions in the charged leptonic sector (LFV). HERA is basically background-free for reactions of the type $eq \to \mu X$ or $eq \to \tau X$, which are predicted by many extensions of the SM, like leptoquarks coupling also to the second and third generation, here taken as a possible model.

ZEUS has recently[2] presented a new analysis in the τ channel, which makes use of a new τ identification technique for the hadronic decay[3]. In this method, variables which are chosen to discriminate between QCD and τ-jets, making use of the fact that taus are low-multiplicity and collimated jets, are combined with a multi-observable discrimination technique. The signal efficiency is therefore improved with respect to previous analyses. The LFV signature is a τ, either decaying to electron, muon or hadrons, aligned with the missing transverse momentum in the event.

No event with such a signature was selected by ZEUS in the 99-00 e^+p data. The result is used to set limits on the production of leptoquarks which decay to $e\tau$. An example, for the two leptoquark types $\tilde{S}_{1/2}^L$ and V_0^R, is shown in Figure 2, where the limit of the Yukawa coupling λ_{eq}, assuming that it is equal to the coupling $\lambda_{\tau q}$, is presented as a function of the LQ mass. For couplings of the order of the electromagnetic interaction, $\lambda = 0.3$, leptoquarks with LFV are excluded in the range $279 \div 299$ GeV.

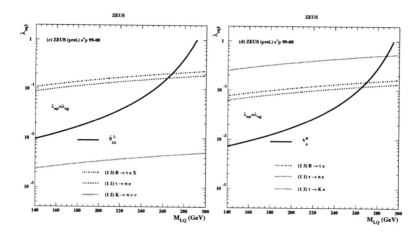

Figure 2. Example of limits on lepton flavour violation production from the ZEUS data, compared also to limits extracted from rare decays.

The limits are also shown in comparison to results obtained indirectly from rare decays of B, K mesons and taus. ZEUS sets the strongest limits in a wide region, especially when a heavy quark in the final state is involved.

4. R−parity violating supersymmetry

The Minimal Supersymmetric Model (MSSM) is widely accepted as a possible extension of the SM. In SUSY models, every particle has its own supersymmetric partner and their spin S differ by $1/2$. A new quantity, the R-parity, defined as $R_p = (-1)^{3B+L+2S}$, is $+1$ for particle and -1 for sparticle. While HERA is not competitive with LEP and Tevatron for reactions which require the conservation of R_p, it can instead look for R_p-violating reactions. In this case the Lagrangian has 3 additional terms:

$$L = \lambda_{ijk} L_i L_j \tilde{E}_k + \lambda'_{ijk} L_i Q_j \tilde{D}_k + \lambda''_{ijk} U_i D_j \tilde{D}_k. \qquad (1)$$

While LEP is more sensitive to the first term and Tevatron to the last, HERA can search for reactions with the term λ', involving the coupling of squarks to an electron and a quark.

An example in $e^+ p$ interaction is the process $e^+ d \to \tilde{u}_L^j$, where the squark $\tilde{u}_L^j$ can decay to: $e^+ d$, giving a signature identical to the one expected by a leptoquark; or via a gauge decay to $q\chi_\alpha^0$, $q\tilde{g}$ or $q\chi_\beta^+$. The neutralinos $\chi^0_{1,2,3,4}$, the charginos $\chi^+_{1,2}$ and the gluino $\tilde{g}$ are unstable and can decay via R−parity violating coupling. For instance the decay $\chi^0 \to e^\pm q \bar{q}$,

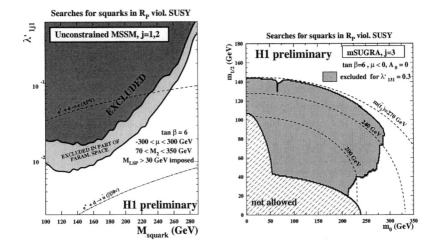

Figure 3. Limits on R_p-violating squark production for $\tilde{u}, \tilde{c}$ squarks in the unconstrained MSSM model (left) and on the stop production in the mSugra model (right).

gives in the final state a total signature of 3 jets and an electron or positron, half of the time of opposite sign compared to the initial beam lepton, which is a background-free process.

The H1 Collaboration has analyzed the 98-00 data and searched for squarks in seven different topologies, according to all possible decays[4]. No evidence was found for a signal, therefore limits on squark production were determined. The limits were derived scanning a wide range of the SUSY parameters μ, M_2 and for two values of the $\tan\beta$ parameters (2,6). However the limits do not depend strongly on these variations, as shown by the grey band in Figure 3 (left). In general $\tilde{u}_L, \tilde{c}_L$ squarks with masses below 240 GeV are ruled out for $\lambda'_{111}, \lambda'_{121} = 0.3$, respectively.

H1 has also calculated limits in the framework of the Minimal Supergravity Model (mSUGRA). An example is shown for the stop quark in Figure 3 (right), where the exclusion region in the two mSUGRA parameters $m_0, m_{1/2}$ is shown for $\lambda'_{131} = 0.3$. The excluded area corresponds approximately to a value of the $\tilde{t}_L$-mass of 270 GeV, as shown by the dashed line in the figure. The present Tevatron limits are around 130 GeV and the sensitivity of Tevatron Run II is around 250 GeV, showing the importance of HERA II for stop searches.

5. Isolated leptons and missing p_T events at HERA

The H1 collaboration has observed regularly, during the HERA I running, an excess of events characterized by an isolated lepton e, μ at high transverse energy, missing transverse momentum p_T and high transverse momentum of the remaining hadronic system p_T^X. Both H1 and ZEUS have published their final results on this study, which are summarized in Table 1[5].

Table 1. Summary of isolated lepton search at H1 and ZEUS.

	Electron obs./exp	Muon obs./exp.	Tau obs./exp.
H1 p_T^X > 25 GeV	4 / 1.49	6 / 1.44	-
H1 p_T^X > 40 GeV	3 / 0.54	3 / 0.55	-
ZEUS p_T^X > 25 GeV	2 / 2.90	5 / 2.75	2 / 0.20
ZEUS p_T^X > 40 GeV	0 / 0.94	0 / 0.95	1 / 0.07

The main SM expectation in the table comes from the production of a W radiated from a quark line, and then decaying into a lepton and a neutrino, giving therefore the observed topology. However, while for the SM expectation the p_T^X distribution peaks at low values, for H1 there is a clear excess of events at $p_T^X > 40$ GeV, where 6 events are observed in the $\mu + e$-channels, while less than 1 event is expected. The ZEUS experiment does not confirm the result and observes no events for $p_T^X > 40$ GeV.

A possible process beyond the SM which could give this signature is the flavour-changing neutral-current (FCNC) reaction $eu \to (e)t \to (e)bW$ (where the scattered electron is seen in the detector only for high Q^2). Top decays could be produced in the SM via CC processes like $eg \to \nu tb$, however the cross-section is less than 1 fb and not measurable with the present integrated luminosity. Any excess of this kind of events could then be due to FCNC reactions. Both ZEUS and H1 have performed a search optimized for single-top production, both in the leptonic and hadronic decay of the W. In the leptonic channel, H1 observes 5 events compatible with a single-top signal, while 1.8 are expected from the SM contribution; no deviation is observed in the hadronic channel, where however the contribution from the SM QCD background is high. ZEUS observes no deviation in both channels, setting therefore strong limits on the FCNC couplings.

ZEUS has calculated the limit taking into account both the $tu\gamma$ and the tuZ contribution at leading order, as shown in Figure 4 (right). H1 has calculated the limit using the NLO order cross-section and neglecting the Z contribution: the excess in the leptonic channel is cause of the slightly

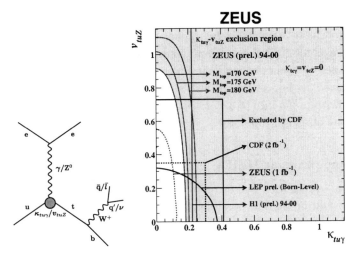

Figure 4. Diagram for FCNC top production at HERA (left) and limit (right) on the FCNC couplings obtained by H1, ZEUS, CDF and the combined LEP preliminary result. The dashed lines show also the prospects on the limits at HERA II and Tevatron II, assuming that no deviation from the SM is observed.

weaker limit in $k_{tu\gamma}$ compared to ZEUS. The limits are compared to the ones derived by the LEP experiments, looking for the decay $e^+e^- \to tu^a$, and to the CDF contraints derived from the branching ratios $t \to q\gamma$, $t \to qZ$. The strongest limit on the γ coupling, $k_{tu\gamma} < 0.174$, comes from ZEUS.

At HERA II, the search for this kind of events will be one of the high priority studies. Even hoping to see still an excess, an absence of this signal will provide very strong limits on the FCNC couplings, as shown in Figure 4 for an expected integrated luminosity of 1 fb^{-1}.

The ZEUS experiment has extended the previous search to the hadronic τ-decays[b]. The same technique described in Section 3 has been used to separate the taus from the QCD jets. The results[3] are summarized in Table 1, where, with some surprise, 2 outstanding events with $p_T^X > 25$ GeV remain after all cuts, with a SM expectation of 0.20 ± 0.05[c]. Note that,

---

[a]The charm contribution has been put to zero in this comparison, as in the HERA high−x regime of interest to this process the charm density is negligible compared to the u-density. The LEP and CDF results have therefore been rescaled by a factor $\sqrt{2}$. Another factor of $\sqrt{2}$, compensating the previous one, derives from the different Lagrangian convention used by HERA and LEP/Tevatron.
[b]The decays $\tau \to e\nu, \tau \to \mu\nu$ are included in the previous search.
[c]The results here reported are an update of what was presented at the Workshop, with the recently published results.

due to the low efficiency of selecting τ-hadronic decays, the cross-section corresponding to these 2 events is far higher than the single-top cross-section already excluded by the e, μ and hadronic-channels. A possible explanation of these τ events beyond the SM requires that there is no W-production involved. It is possible, in SUSY models with large $\tan\beta$, that the stop decays with preference to a τ, i.e. $\tilde{t} \to \tau\tilde{\nu}_\tau b, \tilde{\tau}\nu_\tau b$, with subsequent decays, giving the observed topology.

6. Multi-lepton events at HERA

Also the study of events with two or more leptons (e or μ) at high energy, can provide information on new possible processes beyond the SM. H1 has recently published the final results on this search, ZEUS has presented preliminary results, both use the full HERA I data sets[6].

For the multi-electron search, both experiments require two well-reconstructed *central* electrons[d], well inside the acceptance of the tracking chambers ($20° < \theta_e < 150°$ for H1 and $17° < \theta_e < 164°$ for ZEUS), and with transverse energy $p_T^e > 10$ GeV. At high-Q^2 values, the scattered electron can be observed in the detector, giving a topology of 3-e in the final states: the third electron is allowed in a wider angular region ($5° < \theta_e < 175°$). In general both H1 and ZEUS observe, in all distributions, a good agreement with the SM prediction, which is dominated by the Bethe-Heitler reaction $\gamma\gamma \to l^+l^-$.

Table 2. Summary of multi-electron search at H1 and ZEUS for $M_{12} > 100$ GeV.

	DATA H1	SM H1	DATA ZEUS	SM ZEUS
2e	3	0.30 ± 0.04	2	0.77 ± 0.08
3e	3	0.23 ± 0.04	0	0.37 ± 0.04

H1 observes in total 108 2e events and 17 3e events, in good agreement with the SM predictions of 117.1 ± 8.8 and 20.3 ± 2.1, respectively. However, the invariant mass distribution, M_{12}, of the two highest-p_T^e electrons, shows 6 outstanding events at mass greater than 100 GeV, 3 with 2 electrons and 3 with 3 electrons, while 0.53 events in total are expected. The M_{12} distribution is shown in Figure 5 and the predicted number of events is shown in Table 2. The figure also shows that the events at high mass

---
[d]The term electron here refers to both electrons and positrons, unless specified.

have a different topology in the 2e- and in the 3e-case: while in the 2e-case the two electrons with high invariant mass have also high transverse momentum, this is not the case for the 3e-events, suggesting that, if it is new physics, it could be of different origin.

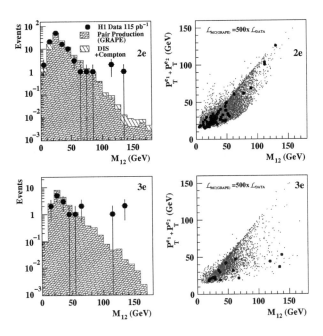

Figure 5. Invariant mass distribution of the two highest-p_T^e electrons in the H1 data, for the 2e- and 3e- samples separately, compared to the sum of the SM expectations. The two plots on the left show the correlation between the mass and the sum of the transverse momentum of the two highest-p_T^e electrons.

The ZEUS collaboration has looked for a similar topology in the HERA I data. In total 191 2e-events and 26 3e-events are selected, in broad agreement with the SM prediction of 213.9 ± 3.9 and 34.7 ± 0.5, respectively. However no deviation is observed at high mass, where two 2e-events have $M_{12} > 100$ GeV, while 1.2 are expected in total (see Table 2). In addition, no deviation from the SM is observed in the muon channel by both experiments.

A possible explanation of the 6 outstanding events observed by H1 is the production of doubly-charged Higgs. A $H^{++}(H^{--})$ can be produced at HERA, radiated from the positron(electron) line. It decays then in two leptons of the same positive (negative) sign. Although there are already very strong limits from OPAL, a dedicated search was done by H1, optimizing the selection cuts for this signal. Only one of the 6 events survive these criteria, making unlikely that these events can be explained by doubly-charged Higgs production. The analysis is instead used to set limits on this process, excluding H^{++}/H^{--} masses of 131 GeV for a value of the Yukawa coupling of the Higgs to the electrons of 0.3. Also this search will be an important chapter of the HERA II program, with expected sensitivities that compete with the present limits from LEP.

7. Summary

The H1 and ZEUS Collaborations have looked for signature beyond the SM model in many processes, ranging from s-channel resonances like leptoquarks, squarks and excited fermions, to contact interactions and to special signatures in the final state like those with high-transverse-momentum leptons. It is natural to ask if anything was forgotten: for this purpose H1 has performed a model-independent search[7], selecting all events with at least two objects (electrons, photons, muons, jets, and neutrinos $i.e.$ missing p_T) at high transverse energy ($E_T > 20$ GeV). The number of selected events has been compared to the expectation, calculated from the sum of the Monte Carlo prediction of various models. The largest deviation is observed in the $\mu - jet - \nu$ channel, corresponding to the events with isolated muons, high p_T^X and high missing p_T, as described in Section 5. All other observations are compatible with all previous analyses and no other anomaly is found. With the higher integrated luminosity promised for HERA II, it will be exciting to confirm or solve the remaining puzzles from the 1994-2000 running period.

References

1. ZEUS Coll., *Phys. Rev.* **D68** 052004 (2003), and references therein; H1 Coll., paper 105 contributed to the EPS03 conference, Amsterdam.
2. ZEUS Coll., EPS03 paper 498.
3. ZEUS Coll., DESY-03-182.
4. H1 Coll., EPS03 paper 101.
5. H1 Coll., *Phys. Lett.* **B561** 241 (2003) and EPS03 paper 100; ZEUS Coll., *Phys. Lett.* **B559** 153 (2003) and EPS03 paper 495.
6. H1 Coll., hep-ex/0307015 and EPS03 paper 104; ZEUS Coll., ICHEP02 paper 910.
7. H1 Coll., EPS03 paper 118.

7
Plans for HERA-II and Beyond

PLANS FOR THE HERMES EXPERIMENT AT HERA II

E.C. ASCHENAUER

ON BEHALF OF THE HERMES COLLABORATION

DESY Zeuthen, Platanenallee 6, 15738 Zeuthen, Germany

Abstract

The HERMES experiment [1] has been operational at the HERA accelerator of DESY since 1995.

In HERMES Run-1 the focus was on a first accurate determination of the flavour separated quark helicity distributions [2]. This was done on the basis of a large data set on both a longitudinally polarized hydrogen and a longitudinally polarized deuteron target. For the deuteron data HERMES could substantially improve the accuracy because of a RICH detector [3], which was installed in 1998. This enabled the identification of pions and kaons over most of the momentum range. For the earlier hydrogen data only pion identification was possible using the previously installed threshold Cherenkov detector. The valence polarizations are well determined, but the much smaller sea polarizations are essentially compatible with 0. The accuracy of these results could substantially be improved by taking an additional 4 Mio DIS events with a longitudinally polarized hydrogen target and the RICH detector. This improvement in statistical accuracy would be particularly important for the sea polarizations. With the present statistics no such asymmetry is evident. Various theoretical calculations exist (again largely prompted by the data taking of HERMES) that predict a non-zero asymmetry. The additional 4 Mio DIS events would improve the statistics enough to allow verification of several models.

In semi-inclusive reactions, when not only the scattered positron but also some of the produced hadrons are detected, a lot of interest has been generated by the appearance of single-spin asymmetries. These offer the best road towards a measurement of the unknown transversity distribution of quarks. HERMES has started Run-2, where a first determination of the transversity through the Collins mechanism is the prime objective for the next years. It was partly driven by the fact that for this the beam does not need to be polarized. Up to February 2003 about 900'000 events have been recorded with a transversely polarized hydrogen target. From estimates reported in e.g. the HERMES Long Range Plan it is known that about 7 Mio DIS events are needed for a statistically satisfactory measurement. This is necessary to discriminate between different theoretical models for the transversity distribution. In fact, the results published by HERMES on Run-1 [4] with a *longitudinally* polarized H and D targets, and the fact that HERMES now runs with a transversely polarized target have led to a spate of new calculations by several theory groups around the globe.

As is well known a measurement of exclusive reactions gives access to the newly introduced Generalized Parton Distributions (GPD), which contain more detailed information on the internal quark structure of the nucleon than the standard Parton Distribution Functions. The cleanest way to arrive at the GPDs is through the Deeply Virtual Compton Scattering (DVCS) process, which was already measured and published in HERMES Run-1 [5]. However, it is obvious that HERMES in its present configuration has insufficient energy resolution to allow an event-by-event separation of exclusive reactions from non-ground state transitions. The easiest way to improve on this situation is by the installation of a recoil detector around the target to detect the recoiling target proton. Such a recoil detector is presently under construction for HERMES and will be ready for installation for the last two years of HERA operation.

Another essential improvement brought by the recoil detector is the much better resolution in momentum transfer $-t$. A full programme of DVCS measurements with the recoil detector needs about 2 years of beam time. Since the installation of the recoil detector necessitates the removal of the present polarized target system, the most efficient planning requires these 2 years to be the last 2 years of data taking with HERMES. An additional condition is that these 2 years of data should consist of roughly equal data sets with opposite beam charge. HERA is the only facility worldwide where beam-charge asymmetries in the interference of DVCS and Bethe-Heitler can be exploited to get to the DVCS amplitudes (rather than cross sections). 1 year of data with electrons and 1 year of data with positrons (both polarized!) will yield unprecedented accuracy for this asymmetry.

The physics case for the next phases in the HERMES running will be presented and an overview of the challenges ahead for the experiment will be given.

References

1. K. Ackerstaff et al., Nucl. Instr. and Meth. A **417** (1998) 230
2. K. Ackerstaff et al., Phys. Lett. B **464** (1999) 123
   A. Airapetian et al., Phys. Rev. Lett. **92** (2004) 012005
3. N. Akopov et al. Nucl. Instr. and Meth. A **479** (2002) 511
4. A. Airapetian et al., Phys. Rev. Lett. **84** (2000) 4047
   A. Airapetian et al., Phys. Rev. D **64** (2001) 097101
   A. Airapetian et al., Phys. Lett. B **562** (2003) 182
5. A. Airapetian et al., Phys. Lett. B **535** (2002) 85
   A. Airapetian et al., Phys. Rev. Lett. **87** (2001) 182001
   F. Ellinghaus et al., hep-ex/0212019
   F. Ellinghaus et al., hep-ex/0207029

PHYSICS WITH H1 AT HERA II

M. KLEIN

DESY, Platanenallee 6, 15738 Zeuthen, Germany
E-mail: klein@ifh.de

A sketch is given of the main results obtained with the H1 apparatus at HERA as is a summary of where progress is to be expected as a result of increased HERA luminosity, the measurement of beam charge and polarisation asymmetries, the variation of beam energies, improved systematics and extended kinematic range.

1. The Past

Before the advent of the world's first electron-proton collider HERA, two decades of experiments on deep inelastic lepton-nucleon scattering (DIS) using stationary targets established what is now the conventional picture of nucleon structure: The proton and the neutron contain valence and sea quarks which are dominant at large Bjorken $x > 0.3$ and low $x < 0.05$, respectively. Here, x describes the momentum fraction of the proton carried by the quarks in the infinite momentum frame. In the Quark Parton Model (QPM), the distribution of quarks and anti-quarks in the nucleon is described by momentum density functions, $q(x)$ and $\bar{q}(x)$, which apply universally to elementary fermion interactions. The DIS cross section is largely determined by the proton structure function $F_2 = x \sum e_q^2 (q + \bar{q})$, where e_q are the quark electric charges. The x dependence of the various parton distributions is to be determined from experiment.

DIS was crucial in establishing Quantum Chromodynamics (QCD) as the correct field theory of the strong interactions which describes parton dynamics in terms of the exchange of gluons and includes gluon-gluon interactions due to its non-Abelian character. In QCD, the structure functions are a convolution of calculable coefficient functions with the parton distributions which due to gluon radiation and quark- antiquark pair production depend both on x and Q^2, the square of the four-momentum transferred by the exchanged virtual photon, $W^\pm$ or Z boson. The strong coupling constant α_s was found to decrease with Q^2, presumably leading to asymptotically free quarks, i.e. to the restoration of the QPM as a limit of QCD. In DIS, secondary particles are emitted which allow the parton dynamics in the formation of the final state to be investigated in detail. Polarised eD scattering was instrumental in establishing the electroweak sector of the Standard $SU(2)_L \times U(1) \times SU(3)_c$ Model (SM).

2. HERA and H1

HERA was built in the eighties to carry these investigations much further. The H1 detector [1] is a large, multipurpose device of nearly 4π acceptance, comprising tracking and calorimetric detectors. The range of x and Q^2 accessible is much wider than that available to fixed target experiments. The kinematics is reconstructed both from the scattered electron and from the final state. This results in a superior accuracy for the HERA DIS measurements which need to be precise to reveal the subtleties of strong interactions. Moreover, due to the high collision energy, HERA provides much richer information about the final state than is accessible at fixed target experiments.

During its first phase of operation, between 1992 and 2000, H1 recorded an integrated luminosity of about 110pb^{-1} of e^+p and 15pb^{-1} of e^-p data with both beams unpolarised. In its second phase, termed HERA II, the aim is to record an increased integrated luminosity of about 1 fb^{-1} with longitudinally polarised leptons at highest energies, and also to provide data at lower proton beam energy. The increase of the luminosity is achieved due to better focussing of the beams in the interaction region (IR). Thus superconducting quadrupoles were put close to the ep interaction point and detectors close to the beam had to be modified. Bending the positron beam nearer to H1 caused stronger synchrotron radiation than in HERA I. This and the modified IR caused beam induced backgrounds, mainly from the dynamic interplay of synchrotron radiation with the proton beam, which initially prevented H1 and the other collider experiment, ZEUS, from collecting data. While this article is being written, early 2004, HERA is delivering beams with increasing intensity yet a factor of $2-3$ below the design value. The longitudinal positron polarisation is typically about 40%. It remains a challenge to achieve routine operation at the highest luminosities but the prospects are good.

The H1 detector underwent a major upgrade for HERA II. A new finely segmented inner proportional chamber will improve the vertex trigger capabilities. A dedicated trigger uses central drift chamber hits to reconstruct tracks, e.g. allowing the online identification of charm in photoproduction events. New forward and backward silicon detectors extend the tracking and thus the heavy flavour physics capabilities to larger and to lower x, respectively. A new forward drift chamber system improves tracking at large x and high Q^2. Downstream along the proton beam line, a new Roman pot fibre detector extends the acceptance for diffractive scattering and measures t dependent cross sections. A new Compton polarimeter will measure the beam polarisation with increased precision and for each bunch separately. These and further improvements represent significant investments and form, together with the standard components of the H1 apparatus, calorimeters and chambers, a solid basis for high precision physics and searches at HERA II.

This paper provides a brief account of the results [2] obtained by the H1 Collaboration from the first phase of HERA operation. Subsequently, selected topics are described which are central to physics with HERA II. With much increased luminosity and improved systematics genuine surprises may be encountered. New developments, such as unintegrated parton distributions or parton correlation functions, may well change our view on proton structure in a qualitative way. QCD predicts that instantons exist and that instanton induced cross sections are accessible to DIS experiments at HERA. Given that interactions mediated by two gluon, or Pomeron, exchange have been observed, should odderons exist, i.e. three gluon exchange? The excitement of HERA physics is largely due to the unprecedented richness of Quantum Chromodynamics.

3. Results of the H1 Experiment at HERA

With beam energies of $E_e = 27.5$ GeV and $E_p = 920$ GeV, corresponding to a centre of mass energy squared of $s \simeq 10^5$ GeV^2, HERA allows the regions of very high $Q^2 < s$, and of very small Bjorken $x = Q^2/sy > 10^{-5}$ to be explored in DIS for the first time. Here y is the inelasticity, which, in the proton rest frame, corresponds to the relative energy transferred by the exchanged boson. At large Q^2, the standard model predicts that the cross sections for exchanging neutral (γ, Z_0) and charged bosons $(W^\pm)$ are of similar strength. A major triumph of the electroweak theory and of HERA was that this indeed could be observed by the H1 [2] and ZEUS [3] experiments, see Fig.1. The detailed exploration of weak neutral and charged current scattering at high Q^2 and large x using high statistics and polarised $e^\pm$ beams is a central issue for the future HERA programme, as described

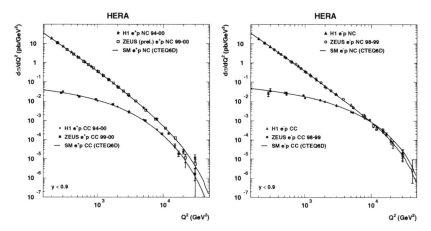

Figure 1. Measurements by H1 and ZEUS of the positron (left) and the electron (right) proton scattering cross sections at large Q^2 in neutral and charged current scattering. The curves represent calculations of the DIS cross section using standard model electroweak couplings and a set of parton distributions determined by the CTEQ Collaboration.

below.

Soon after the start of HERA operation, H1 and ZEUS discovered new features of the strong interactions and of proton structure. A striking observation from the first 1992 data was that the proton structure function $F_2(x, Q^2)$ rises strongly towards low x, at fixed Q^2. This observation was later confirmed with much improved accuracy [2,3], see Fig.2. The rise of F_2 in the DIS region implies that the sea quark density at low x is high while the strong coupling constant, $\alpha_s(Q^2)$, is rather small. This defines a new regime of parton dynamics, the theoretical understanding of which is not complete despite impressive theoretical attempts over the past decades [4]. The x dependence of $F_2(x, Q^2)$ is observed to change [2] at momentum transfers $Q^2 \sim 1\,\text{GeV}^2$, corresponding to a distance of 0.3 fm which is where partonic interactions seem to emerge. As striking as the x dependence, is the strong rise of $F_2(x, Q^2)$ with Q^2 at low, fixed x. In the classic evolution equations of perturbative Quantum Chromodynamics (pQCD), the Q^2 dependence, characterised by the derivative $\partial F_2/\partial \log Q^2$, is related to the gluon density. It was soon realised that, at low $x < 0.01$, the quark contribution to the F_2 scaling violations is small for $Q^2 > 3\,\text{GeV}^2$. Therefore, the measurement of a large derivative $\partial F_2/\partial \log Q^2$, see Fig.2, indicates that the gluon density is large at low x. The experimental investigation of low x physics will be extended at HERA II as part of the effort to establish

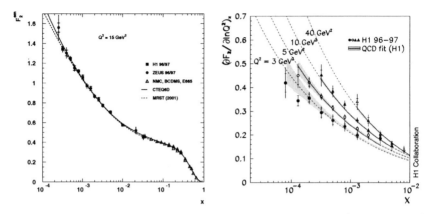

Figure 2. Left: Measurements of the proton structure function $F_2(x, Q^2)$ at medium Q^2 as a function of the proton momentum fraction x carried by the struck quark. HERA has significantly extended the kinematic range of these measurements and discovered a new regime of high sea quark densities. This can be attributed to a large gluon density as can be inferred from the measured scaling violations of $F_2(x, Q^2)$, right. Further improvements of the measurement accuracy and new data are envisaged to accurately determine xg and the strong coupling constant.

the theory of high density parton dynamics.

Low x physics soon turned out to be much richer than one could deduce from the inclusive DIS measurements. A further outstanding result from HERA in the early nineties was the observation [2,3] that in about 10% of all DIS events the proton does not dissociate but remains intact [a]. Since then deep inelastic diffraction has been investigated in much more detail. Diffraction, which may be related to colour confinement, occurs in a simplified view, as a two-gluon exchange which is both a fascinating theoretical problem and may prove to be a means of finding new particles at hadron-hadron colliders as it gives rise to particularly clean events, if both hadrons stay intact.

Because of the high energy available for particle production and since the gluon density is large, HERA is also a laboratory for heavy quark physics allowing the production mechanisms of charm and beauty particles to be studied. This has led to the development of heavy flavour theory [6] in next to leading order (NLO) QCD, as it turned out that much can be learned about parton dynamics from a 20-40% charm event fraction. No consistent view on strong interactions can be obtained, and no reliable extraction of

---

[a]Note that the elastic $ep \to ep$ scattering cross section decreases like Q^{-12} and is thus negligible at HERA [5].

$\alpha_s(M_Z^2)$ is possible, without accurately understanding heavy flavour physics at HERA. Heavy quarks may be a clue to finding exotic multiquark bound states [7].

The production of final state particles became a further testing ground for QCD. Going much beyond DIS fixed target final state physics, at HERA jet physics became a central issue. Fig.3 shows, as an example, a recent H1 measurement [2] of the jet production cross section at low $Q^2 < 1\,\text{GeV}^2$, termed the photoproduction region. The cross section, measured as a function of the transverse energy of the jet, is seen to decrease by 7 orders of magnitude, a behaviour which is consistently described by NLO QCD calculations. Jet measurements are used [2] to measure $\alpha_s(M_Z^2)$ and jet shapes are investigated [2] in resummed NLO QCD calculations which also include power corrections.

A most exciting aspect of HERA physics is the search for new interactions and the new particles with which they are associated, such as leptoquarks and squarks. Outstanding events with high mass multi-electrons or with large missing transverse momentum, including flavour changing neutral current events, have been observed by H1 [2]; somewhat more than expected in the Standard Model. At high Q^2 and large x, a rich field of physics has been opened, allowing, for example, the search for quark-substructure [2] and measurements of parton distributions [2] in the valence quark region, at $Q^2 \sim M_Z^2$. The physics at large x and high Q^2 cannot be explored thoroughly with the luminosity obtained so far, since the cross section decreases with x as $(1-x)^3$ at large x and, for photon-exchange, with Q^2 as $1/Q^4$. High luminosity will lead to new insight and perhaps discoveries at HERA. In this regard, lepton beam polarisation is an interesting feature, as will be discussed below.

4. The Programme for HERA II

4.1. *High Precision Inclusive DIS*

In QCD, variations are logarithmic, often involving factors of $\ln Q^2/\Lambda^2$ or $\ln 1/x$ where Λ is the scale parameter in QCD. Tests of QCD therefore need to reach a high level of accuracy, typically 1%, if they are to be decisive and if the data are to be of use for predicting cross sections at other machines, such as the LHC or for neutrino astrophysics. An important example in this regard constitutes the measurement of α_s in DIS. The strong coupling $\alpha_s(M_Z^2)$ is the least well known of all the coupling constants, but is of great importance for testing QCD and for the unification pattern of the funda-

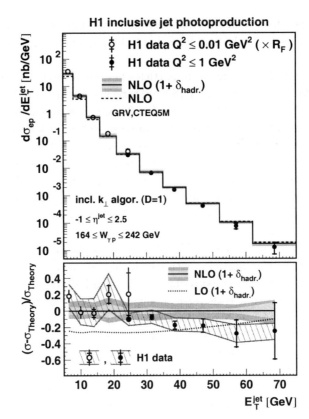

Figure 3. Top: Differential cross section for inclusive jet production as a function of the jet transverse energy. Bottom: relative difference between the data, the LO QCD prediction and the NLO calculation including hadronisation corrections.

mental interactions at extremely high energy. Inclusive deep inelastic scattering is a process particularly well suited to the measurement of α_s because it is calculable to high orders, in perturbation theory, and independent of the non-perturbative corrections that arise in jet and other analyses from final state effects. An accurate measurement of $\alpha_s(M_Z^2)$ in DIS was obtained by H1, the result being $\alpha_s(M_Z^2) = 0.1150 \pm 0.0017(exp) \pm 0.0008(model)$. This measurement is based on a systematic analysis of the BCDMS μp and the H1 ep inclusive cross section data. Contrary to other determinations of $\alpha_s(M_Z^2)$, the correlation of $\alpha_s(M_Z^2)$ with the gluon distribution is resolved, leading to competitive, consistent contributions of the data of both experiments to the final result, see Fig.4. The HERA II programme foresees further progress with this analysis which can be achieved by still improving the measurement of the ep cross section, using low E_p HERA data to access large x, a better understanding of the model uncertainties like the charm mass, possibly the use of data with deuterons [8] and, last but not least, from the calculation of the splitting and coefficient functions in NNLO [9] or their modification [10]. In an *ad hoc* procedure, the variation of the renormalisation scale μ_r^2 between $Q^2/4$ and $4Q^2$ leads in NLO to an estimated uncertainty of about ± 0.005 on $\alpha_s(M_Z^2)$. It is expected that this will be halved using forthcoming NNLO calculations. The prescription of varying μ_r by factors of $1/2$ and 2 must be reexamined since the data exclude the possibility of such big variations [2,12].

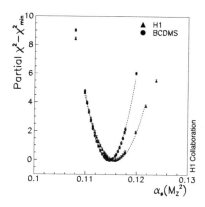

Figure 4. Contributions from the H1 and the BCDMS data to the χ^2 in the minimisation procedure using NLO QCD theory as a function of $\alpha_s(M_Z^2)$.

A further example of the importance of large data samples at HERA regards the flavour decomposition of the partonic contents of the proton. Not much is known about the strange quark and its possible asymmetry $s-\bar{s}$ at high Q^2 and x which may be important to the understanding of the electroweak mixing angle measurements [13]. The problem of the ratio of down and up quark distributions at large x is important for discoveries of high mass particles [b]. HERA II can contribute to this through improved measurements of the neutral and charged current cross sections at very high

---

[b] As was pointed out recently [8], the H1 detector offers the opportunity of tagging spectator protons in electron-deuteron scattering, thereby removing the binding corrections otherwise necessary in the unfolding of the neutron structure. This allows both the sea quark asymmetry $\bar{u}-\bar{d}$ to be extracted accurately at low x, which is not accessible in ep scattering at HERA II, and the ratio d/u to be measured at large x.

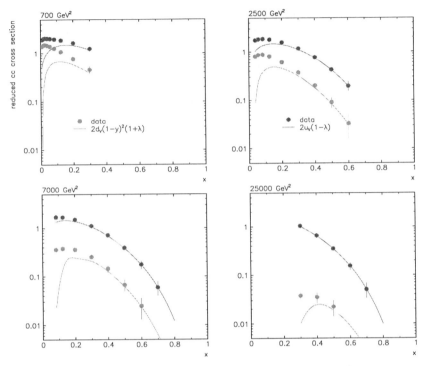

Figure 5. Simulation of charged current measurements with 250pb^{-1} luminosity for 50% polarisation showing statistical errors only for e^+p (grey points) and for e^-p (black points). The curves represent the valence quark contributions to the CC cross sections.

Q^2, see [14] Fig.5.

4.2. Physics with Polarised $e^{\pm}$ Beams

4.2.1. Parity Violation and Valence Quarks

The increased luminosity and the use of longitudinal lepton polarisation will allow the electroweak structure functions [15] to be measured. For NC interactions, in addition to F_2, this applies to the lepton beam charge asymmetry which essentially determines xG_3, the $\gamma - Z$ interference part of the generalised structure function $\mathbf{xF_3}$. Variation of the lepton beam helicity determines the structure function, $G_2 = 2x\sum e_q v_q(q + \bar{q})$, so far unmeasured, which is the $\gamma - Z$ interference part of the generalised structure function $\mathbf{F_2}$. Here, v_q is the weak vector coupling of the quark q. At high x, $G_2(x, Q^2)$ is proportional to the sum $u_v + d_v$ while $F_2 \propto 4u_v + d_v$. Both G_2 and F_2 will thus provide complementary information on the valence-quark distributions, independent of nuclear binding corrections. A simulated measurement of G_2 was presented in [16]. Unique information on the valence quarks at rather low $x \geq 0.01$ can be obtained from beam charge asymmetry data determining the non-singlet structure function $xG_3 \propto 2u_v + d_v$. Since the underlying asymmetries rise with Q^2, the charge asymmetry is proportional to $1-(1-y)^2$ and the valence quark distributions drop like a power of $(1-x)$, it would be desirable to significantly exceed the current goal of 1 fb^{-1} with the upgraded HERA collider.

4.2.2. Right Handed Currents

The determination of the helicity structure in neutrino-nucleon scattering has long been considered an interesting problem. It can finally be solved using polarised lepton beams at HERA, which provides the equivalent of neutrino and antineutrino beams on fixed targets of 54 TeV energy. The charged current scattering cross section is predicted to be $\sigma_{CC}^{\pm}(\lambda) \propto (1\pm\lambda)$. A departure from this expectation would hint at the existence of right handed currents. The accuracy of this measurement depends mainly on the maximum degree of beam polarisation achievable.

4.2.3. Generalised Parton Distributions

Deeply virtual Compton scattering (DVCS), the hard diffractive scattering of the virtual photon off the proton, $ep \to e\gamma p$, interferes with Bethe-Heitler (BH) scattering. The interest in this process arises from the possibility it presents to measure parton correlations, i.e. to measure generalised, or non-diagonal parton distributions [17]. It is in this respect similar to vector meson (VM) production, but avoids the complications due to the VM wave function. Large asymmetries are predicted [18], both in lepton beam charge and polarisation asymmetry measurements, which access the real and the imaginary part of the $BH \cdot DVCS$ interference amplitude, respectively. Improvements in the accuracy of the measurements [2,3], see Fig.6, is likely to shed new light on the partonic structure of the proton and on the nature of the diffractive interaction. Exploitation of this process requires maximum luminosity.

4.3. Physics with Varied Proton Beam Energy

Apart from one adjustment of the proton beam energy, from 820 GeV to 920 GeV, HERA has been operated for 11 years at fixed beam energies.

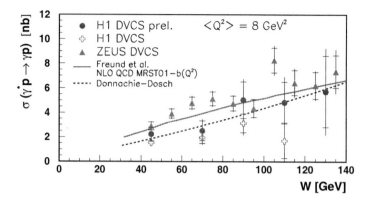

Figure 6. The $\gamma^*p \to \gamma p$ DVCS cross section as a function of W measured by H1 and ZEUS. The data are compared with NLO QCD using a GPD parameterisation (solid curve) and with a Colour Dipole model prediction (dashed curve).

However, there are good reasons to vary the beam energies. Increasing the proton beam energy [c], E_p, is most desirable when searching for new phenomena, since all search limits are set by the available energy, rather than by the luminosity, and are given, e.g. in the mass range for new particles, roughly by $M \leq \sqrt{s} = 2\sqrt{E_e E_p} = 318\,\mathrm{GeV}$ at the present energy settings. An increase in proton beam energy involves certain risk, or investment, regarding the quench protection of the proton machine dipole magnets.

Lowering the beam energy up to about half the present values is possible at the expense of reducing the luminosity. It makes possible the extension of the kinematic range for many HERA measurements, such as the total γp and vector meson cross sections [2]. The study of two further, specific questions requires E_p to be lowered. The first regards the understanding of the gluon density and pQCD at low x. Investigations based on $\partial F_2 / \partial \log Q^2$ only cannot uniquely pin down the behaviour of $xg(x, Q^2)$ at low x. More generally, an additional constraint is necessary if QCD is to be tested to higher order at low x. This requires an accurate determination of the longitudinal structure function $F_L(x, Q^2)$. Such a measurement [19], accurate to about 5%, can be made with the present H1 apparatus for $Q^2 > 3\,\mathrm{GeV}^2$ in a sequence of runs at low E_p, for example at 460 and 570

---

[c]The electron beam energy of 27.6 GeV is chosen to achieve maximum lepton beam polarisation. HERA could be operated at somewhat larger E_e depending on the rf. power available.

GeV, with luminosities of about 10 and 5 pb^{-1}.

The second question concerns access to the medium Q^2, large x region with HERA. This is of great interest for a precision measurement of $\alpha_s(M_Z^2)$, for the determination of parton distributions at large x and perhaps also for the search for an intrinsic heavy flavour contribution to proton structure. Since the luminosity is expected to degrade roughly like E_p^2, it is presently not realistic to envisage a luminosity larger than about 30 pb^{-1}. As is demonstrated in Fig.7, such data would be of considerable use in extending the kinematic range of the HERA data and achieving a larger overlap with the data from the BCDMS experiment. The latter have the peculiar feature of forcing $\alpha_s(M_Z^2)$ in DIS QCD analyses to small values unless the BCDMS high y data are excluded, see [11].

4.4. *Final States, Diffraction and Heavy Flavour*

Final state physics at HERA allows the parton dynamics to be studied in detail. At low x it has been realised that, while collinear factorisation, as is inherent in DGLAP evolution, seems to be able to account for the rise of F_2 towards low x, data on final state production may be better described in k_t factorisation approaches. The phenomenology based on different evolution equations (DGLAP, BFKL, CCFM) applied to data on forward particle production [2] or on azimuthal correlations between dijets [2], for example, promises to yield new insight [4] into parton dynamics at low x. Larger data samples of increased systematic accuracy and extended kinematic range will be extremely valuable in the pursuit of these studies. It is noteworthy that currently a major step is being made in these areas. While previously final state physics, diffraction and heavy flavour were, and will of course continue to be, considered separately, joint studies of all these fields and their interdependence using data of higher luminosity are now being made. This concerns, for example, the investigations of how NLO parton distributions in the pomeron, determined in inclusive diffraction, are applicable to diffractive charm [2] or jet production [2], see Fig.8. The study of diffraction with H1 will be further pursued using an upgraded system of proton tagging spectrometers and high luminosity data permitting the t dependence to be employed.

Jet physics is a major testing ground for QCD. Recently progress was made in describing consistently event shapes in jet production [2] using resummed NLO QCD calculations. With high luminosity, the measurement of $\alpha_s(M_Z^2)$ from jet rates will become more precise and joint analyses of

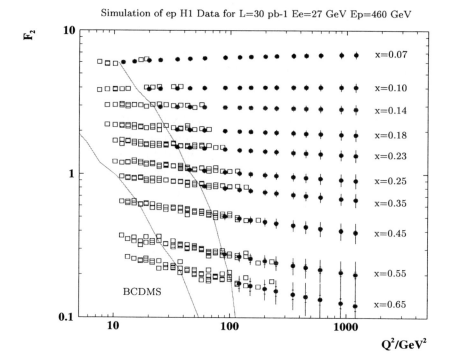

Figure 7. Measurement of the proton structure function $F_2(x,Q^2)$ by the BCDMS Collaboration using 4 different muon beam energies, between 120 GeV and 280 GeV (squares). Simulation of precision low energy run data (points), for $E_p = 460$ GeV, $E_e = 27.6$ GeV and 30 pb^{-1} luminosity, with a full simulation of the systematic accuracy (inner error bars). The lines mark the kinematic range limit $Q^2 = 2ME_\mu xy$ for $y = 0.3$ and 120 GeV and 280 GeV enforced in the H1 QCD fit to the BCDMS data. Precision H1 data from extended low energy running will have substantial overlap with the BCDMS data.

inclusive DIS with jet production may help in the determination of the gluon density at large x, that remains unresolved in DIS. The description of multi-jet final states will be much improved in following the development of parton shower simulation programs at higher orders [20] (MC@NLO).

A consistent theory is necessary to describe charm and beauty production in DIS. Describing correctly the effects that occur close to charm threshold is a nontrivial task in the DGLAP framework which considers the charm quark to be light at high $Q^2 >> m_c^2$. The amount and accuracy of heavy flavour data [21] from HERA I, on charm mesons [2], fragmentation functions [2], J/Ψ production [2] etc., is most remarkable and work is ongoing. Based on the first observations, see Fig. 8, one may well speculate that beauty at HERA II will play the role of charm at HERA I. Accurate beauty data, based on the large transverse momentum of the b-quarks and on their measurable lifetime, corresponding to about 0.5 mm decay lengths, promise to open an important new testing ground for QCD. New reactions like the production of strange quarks in CC scattering will be also studied.

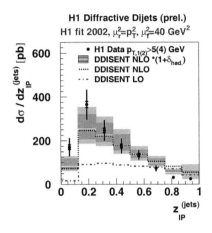
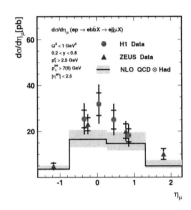

Figure 8. Left: Diffractive dijet production cross section in DIS as a function of the longitudinal momentum fraction of the Pomeron. The data are compared with predictions, within an uncertainty dominated by the renormalisation scale choice, using (N)LO diffractive parton distributions determined from inclusive diffractive H1 data. Right: Measurements of the beauty jet production cross section in photoproduction as a function of the muon rapidity compared with NLO QCD calculations including hadronisation corrections. A considerable increase of the luminosity is required to perform precision measurements of such quantities at HERA II.

4.5. Searches for Exotic Physics

Dedicated analyses of the H1 data from HERA I, using about 15 pb^{-1} of e^-p and 110 pb^{-1} of e^+p data, have discovered a few peculiar events of a type or a rate beyond the expectations of the Standard Model. Isolated electrons and muons with large missing transverse momentum, 10 events with $p_t > 25$ GeV, are observed and 2.9 ± 0.5 are expected. Furthermore, H1 saw multi-electron production at a rate above the Standard Model expectations. Three $2e$ and three $3e$ events were observed which is to be compared with an expectation of 0.30 ± 0.04 and 0.23 ± 0.04, respectively. Based on a considerable simulation effort, new effects were also searched for *generically* in 28 event classes, e.g. of the type ej or $e\mu\mu$, using electrons, muons, jets and missing energy (neutrinos) to characterise the classes. The preliminary H1 result [2], using all available HERA I data, is shown in Fig.9. It confirms the excesses in the aforementioned channels but no further significant departure from the SM expectation is found. It is obvious that a larger data sample is required to put such generic studies and dedicated searches on a firmer ground and possibly to discover new particles or interactions, a major goal of HERA. If, for example, leptoquarks were seen, charge and lepton beam polarisation would enable their spectroscopy to be studied. New ways could be found to further increase the luminosity beyond the HERA II values. Moreover, new, LHC type, magnets could be used to double the proton beam energy [22] if findings at HERA, the TeVatron or the LHC would require *ep* data in an extended range of energy.

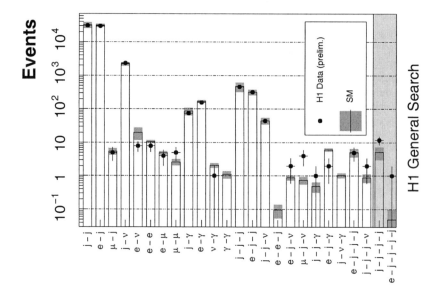

Figure 9. Event yields, measured and calculated in the standard model using various simulation programs, for all event configurations with an expectation larger than 0.1. Four jet simulations are not yet reliable and kept separate in the generic analysis.

5. Future

As was sketched here, the HERA programme is of fundamental importance in its own right. Seen from the perspective of the LHC, which will take particle physics into the TeV energy region, it is becoming clear that the results of HERA are of interest in all areas: high precision measurements will pin down the SM expectations and may be necessary for the determination of the LHC luminosity; high x parton distributions are of crucial importance for the understanding of any state possibly discovered at the LHC; understanding low x and diffractive phenomena appears also to be important for the LHC, because of the significance of the quark and gluon distributions in understanding pp collisions and the possibilities of detecting Higgs production in double diffractive scattering [23]. It is desirable to operate HERA when the LHC runs and new questions may be posed which HERA could answer.

The past and the current programmes have not exhausted HERA's potential. Precision studies of high parton density phenomena near the transition from parton-like to hadron-like behaviour, at $Q^2 \sim 1\,\text{GeV}^2$ and smallest x, the investigation of neutron structure in the range opened up by HERA, the study of nuclear and thus high density gluon structure with high energy electron probes and the exploration of the completely new kinematic region accessible to a polarised eN collider are all scientifically most attractive elements of a programme to which particle physicists will have to return to if the study of QCD and nucleon structure is to be brought to a satisfactory conclusion: they are all topics which HERA allows to be studied uniquely.

Acknowledgments

I would like to thank the organisers of the 2003 Ringberg workshop for the invitation to attend what has been an excellent discussion meeting of physicists interested in deep inelastic scattering at HERA. The H1 results presented here, and those I had no time to discuss, rely on the efforts of a large collaboration of physicists and engineers to whom I want to express my deep gratitude. In particular, I like to thank Allen Caldwell, Tim Greenshaw, Peter Kostka, Tomas Lastovicka, Paul Newman, Carsten Niebuhr, Daniel Pitzl, Vladimir Shekelyan and Ferdinand Willeke for a most faithful and enjoyable collaboration in recent times.

References

1. H1 Collaboration, I. Abt et al., *Nucl. Inst. Meth.* **A386** (1997) 310 and 348.
2. Results of the H1 Collaboration presented here are either published or were submitted to recent conferences, see the H1 paper collection on the web http://www-h1.desy.de. For a recent review see P. Newman, hep-ex/0312018 (2003).
3. For results of the ZEUS Collaboration see http://www-zeus.desy.de and R. Yoshida, these Proceedings.
4. For reviews see e.g.: The Small-x Collaboration, J. Andersen et al., hep-ph/0312333 (2003) and references cited therein; E. Iancu, A. Leonidov and L. McLerran, hep-ph/0202270 (2002).
5. M. Klein and T. Riemann, *Z. Phys.* **C8**, 239 (1981).
6. For a review see W. K. Tung et al., Procs.: New trends in HERA Physics, Ringberg, Germany, June 2001, *J. Phys.* **G28**, 983 (2001).
7. P. Z. Huang et al., hep-ph/0401191 (2004).
8. H. Abramowicz et al., MPI-PhE/2003/06 (2003); T. Alexopoulos et al., DESY/03-194 (2003).
9. S. Moch, J. A. M. Vermaseren and A. Vogt, hep-ph/0309056 (2003).
10. G. Altarelli, R. D. Ball and S. Forte, *Nucl. Phys.* **B674**, 459 (2003) [hep-ph/0306156].
11. H1 Collaboration, C. Adloff et al., *Eur. Phys. J.* **C21**, 33 (2001) [hep-ex/0012053].
12. S. Chekanov *et al.* [ZEUS Collaboration], *Phys. Rev.* **D67**, 012007 (2003) [hep-ex/0208023].
13. S. Kretzer et al., hep-ph/0312322 (2003). B. Portheault, Proceedings DIS03 Workshop, to be published.
14. M. Klein, Inv. Talk at HiX2000, Philadelphia, USA, *http://ba323.scitech.temple.edu/hix2000*, unpublished.
15. M. Klein and T. Riemann, *Z. Phys.* **C24**, 151 (1984)
16. M. Klein, *Proceedings of the 9th International Workshop on Deep Inelastic Scattering (DIS 2001)*, Bologna, Italy, April 2001, p.409 (2002)
17. For a review see: M. Diehl, *Phys. Rept.* **388**, 41 (2003) [hep-ph/0307382].
18. A. Freund, *Phys. Rev.* **D68**, 096006 (2003) [hep-ph/0306012].
19. L. A. T. Bauerdick, A. Glazov and M. Klein, hep-ex/9609017 (1996).
20. S. Frixione and B. R. Webber, hep-ph/0309186 (2003).
21. For a recent review see: R. Gerhards, *Nucl. Phys. Proc. Suppl.* **115**, 126 (2003).
22. D. Pitzl, Energy upgrade of HERA, DESY 2003, private communication.
23. A. B. Kaidalov, V. A. Khoze, A. D. Martin and M. G. Ryskin, hep-ph/0311023 (2003).

PROSPECTS AND STATUS OF ZEUS AT HERA II

R. YOSHIDA

Argonne National Laboratory
9700 S Cass Ave.
Argonne, IL, 60439 USA
E-mail: rik.yoshida@anl.gov

Selected physics topics of the HERA II program are briefly reviewed. An outlook for data taking as of October 2003 and estimated luminosity requirements for some physics topics are given.

1. Physics of HERA II

Some physics topics of HERA II were covered by the previous speaker[1]. Here I briefly outline some of the remaining topics.

1.1. *Structure Functions*

The HERA Collider experiments H1 and ZEUS have made measurements[2] of the proton lstructure function F_2 spanning five orders of magnitude in both the Bjorken variable x and the virtuality of the photon Q^2, from the data taken in the period 1992-2000. Below Q^2 of 500 GeV^2, these measurements have a 2-3% precision, which is dominated by systematic effects. In this kinematic region, these measurements will most likely be the most precise for the foreseeable future.

For higher Q^2, and so for high x ($x > 0.1$), the statistical uncertainties dominate the current F_2 measurements. Studies[3] have shown that with an integrated luminosity of 1 fb^{-1} planned at HERA II, the precision measurements at high x will be such that, for example, the strong coupling constant $\alpha_s(M_Z)$ can be constrained to 0.002, in experimental uncertainties, and the gluon to 3% when these data are used in a QCD fit.

1.2. *Electroweak Studies*

The sensitivity of polarized Neutral Current (NC) and Charged Current (CC) DIS cross-sections to electroweak parameters means that there will be a rich spectrum of Electroweak results from HERA II. Again most of these are discussed in detail elsewhere[4].

An example of such a result is the determination of the NC couplings of the light quarks[5]. Depending on the polarization and integrated luminosity achieved, these can be determined to approximately 10%, and provide a result complementary to the heavy quark couplings at LEP.

Another example is the use of measured electron-proton cross-sections, both NC and CC, to determine the consistency of the electroweak description[6]. The results can be expressed in terms of the W mass and the top mass. A precise measurement of the top mass at the Tevatron or the LHC at the 5 GeV level would enable the HERA W mass to be constrained at the 50 MeV level. This would be a stringent test of the electroweak sector.

1.3. Exotics and Anomalies

There are intriguing anomalies observed in HERA I. H1 has observed excess of isolated electrons(and positrons) and muons accompanied by missing P_T; also an excess of di- and tri-electrons at high masses was seen[7]. While ZEUS does not see these excesses[8], it observes more than the expected number of isolated τ leptons accompanied by missing P_T, not associated with the τ decay[9]. All of these anomalies are at a level where more data are needed to study these effects. Since the HERA I data are based on an integrated luminosity of about 100 pb^{-1}, we would need something in the same order of magnitude in luminosity to start to make progress.

Also, there are new-physics channels, including R-parity violating SUSY and leptoquarks where HERA II remains competitive with the Tevatron[10]. In any case, the sign of the lepton beam and polarization are important for understanding any new physics that may emerge.

2. Prospects for data

2.1. HERA performance

Through 2002, the HERA machine has demonstrated its readiness for the high luminosity operation. The specific luminosity goal for HERA II of $1.8 \times 10^{30} cm^{-2} s^{-1} mA^{-2}$ has already been reached in November 2001. After a period of improvements to background conditions due to synchrotron radiation and aperture limitations, a HERA record luminosity of $2.7 \times 10^{31} cm^{-2} s^{-1}$ was reached in February 2003. This is to be compared to the design goal of $7.0 \times 10^{31} cm^{-2}s^{-1}$. Polarization of 50% during luminosity runs was also achieved in February 2003. However, the maximum beam current storable in HERA was limited by the severe backgrounds encountered at H1 and ZEUS; this has severely limited the luminosity at which the machine could be operated practically.

2.2. *Background*

Throughout 2002 and early 2003, H1 and ZEUS could not effectively take data due to the background rates. To a very good approximation, they can be equated with the current being drawn in the central tracking detector, which is too high so that it cannot be operated safely. Many measures against the background were taken during the HERA shutdown of 2003 in time for beam operation in September 2003.

There are three components to the background:

- Backscattered synchrotron radiation. This background, up to early 2003, was worse in ZEUS than in H1. This was reduced with improved synchrotron masks at the experiments. At ZEUS, the reduction is estimated to amount to a factor of ten.
- Lepton-beam-related particle background. This was reduced by reducing the thickness of some collimators and making provisions for better vacuum such as additional pumps or improvement in existing pumps. A reduction of a factor of two in this background is expected.
- Proton-beam-related particle background. This is the dominant background after the improvements above. This is also the background that is the least understood quantitatively. The vacuum improvements, better vacuum conditioning and collimators re-designed to reduce higher-order-mode heating will certainly reduce this background. The aimed-for reduction for this background is a factor of two.

As of the date of this report, only sizable positron together with small proton currents hadve been stored in HERA. However, studies at ZEUS of the synchrotron background showed that the improvement was as expected. Also the positron background showed steady improvement over several days.

Since no significant proton currents have been stored, the situation with respect to the proton background was not known[a].

3. Luminosity requirements and constraints

An estimate of the integrated luminosity required for the main physics topics for HERA II is shown in Figure 1. It is clear that the expected integrated luminosity for HERA II of 1 fb^{-1} is enough to cover a large portion, but not all, of the topics listed. The default plan of 250 pb^{-1} appears to be a relatively good compromise. Obviously, the actual operation of the machine and developing physics interests as the run progresses will determine the character of the data actually taken.

Physics	Required Luminosity (pb^{-1})				Total	E_p<920GeV
	e^+_L	e^+_R	e^-_L	e^-_R		
EW: EW parameters	250	250	250	250		
$\delta M_W \sim 50$ MeV			1000			
Large-x:F_2 for x>0.7					1000	
or						100
xF_3		500	250			
d_V from CC		500				
Med-x:F_2^b					500	
strange quark		250	250			
Small-x:High Q^2 VM					500	
Extend W coverage						50
F_L						30
Exo: rule out anomalies		200				
study anomalies	250	250	250	250		

Figure 1. A table of estimated integrated luminosities needed for HERA II physics topics.

References

1. M.Klein in these proceedings.
2. H1 Coll., C.Adloff et al., Nucl. Phys. B **470**, 3 (1996).
   H1 Coll., C.Adloff et al., Nucl. Phys. B **497**, 3 (1997).
   ZEUS Coll., J.Breitweg et al., Eur. Phys. J. C **11**, 427 (1999).
   H1 Coll., C.Adloff et al., Eur. Phys. J. C **13**, 609 (2000).
   H1 Coll., C.Adloff et al., Eur. Phys. J. C **19**, 269 (2001).
   H1 Coll., C.Adloff et al., Eur. Phys. J. C **21**, 33 (2001).

---

[a]As of January 2004, the background had been reduced to such an extent that it no longer limited the performance of the machine.

ZEUS Coll., S.Chekanov et al., Eur. Phys. J. C **21**, 443 (2001).
ZEUS Coll., S.Chekanov et al., Eur. Phys. J. C **28**, 175 (2003).
H1 Coll., C.Adloff et al., Eur. Phys. J. C **30**, 1 (2003).
ZEUS Coll., S.Chekanov et al., DESY-03-218, submitted to Phys. Rev. D.
3. M.Botje, M.Klein and C.Pascaud, *Proceedings of the Workshop on Future Physics at HERA*, G.Ingelman, A.DeRoeck and R.Klanner (eds.), Vol. 1, p. 33. Hamburg, Germany, DESY (1996).
4. R.Cashmore et al., ibid., Vol. 1, p. 129.
5. R.Cashmore et al., ibid., Vol. 1, p. 163.
6. S.Beyer et al., ibid., Vol. 1, p. 140.
7. H1 Coll., C.Adloff et al., Eur. Phys. J. C **5**, 575 (1998).
   H1 Coll., C.Adloff et al., Phys. Lett. B **561**, 241 (2003).
   H1 Coll., C.Adloff et al., Eur. Phys. J. C **31**, 17 (2003).
8. ZEUS Coll., S.Chekanov et al., Phys. Lett. B **471**, 411 (2000).
   ZEUS Coll., S.Chekanov et al., Phys. Lett. B **559**, 153 (2003).
   ZEUS Coll., S.Chekanov et al., contributed paper to XXXIst International Conference on High Energy Physics, Amsterdam, (2000). Abs. no. 910.
9. ZEUS Coll., S. Chekanov et al., DESY-03-182, accepted by Phy. Lett. B.
10. M.Dreiner et al., *Proceedings of the Workshop on Future Physics at HERA*, G.Ingelman, A.DeRoeck and R.Klanner (eds.), Vol. 1, p. 239. Hamburg, Germany, DESY (1996).

A NEW ROUND OF EXPERIMENTS FOR HERA

ALLEN C. CALDWELL

Max-Planck-Institut für Physik
Munich, Germany   and
Columbia University
New York, NY, USA
E-mail: caldwell@mppmu.mpg.de

A new round of experimentation with the HERA accelerator is discussed.

1. Introduction

QCD is the most complex of the known forces operating in the microworld. Myriad effects are seen in condensed matter physics resulting from simple non-relativistic quantum mechanical interactions between electrons. Imagine the possibilities resulting from strong interactions of large numbers of gluons and quarks. Further studies are bound to reveal new and exciting results which will challenge our existing paradigms.

QCD is fundamental to the understanding of our universe. The bulk of the known mass in the universe arises via the strong interactions - nearly massless quarks become massive baryons and mesons in the QCD potential. One can therefore speculate on deep connections between gravity and QCD. What is clear is that we need to understand the strong force at short and long distance scales. We believe we can make predictions for strong interaction effects at small distance scales where the coupling is weak. Can we do this at high energies ? Data from HERA on forward jet production already show weaknesses in our ability to make accurate calculations. As the distance scale increases, the strong interactions increase dramatically in strength, radiation is suppressed, and confinement sets in. How this comes about is not well understood, but HERA data clearly show this transition, e.g. in the change of the energy dependence of the total cross section around $Q^2 = 0.5$ GeV^2. Studying this transition from a partonic to a hadronic behavior of matter is of the highest importance, and would be a highlight of further HERA running.

The goal of further experimentation at HERA would be to follow up on the interesting effects already seen at HERA:

- disappearing gluons at small Q^2;
- transition from partonic to hadronic behavior of cross sections;
- forward particle (jet) production and disagreement with NLO DGLAP calculations;
- energy dependence of diffractive cross sections,

with new and/or improved detectors, and make new measurements which will qualitatively change our understanding:

- the longitudinal structure function, F_L;
- a full program of electron-ion collisions;
- scattering on polarized hadronic beams.

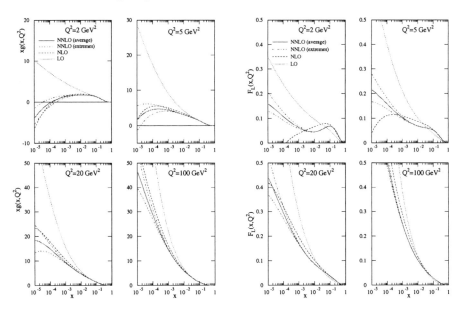

Figure 1. Solution of the DGLAP evolution equation for gluons (left) at different orders of perturbative expansion and the corresponding expectations for F_L (right), as a function of x for fixed Q^2 [3].

Two letters of intent were submitted to the DESY PRC in May 2003 to pursue this program [1,2]. The first letter of intent focuses on measuring eD interactions with an improved H1 detector, while the second letter of intent proposes a new detector optimized to study the physics mentioned above. The LoI's will be described in more detail below.

It should be pointed out that the research program discussed in the LoI's cannot be carried out with the ongoing HERA running (HERA II). This program is intended to study high Q^2 physics with maximum luminosity, and is made possible by the insertion of accelerator elements near the interaction points, thereby limiting access to the small-x physics.

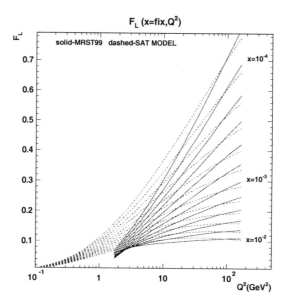

Figure 2. Comparison of the expectations for the Q^2 dependence at fixed x of F_L in the DGLAP approach (MRST99) and the saturation dipole model.

2. HERA III Physics Program

The proposed program goes by the name HERA III. The main research topics discussed in the LoI's are outlined here.

2.1. Precision Structure Function Measurements

It is proposed to measure the F_2, F_2^D, F_L, F_L^D structure functions in the range $0.1 < Q^2 < 100$ GeV^2. The gluon densities extracted at HERA show non-intuitive behavior, such as the tendency to go negative at small x (see Fig. 1). While this is not necessarily a breakdown of the calculational machinery, the fact that these gluon densities also result in negative predictions for F_L clearly shows a breakdown in our understanding. More precise measurements of F_2, and measurements of F_L at small x, will directly test our understanding of the gluon density. The measurement of F_L will also be a strong test of the popular dipole model [4], which makes significantly different predictions than current NLO DGLAP calculations (see Fig. 2).

Due to the lack of theoretical constraints, many parameters are needed to describe the shape of the parton densities. Several assumptions are required in order to limit the number of parameters below about 30. Once these assumptions are relaxed, the uncertainty on the extracted parton densities get much larger. One example is the assumption that $\bar{u} = \bar{d}$ at small x. The effect of removing this constraint is shown in Fig. 3. It is clear that more observables are required to experimentally constrain the parton densities. One such observable is F_L, which will provide a strong constraint on the gluon density at small-x. Running HERA in eD mode would give the information necessary to separate u and d quark distributions.

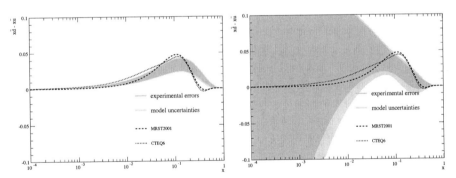

Figure 3. The sea quark difference at $Q^2 = 4$ GeV^2, as determined in a NLO QCD analysis [5]. Left: the constraint $\bar{u} = \bar{d}$ for $x \to 0$ is imposed. Right: the constraint is not imposed.

The structure function F_2 shows a particularly interesting x-dependence near $Q^2 = 0.5$ GeV^2, as shown in Fig. 4. Below this Q^2, the x-dependence is consistent with the energy dependence of hadron-hadron total cross sections, while above this Q^2 the energy dependence becomes significantly steeper. This transition signals a different physics mechanism at work (presumably hadronic degrees of freedom versus partonic degrees of freedom). A new experiment would focus on precision measurements with full acceptance in this Q^2 region.

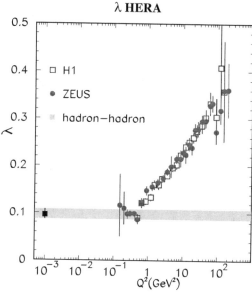

Figure 4. The logarithmic derivative $\lambda_{eff} = -\partial ln F_2/\partial ln x$ as a function of Q^2 (courtesy of A. Levy).

2.2. Exclusive Processes

Exclusive processes allow for a very detailed study of eP interactions. Two additional variables are introduced (4-momentum transfer at the proton vertex and mass of the final state), which allow a many-fold differential study. In the proton rest frame, we can think of the electron emitting a photon far upstream of the interaction point. The photon probes the proton with a transverse resolution which scales as $\hbar c/Q$, and at impact parameter $\sim \hbar c/\sqrt{|t|}$, where t is the square of the 4-momentum transferred at the proton vertex. At small x, the photon interactions are really dipole-proton interactions, with the lifetime of the dipole fluctuation given by

$1/2xM_P$. The dipole converts into a vector meson or real photon (DVCS process) after the interaction. The wavefunction of the produced state constrains the dipole configurations which can take part in the scattering. The extra variables measured in exclusive reactions allow a much more detailed mapping of the hadronically interacting matter. An example of the information which may be extracted is given in Fig. 5, where the dipole-proton scattering amplitude is shown as a function of impact parameter. These processes will allow a 3-dimensional mapping of proton structure.

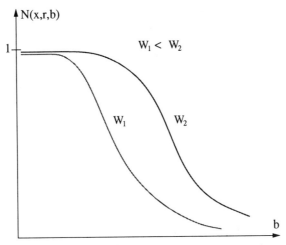

Figure 5. Expectations for the dipole-proton amplitude, $N(x, r, b)$ as a function of impact parameter, b, for different photon-proton energies, W.

2.3. Forward Jet Production

The HERA I results indicate that NLO DGLAP alone is not enough to describe forward jet production. Figure 6 shows the differences predicted [6] in different calculations for different ranges of pseudorapidity, η. It is clear that extending the rapidity coverage of the detectors would greatly enhance the sensitivity. This applies not just to jet production, but also to particle production. In particular, heavy quark production at high rapidities is a very interesting research topic.

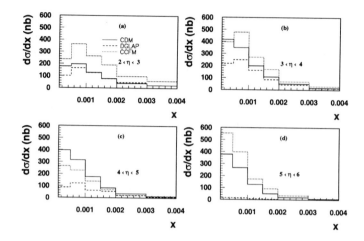

Figure 6. Forward-jet cross section as a function of x obtained from the CDM (Ariadne), DGLAP (Lepto) and CCFM (Cascade) models. The cross sections are shown in different rapidity intervals.

2.4. Precision eA Measurements

Replacing protons with nuclei is widely expected to enhance the striking effects observed at HERA. For example, it is expected that the diffractive cross section will approach 50 % of the total cross section, the maximally allowed value. The black-body limit of QCD will be reached, and new states of matter, such as the color-glass-condensate [7], can be studied. The main argument leading to such conclusions can be stated as follows. The dipole fluctuation from the virtual photon typically lives for distances much larger than the nuclear size at small-x. It is therefore sensitive to all the hadronic matter along a line through the nucleus, whose length scales as $A^{1/3}$. For large nuclei, the extra path length through nuclear matter can lead to saturation of cross sections. The interesting measurements for eA scattering are the same as those for eP scattering - structure function measurements, diffractive cross sections, and exclusive reactions.

2.5. *Spin*

It is not understood how the intrinsic angular momentum of hadrons is built up from the constituents. This flagrant hole in our understanding of subatomic physics demands an explanation. Colliding polarized electrons with polarized protons at HERA would allow the study of the angular momentum carried by small-x partons, and would open a new line of attack on this fundamental problem.

3. H1 Detector Upgrade

The H1 LoI [1] focuses initially on eD scattering, and discusses a second phase which focuses on the forward and backward region. For the first phase, a new spectrometer would be added to measure protons exiting with approximately half the deuteron beam energy. For the second phase, upgrades in the electron direction (new silicon based tracking, a new small-angle calorimeter) and proton direction (instrumented beamline, new forward calorimeters) are envisaged.

As examples of what could be achieved with an eD program of moderate luminosity, the $\bar{d}-\bar{u}$ measurement at small-x and the ratio of valence quark densities, d_v/u_v, were studied. The results are given in Fig. 7. It is clear that the isospin symmetry of the sea can be extremely well tested, and the valence d/u ratio well constrained.

4. A New Detector for HERA

The H1 and ZEUS detectors were optimized for high Q^2 physics, and as such the detectors are optimal for the HERA II program which is now getting under way. For the small-x physics which has been the highlight of HERA so far, and which would be the physics goal of a HERA III program, a new type of detector concentrating on the forward directions is needed. A new detector designed with this physics in mind is sketched in Fig. 8. The main idea is to build a compact detector with tracking and central electromagnetic calorimetry inside a magnetic dipole field. Calorimetric end walls are located outside the dipole. The tracking focuses on forward and backward tracks, while the calorimetry focuses on e/π separation and the central electron and photon measurements. The magnetic field extends from ±4.5 m, and points along the vertical direction. There are 28 tracking planes located along the length of the magnet capable of producing a 3-D coordinate. The barrel calorimeter has an inner radius of 40 cm, and is

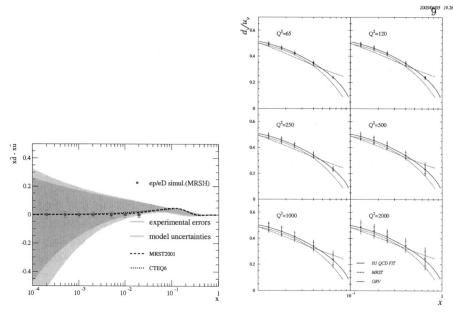

Figure 7. Left: simulation of the difference $\bar{d} - \bar{u}$ using 40 pb^{-1} of eP and 20 pb^{-1} of eD data. The error bar represents the quadratic sum of the statistical and systematic errors. Right: simulation of the measurement of the ratio d_v/u_v with H1 assuming 50 pb^{-1} luminosity for eP and for eD scattering.

confined to a tube of 60 cm. The dipole magnet has an inner radius of 80 cm.

In addition to the detector components shown in the figure, additional detectors located further upstream in the proton and electron directions would also be included. These detectors would be used to tag proton dissociation and initial state radiation, as well as be helpful for luminosity measurements.

The acceptance of the detector for the scattered electron and for forward particles in shown in Fig. 9. Here acceptance is defined as at least three silicon planes crossed. As is clear from the plots, full acceptance is achieved in the transition region near $Q^2 = 0.5$ GeV^2, and the acceptance for forward tracks and jets is vastly increased over what is currently achieved by H1 and ZEUS.

A full GEANT simulation of the detector was performed, including realistic material estimates for the silicon detectors and supports. The EM calorimeters were assumed to be made of Tungsten and silicon, while the forward hadron calorimeter was simulated as the ZEUS FCAL. Momentum

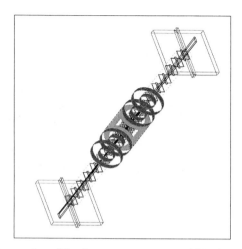

Figure 8. Schematic overview of the detector components within ±6 m of the interaction point. The silicon planes are visible along the beamline. The calorimeter system consists of a central barrel, two catcher rings on each side, and end walls. The calorimeters are all electromagnetic calorimeters, except for the end wall in the proton direction, which also includes a hadronic calorimeter. The dipole magnet enclosing the silicon detectors, central barrel and catcher rings is not shown.

and energy resolutions were studied, as was the e/π separation achieved. As an example of the performance of this new detector, the range and precision of a possible F_L measurement was estimated. For this, three beam energies were assumed ($E_P = 460, 690, 920$ GeV). The x range over which a 10 % measurement would be possible is shown in Fig. 10. The required luminosity for this measurement is given in the figure. At the higher Q^2 values, larger Q^2 intervals would be used to allow for reasonable statistical precision with luminosities of order 100 pb^{-1}.

The performance of the detector was also studied for many other physics processes, as described in the LoI. It is clear from these studies that a detector optimized for forward physics would yield a vast quantity of exciting new results.

5. Conclusions

HERA is a unique facility. With experiments dedicated to strong interaction physics studies, substantial progress can be made in understanding QCD on different distance scales. This is clearly of fundamental importance, as QCD is at the heart of matter. There is a distinct possibility that

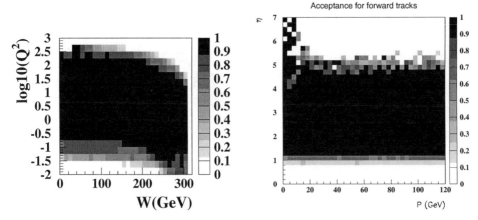

Figure 9. Left: acceptance of the tracking system for the scattered electron vs. W and Q^2 for 3 tracking planes crossed. Right: the acceptance of the tracking system in the proton direction in terms of momentum and pseudorapidity.

a paradigm shift in our conception of nature would result as a consequence of the studies outlined here.

In addition to the fundamental nature of the measurements described above, there are many additional benefits which would be derived from carrying out the program. For example, the parton densities extracted are required for high energy particle, astroparticle and nuclear experiments.

The proposed experiments are of moderate scale compared with the LHC or the Linear Collider efforts, and would offer an attractive alternative to the very large collaborations involved in those efforts. Additionally, the HERA accelerator is an existing facility which still has substantial physics potential. The manpower and financial expenditures already invested in HERA should be exploited fully.

Acknowledgements

I would like to thank all the authors of the two LoI's for their efforts in promoting this important research program. The help of Halina Abramowicz, Iris Abt, Tim Greenshaw, Max Klein, and Uta Stoesslein is particularly appreciated.

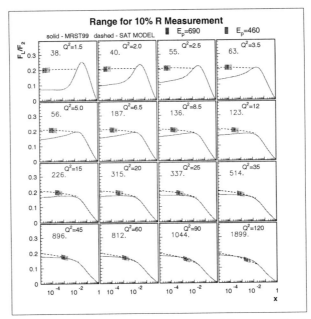

Figure 10. The x-range where 10 % or better measurements of F_L are possible is shown as the shaded band. The required luminosity is given in each bin in the upper left corner in units of pb^{-1}.

References

1. Electron-Deuteron Scattering with HERA, a Letter of Intent for an Experimental Programme with the H1 Detector, H1-04/03-609, http://hep.ph.liv.ac.uk/g̃reen/HERA3/home.html.
2. A New Experiment for the HERA Collider, Expression of Interest, H. Abramowicz et al., MPP-2003-62, MPI-PhE/2003-06, http://wwwhera-b.mppmu.mpg.de/hera-3/
3. A. D. Martin, R. G. Roberts, W. J. Stirling and R. S. Thorne, Phys. Lett. **B531** 216 (2002);
   see also R. S. Thorne, talk at the HERA III workshop, MPI, Munich, Dec 2002, unpublished.
4. For a review, see M. F. McDermott, arXiv:hep-ph/0008260, and references therein;
   K. Golec-Biernat and M. Wüsthoff, Phys. Rev. **D59** 014017 (1999).
5. H1 Collaboration, C. Adloff et al., hep-ex/0304003, Eur. Phys. J. **C30** 1 (2003).
6. H. Jung, private communication.
7. L. McLerran and R. Venugopalan, Phys. Rev. **D50** 225 (1994); Phys. Rev. **D59** 094002 (1999).

List of participants

SERGEY ALEKHIN	IHEP Protvino	alekhin@sirius.ihep.su
ELKE-CAROLINE ASCHENAUER	DESY Hamburg	elke.aschenauer@desy.de
DANIEL BOER	U Amsterdam	dboer@nat.vu.nl
YURAJ BRACINIK	MPI Munich	juraj.bracinik@desy.de
GERD BUSCHHORN	MPI Munich	gwb@mppmu.mpg.de
ALLEN CALDWELL	MPI Munich	caldwell@mppmu.mpg.de
VLADIMIR CHEKELIAN	MPI Munich	shekeln@mail.desy.de
FRANCOIS CORRIVEAU	U McGill	francois@mail.desy.de
JAROSLAV CVACH	Acad. of Sci. Prague	cvach@fzu.cz
KAI DIENER	PSI Villigen	kai.diener@psi.ch
ELISABETTA GALLO	U Florence, INFN	gallo@mail.desy.de
CLAUDIA GLASMAN	U Madrid	claudia@mail.desy.de
GÜNTER GRINDHAMMER	MPI Munich	guenterg@desy.de
GÖSTA GUSTAFSON	U Lund	gosta@thep.lu.se
GUDRUN HEINRICH	Durham, IPPP	gudrun.heinrich@durham.ac.uk
LEIF JÖNSSON	U Lund	leif.jonsson@hep.lu.se
ROSITA JURGELEIT	MPI Munich	roj@mppmu.mpg.de
DORIAN KCIRA	U Wisconsin	kcira@mail.desy.de
CHRISTIAN KIESLING	MPI Munich	cmk@mppmu.mpg.de
MAX KLEIN	DESY Zeuthen	klein@ifh.de
BERND KNIEHL	U Hamburg	kniehl@mail.desy.de
GUSTAV KRAMER	U Hamburg	kramer@mail.desy.de
SERGEY LEVONIAN	DESY Hamburg	levonian@mail.desy.de
MARKOS MANIATIS	U Hamburg	markos@mail.desy.de
SVEN MOCH	DESY Zeuthen	moch@ifh.de
LESZEK MOTYKA	Jagellonian U Cracow	leszekm@th.if.uj.edu.pl
OTTO NACHTMANN	U Heidelberg	o.nachtmann@thphys.uni-heidelberg.de
WOLFGANG OCHS	MPI Munich	wwo@mppmu.mpg.de
SANJAY PADHI	DESY Hamburg	spadhi@mail.desy.de
ROBERT PESCHANSKI	CNRS Saclay	pesch@spht.saclay.cea.fr
RINGAILE PLACAKYTE	MPI Munich	ringaile@mppmu.mpg.de
BENJAMIN PORTHEAULT	LAL Orsay	portheau@lal.in2p3.fr
JON PUMPLIN	Michigan State U	pumplin@pa.msu.edu
ZUZANA RURIKOVA	MPI Hamburg	zuzana.rurikova@desy.de
INGO SCHIENBEIN	DESY Hamburg	schien@mail.desy.de
FRIDGER SCHREMPP	DESY Hamburg	fridger.schrempp@desy.de
UTA STÖSSLEIN	DESY Hamburg	uta.stoesslein@desy.de
PAUL THOMPSON	U Birmingham	thompspd@mail.desy.de
MICHAEL TYTGAT	U Gent	michael@inwfsun1.UGent.be
NIKOLAI VLASOV	U Freiburg	vlasov@mail.desy.de
BILJANA VUJICIC	MPI Munich	bivuj@mppmu.mpg.de
RIK YOSHIDA	Argonne	rik.yoshida@desy.de